Springer Texts in Business and Economics

More information about this series at
http://www.springer.com/series/10099

Henrik Gudmundsson • Ralph P. Hall •
Greg Marsden • Josias Zietsman

# Sustainable Transportation

## Indicators, Frameworks, and Performance Management

Springer

Henrik Gudmundsson
Department of Transport
Technical University of Denmark
Kgs. Lyngby
Denmark

Ralph P. Hall
School of Public and International Affairs
Virginia Tech
Blacksburg
Virginia
USA

Greg Marsden
Institute for Transport Studies
University of Leeds
Leeds
UK

Josias Zietsman
Texas A&M University System
Texas A&M Transportation Institute
College Station
Texas
USA

Co-publication by Springer-Verlag, Heidelberg Germany
and Samfundslitteratur, Frederiksberg Denmark.
The first edition has to be published at the same time, with
the same title, and by the same authors.

This edition is for countries outside of Scandinavia.
Please order the book from Springer Verlag: www.springer.com

For orders from Scandinavia please order the book from
Samfundslitteratur Press: www.samfundslitteratur.dk

ISSN 2192-4333 ISSN 2192-4341 (electronic)
Springer Texts in Business and Economics
ISBN 978-3-662-46923-1 ISBN 978-3-662-46924-8 (eBook)
DOI 10.1007/978-3-662-46924-8

Library of Congress Control Number: 2015939346

Springer Heidelberg New York Dordrecht London


Printed on acid-free paper

Springer-Verlag GmbH Berlin Heidelberg is part of Springer Science+Business Media (www.springer.com)

# Preface

This book is the result of 6 years of collaboration between four scholars from leading universities in Europe and North America. It is based on decades of collective experience in the areas of sustainable development and sustainable transportation, focused on research, teaching, and practice. The book's development was driven by the need for a comprehensive text for students, academics, and practitioners interested in the broad area of sustainable transportation.

The book provides a rich text for advanced undergraduate and graduate students, academics, researchers, and transportation practitioners. It will provide readers with a deep understanding of the basic concepts of sustainability as well as a coherent framework for how to apply the concepts consistently within the context of transportation planning, management, and decision-making.

The book contains 12 chapters and is organized into two main parts connecting theory and methodology to practical examples of sustainable transportation indicator systems followed by our concluding reflections. It is intended to be both a valuable reference on the subject and a source of ideas for how to approach the development of sustainable transportation indicator systems. The book is grounded in the belief that there is no one right way to develop such a system; however, there is a set of ideas and tools that should be applied to ensure that any system developed is informed by sustainability principles, is effective, and is used by all participants and stakeholders.

In the academic realm, the book is designed for use in courses involving the application of sustainability to decision-making in transportation. The structure of the chapters in Part I (Chaps. 2–7) was designed from a pedagogical/learning perspective. Each chapter builds on the previous set of ideas to enable students to develop a broad and interconnected understanding of the material and how it can be applied in a real-world setting. Where relevant, the text provides key terms, important references, and discussion questions to facilitate in-class discussions. The book can also be used in a range of existing courses on transportation planning, policy analysis, or performance management in general.

In the practitioner realm, the book will support planners, managers, consultants, and other professionals who are challenged with transitioning their transportation systems toward sustainability. It offers a frame of reference on what sustainability is and how a measurement system can be developed to make informed decisions. It provides clear guidance on what we should measure, how we should measure, and what we should report. We have endeavored to make the text accessible, while not undermining the importance of using consistent and accurate terminology throughout. We hope to empower the reader with the correct terminology to facilitate effective communication. The descriptions, examples, and case studies in Part II of the book (Chaps. 8–11) in particular are intended to enable practitioners to develop sustainable transportation performance measurement systems that are well conceived and, hence, valuable to their organization.

In this book, we show that transportation plays a key role in addressing the broader topic of sustainability, while at the same time recognizing that transportation has to become more sustainable to make progress toward sustainable development. The message that there are multiple ways to implement a sustainable transportation performance measurement system may frustrate some who are looking for an "off-the-shelf" answer. Similar to sustainable development, moving toward sustainable transportation is a *process* of change that thrives when supported by a flexible and learning-oriented approach. We hope that the performance measurement frameworks and best practices discussed in this book provide useful starting points for those looking to embark on transforming their transportation systems toward sustainability. We provide our reflections on what we have learnt in the conclusion of the book (Chap. 12). Our intention is to continue developing and sharing ideas through a website associated with the book at https://sustransindicators.com/ and we would encourage readers to engage with us in this enterprise.

Kgs. Lyngby, Denmark — Henrik Gudmundsson
Blacksburg, VA — Ralph P. Hall
Leeds, UK — Greg Marsden
College Station, TX — Josias Zietsman

# Acknowledgments

The preparation of this book would not have been possible without the financial support we received from the *2010 Lærebogsprisen Award*. We would like to thank the publisher Samfundslitteratur in Denmark for these funds that enabled the team to meet and work together on the structure and content of this book over the past several years.

We would also like to thank Tara Ramani for coauthoring Chap. 6 and for general writing support, as well as Erin Puckett and Amanda McGee for providing research support.

We are indebted to the coauthors of the case studies for providing invaluable local information and reviewing our draft material and answering numerous questions—Daisuke Fukuda (Tokyo Institute of Technology) and Paul Krekeler and Debra Nelson (New York State Department of Transportation).

We would also like to thank Terry Hall and Gary Lobaugh for editorial assistance and John Henry for support with the graphics.

We thank our respective institutions (Technical University of Denmark, University of Leeds, Virginia Tech, and the Texas A&M Transportation Institute) for enabling us to enrich this text through our teaching and research activities. We would also like to thank our students for the feedback they provided on early drafts of this text. We believe the flow of the book and delivery of information has been greatly improved as a result of their insights.

We also wish to thank our numerous friends and colleagues in the sustainable transportation research community for their work, ideas, and encouragement. While not directly involved in the writing of this book, we have benefited significantly from their intellectual contributions that continue to shape and advance the fields of sustainable transportation and performance assessment.

Finally, we owe a great debt of thanks to all of our families who have supported us through the visits, exchanges, long nights, and weekends which have made this book possible.

# Contents

# Introduction

# 1

Sustainability has become an overarching concern for transportation policy and planning around the world. Like sustainable development, the concept of sustainable transportation is broadly defined, which permits policies and practices to be labeled as "sustainable" while pursuing business-as-usual approaches. Thus, there is a pressing need to better integrate and apply sustainability principles to transportation. Performance measurement frameworks offer an effective way to do this.

Over the past two decades, much effort has been made on understanding and applying the concept of sustainable development to transportation.[1] In this regard, there is a wealth of research and experience that we can learn from. Yet, substantive progress on realizing more sustainable forms of transportation remains limited. In many regions, the negative impacts from transportation are likely to worsen in the face of increasing demand for mobility and infrastructure (Dulca 2013; AfDB et al. 2012). For example, it is estimated that around 25 million paved road lane kilometers and 335,000 rail track kilometers will be needed globally by 2050 in response to growth in passenger and freight travel, primarily in emerging economies (Dulca 2013). To put this in perspective, this would be a 60 % increase in the combined length of all road and railway networks around the world (*ibid.*). These predictions are accompanied by an expected upward trend in oil consumption in 2035, driven primarily by demand in China and India, with oil consumption declining in Organisation for Economic Co-operation and Development (OECD) countries (IEA 2013b).

The demands for higher levels of mobility and infrastructure expansion reflect that transportation delivers beneficial and often essential services to local as well as global economies. In many areas of the world, investing in better transportation

[1] For example, see Replogle (1991), Black (1996, 2010), Gudmundsson and Hojer (1996), Button and Nijkamp (1997), UKRTSD (1996), Greene and Wegener (1997), Whitelegg (1997), Black and Nijkamp (2002), Hoogma et al. (2002), Steg and Gifford (2005), RAE (2005), Banister (2005), Hall (2006), Barrella et al. (2010), Amekudzi et al. (2011), Zietsman et al. (2011), Holden et al. (2013), and Booz Allen Hamilton (2014).



H. Gudmundsson et al., *Sustainable Transportation*, Springer Texts in Business and Economics, DOI 10.1007/978-3-662-46924-8_1

systems to move people and freight is seen as one of the most effective ways to improve economic well-being and performance.[2] However, the negative impacts of adding additional passengers and traffic to existing transportation systems and/or expanding these systems could begin to undermine the benefits realized from this growth. The key focus of most existing methods used for transportation planning and decision-making is to quantify the net balance of economic benefits. We argue in this book that there is a pressing need to expand the scope and scale of transportation decision support to encompass the full vision of sustainable development of which net present economic benefits are only one element.

In terms of the negative impacts of transportation, the sector's reliance on oil has long been a major indicator of its unsustainability, although some modest inroads in reducing fuel consumption are beginning to be made through the sale of hybrid, plug-in hybrid, and all-electric vehicles. In 2011, the concentration of carbon dioxide in the atmosphere reached 391 ppm (parts per million), an increase of 40 % above pre-industrial levels and close to the 400 ppm level that is predicted to increase the Earth's global average surface temperature by 2 °C (3.6 °F) (IPCC 2013). That same year, over 31,000 million tonnes of $CO_2$ were emitted from fuel combustion, 22 % of which came from the transportation sector, with 17 % attributed to road transportation (IEA 2013a). Similarly, transportation also remains a major contributor to the emissions of other air pollutants including diesel particulate matter, hydrocarbons, and oxides of nitrogen. Within the context of increasing global demand for transportation infrastructure and services, the transportation sector will come under growing scrutiny as efforts to address climate change and other global and national environmental concerns intensify.

Beyond climate change, the transportation sector is also responsible for a wide range of impacts that affect ecosystem integrity and biological diversity and directly affect human health and well-being. The growing field of "road ecology" provides a good example of the concerns that researchers (across a wide range of disciplines) have with the physical, chemical, and noise impacts of the road network and traffic on vegetation, wildlife, aquatic systems, etc. (Forman et al. 2003; van der Ree et al. 2011). In terms of human health, mobile source air toxics—such as benzene, 1,3-butadiene, formaldehyde, acetaldehyde, acrolein, polycyclic organic matter (POM), naphthalene, and diesel particulate matter—remain closely monitored and regulated due to the potential health risks they pose (Carr et al. 2007; Shrouds 2009; Milojevic et al. 2014). It is estimated that road traffic accidents typically cost between 1 and 2 % of GDP for both developed and developing countries (WHO 2004, 2013). The total global cost of accidents was estimated in 2000 to be almost US$518 billion per year. Finally, the design and layout of transportation infrastructure can directly shape the livability and quality of neighborhoods (Wheeler 2013).

[2] This is especially the case in emerging economies where the development of rural roads is considered to be essential for connectivity and economic development (Faiz 2012).

As transportation systems continue to develop and expand across the world, these types of environmental and social impacts are likely to remain or in some cases, such as climate change, become an increasingly important driver of the need for sustainable development. At the 2012 Rio+20 conference, the subject of sustainable transportation was highlighted as central to sustainable development (UN 2012), and many nations, regions (such as the European Union[3]), and multilateral development banks are responding to this call for action—see, for example, the "Commitment to Sustainable Transport" made by eight development banks (AfDB et al. 2012, 2013). Such activities are increasing the global demand for professionals who are well-equipped to manage the transition toward more sustainable transportation systems.

The science (and social science) of understanding sustainable development issues is well documented and continues to evolve (Black 2010; Cox 2010; Alonso et al. 2015). This has been matched by neither progress in policy making nor in how to approach policy making to tackle such cross-cutting problems. A goal of this book, therefore, is to provide, through theory and case study analysis, some generic principles to advance the capability of the transportation profession to promote sustainable development, while the debate on what this means in different contexts continues to unfold. It also asks questions that highlight the types of research that academics might do to improve decision-support tools and techniques.

This book focuses on the role that indicators and performance measurement frameworks (or systems) have in making sustainability count for decision-makers, planners, operators, and other stakeholders within and beyond the field of transportation. It does so because "what is measured is what matters." This often used statement hides some very important yet under-discussed issues relating to the politics and practice of decision-making. What gets measured and how it gets used are part of the political process. The information which is considered to be important in a debate and how it is interpreted is a reflection of the framing of the problems to be tackled by government and nongovernmental actors. This means that, far from being about what to measure and how, this book addresses more fundamental questions about how the different actors in the transportation system see their role in the broader sustainable development debate. This is discussed further in Chap. 5.

For sustainability to matter, the concept needs to be made a priority and then effectively operationalized in our decision-making frameworks. Once this occurs, what is measured will also matter for sustainability. In the next four chapters of the book, attention is given to clearly describe sustainable development and sustainable transportation and to draw a boundary around the transportation system and its governance to which these concepts are applied. This provides a critical baseline position to understand and argue for the application of sustainability procedures to improve decision support.

[3] See the European Union's efforts to promote Sustainable Urban Transport Plans, http://ec.europa.eu/environment/urban/urban_transport.htm (accessed 6/2/2014).

Sustainability is a complex and multidimensional issue that we argue cannot be made operational without the use of indicators. Since the term "indicator" has different meanings across multiple disciplines (such as engineering, law, finance, and policy), a consistent definition of an indicator is provided (in Chap. 6) and applied throughout the book. While there is a growing body of literature on sustainable transportation indicators—for example, see Bongardt et al. (2011), Holden (2013), Holden et al. (2013), Jeon et al. (2013), and Zietsman et al. (2011)—we argue that there is no "predefined" set of indicators that can be applied to measure the performance of sustainable transportation policies or programs. What works in one place cannot simply be taken and reapplied elsewhere. Predefined sets of sustainable transportation indicators can be informative, but they should not be accepted wholesale, without consideration to the political, organizational, and economic environment in which they are being applied.

During the preparation of this book, the authors supported a National Cooperative Highway Research Program (NCHRP) research project to develop NCHRP Report 708, "Guidebook for Sustainability Performance Measurement for Transportation Agencies" (Zietsman et al. 2011). This project led to the identification of hundreds of potential transportation indicators that could be used within a sustainable transportation performance measurement system.[4] The research also supported a number of interactive practitioner workshops to test the application of the performance measurement framework developed by the research team. These workshops revealed how an individual's or group's *perspective* plays a critical role in shaping which indicators were selected for what purpose. For example, a group of planners from a transportation agency might be interested in using indicators to *describe* a perceived problem or *communicate* progress that has been made on addressing a problem. The indicators selected for each type of application may be quite different. The same indicator may also be selected, but it could be viewed quite differently due to the *framework* through which the indicator is viewed (see Chap. 7). As a result of this experience, and from supporting research undertaken in the EU (Joumard and Gudmundsson 2010), the importance of clearly specifying the context in which indicators are "applied" became apparent. Every institution operates at a different scale, with a different purpose, and to a different set of stakeholders. This context-specific nature of indicator application means that it is not possible (or perhaps wise) to provide a single set of metrics that will be applicable across different transportation agencies. It is instead necessary to focus on how indicators can support the alignment of purpose to sustainability goals across the system. While we do not dispute the value of standardizing indicators for comparative purposes, our focus is on the development of performance measurement systems that are fine-tuned to the specific needs of a transportation entity wherever they sit in the system.

The real-world dynamic environment in which indicators are applied is probably one reason why there has been limited progress in realizing more sustainable forms

[4] These indicators are provided as an appendix to the main NCHRP report.

of transportation. As will be discussed in this book, and highlighted by several of the case studies, the implementation of an effective sustainable transportation performance measurement system is a significant undertaking that requires leadership and a sustained effort, along with space and time to adapt the framework as learning occurs. In this book, we deliberately stopped short of presenting an idealized performance measurement framework, since we recognize the importance and persistence of existing decision-making practice and the need to make sustainable transportation work within existing governance structures. This is not to say we are content with the status quo. Rather, it is a recognition that institutions change slowly and therefore we need to start the task now, demonstrate that planning for sustainable transportation leads to better decisions and outcomes, and work to transform systems over time.

## 1.1 The Structure of the Book

This book is structured into two distinct parts which are preceded by this introduction (Chap. 1) and followed by our concluding remarks in Chap. 12.

**Part I—Conceptual Foundations**

- Chapter 2: Sustainable Development
- Chapter 3: Planning for Transportation
- Chapter 4: Transportation and Sustainability
- Chapter 5: Governance and Decision-Making in Transportation
- Chapter 6: Indicators
- Chapter 7: Frameworks

**Part II—Case Studies**

- Chapter 8: European Union Transport White Paper
- Chapter 9: High Speed Rail in England
- Chapter 10: New York's GreenLITES Rating Systems
- Chapter 11: Japan's "Eco-City Model" Program

Part I of the book develops several conceptual foundations on which the remainder of the book rests. It begins by exploring the emergence of the concept of sustainable development through the lens of key international conferences and publications (Chap. 2). By tracking the historical evolution of the concept, the compromises that were made when crafting key foundational texts such as *Our Common Future* and the Rio Declaration are revealed. One of the main objectives of the chapter is to clarify the type of development that is being promoted or endorsed when the Brundtland formulation of sustainable development is invoked. Since most efforts to promote sustainable transportation are linked with the Brundtland definition, having a clear grasp of the strengths and weaknesses of this definition is important. Chapter 2 also outlines the "weak" and "strong" forms of sustainability to create a continuum along which sustainable transportation initiatives and programs could be placed. Weak sustainability can be described as an environmentally oriented business-as-usual approach to development, whereas strong sustainability is a more radical reformulation where human activity has to be kept within macroecological limits. Regardless of one's perspective on sustainable development, it is helpful to be able to articulate how policies and initiatives designed to promote sustainable transportation align with established theoretical frameworks. Knowing where, in principle, an organization stands on the weak to strong sustainability continuum is likely to promote learning and a deeper appreciation for what is or is not likely to be achieved through the organization's actions. Chapter 2 concludes by highlighting the need to adopt a holistic and integrative perspective to the design of policies, programs, or initiatives targeted at addressing unsustainability.

Having established a common understanding of sustainable development, Chap. 3 provides a comprehensive definition of a transportation system and attempts to draw a boundary around what we mean by "transportation." In particular, it considers how the components of the different transportation modes and networks fit together and are organized within a societal and environmental context. This broad description of a complex socio-technical system highlights the connections that exist between societal demand for travel and the consequences of this travel on communities and the natural environment. The chapter also discusses how the transportation system is continually shaped by political-economic actors/stakeholders, as well as the availability of financial resources and the capacity to develop/deliver transportation services—topics explored in more depth in Chap. 5.

In Chap. 4, we revisit the material introduced in Chap. 2 through the lens of transportation. The chapter follows the evolution of the definitions and principles of sustainable transportation since the early 1990s and argues that the current focus on the concept might be too narrow and constraining. By positioning the transportation system as one of many systems (or sectors) contributing to development, a holistic perspective is presented that considers the transportation system through the lens of sustainable development. Thus, the transportation system is conceived as one of several interconnected systems, which raises the importance of developing integrated, multi-sectoral solutions to the sustainability challenges ahead.

Having defined sustainable development and sustainable transportation and drawn a boundary around what we consider to be a transportation system, Chap. 5 explores the governance of this system. In particular, the chapter discusses how the transportation system is governed by a range of state and non-state actors that operate at the local to national/international level. The chapter makes the case for the state's intervention in shaping the transportation system—acknowledging that this task in itself is a highly complex endeavor—and argues for the coordination (or better still, integration) of policy within any given level of government to promote sustainable development/transportation. Several characteristics of governance systems that are considered to promote planning for sustainable transportation are also discussed. Finally, the chapter defines two broad domains of the transportation planning process—"planning" and "delivery." Within each domain, a series of functional areas are identified from long-term strategic planning to operation and maintenance. The provision of transportation services within and across these different functions is where the alignment of transportation and sustainability happens in practice.

Chapter 6 defines, in depth, the various types of indicators that inform transportation planning and management. In addition, other key concepts such as performance measures, indices, and benchmarks are discussed. Particular attention is paid to identify ways to distinguish "good" from "bad" indicators using a set of well-established criteria. The chapter also explores three indicator typologies by looking at the *dimensions* of indicators (e.g., time and space), the *messages* they convey, and their *positions* (i.e., whether they lead or lag a phenomenon of interest). The discussion of indicators is not an indulgent technocratic activity. As we explore in Part II of the book, indicators are everywhere in our decision-making processes so we need to understand their purposes. The final section of the chapter therefore focuses on eight *indicator application* areas—i.e., describe, forecast, review, diagnose, decide, account, learn, and communicate—that link the use of indicators to a wide variety of planning, decision support, and operational tasks. Each of these tasks may utilize different types of indicators or apply the same indicator in a different context. Thus, having a clear understanding of the indented *use* of an indicator is an essential first step in knowing whether the most appropriate type of indicator has been selected. Further, explicitly focusing on whether a performance measurement system is intended to *diagnose* a problem or help decision-makers *decide* on a future course of action will help clarify the system's purpose and improve its relevance and value to an organization. This chapter begins therefore to build the bridge between the institutional settings and the application of information to real decisions.

Chapter 7 moves beyond individual measures to *frameworks* that connect variables together in systems for planning and policy making in the area of sustainability and transportation. The chapter begins by asking the basic question of what is a framework and how can it provide a useful context for making a set of indicators count? This simple question leads to a rich discussion of ways to "frame" indicators, from broad ideas and paradigms to more specific systems adopted by organizations. Both generic and practical frameworks that emphasize relevant areas such as transportation appraisal, environmental planning, sustainability assessment, and performance measurement are discussed. A key objective of the chapter is to highlight how to identify the strengths and weaknesses of frameworks in terms of how well they support thinking and acting on sustainability in transportation and how well they support the different indicator applications introduced in Chap. 6. An important conclusion from Chap. 7 is that efforts targeted at creating one single performance measurement framework that works effectively everywhere is likely to be a futile endeavor. This conclusion motivated the development of several detailed case studies to explore how governments and transportation agencies are framing and applying indicators to measure progress toward sustainable transportation. These case studies are presented in Part II of this book.

In an effort to internationalize the scope to this book, four case studies were selected from different developed regions of the world. The case studies were selected because each, to some respect, captures something of what we consider to be a globally interesting practice. They each exemplify different interpretations of sustainability, different governance issues, different frameworks, and different applications of indicators. They each exhibit strengths and weaknesses. Taken together, they illustrate the scope of the challenge of, and the opportunities for, joining up decision-making for sustainability. Part II of the book begins with an introduction to the case studies that links the case studies to the material introduced in Part I of the book.

Chapter 8 examines the 2011 European Transport White Paper as an example of the application of indicators in the planning domain. The White Paper is unique in that it emphasizes sustainability as an overarching goal for all areas of European policy making and is perhaps the largest scale attempt to provide a guiding framework for sustainable transportation that exists. The large geographic scale and significant variations in baseline conditions in each of the countries that comprise the EU provide for a useful reflection on the importance of frameworks in providing a meta-environment within which other actors then operate.

Chapter 9 provides an ex-ante assessment of the case for the development of a High-Speed Rail network in England, which would connect London and cities to the north including Birmingham, Manchester, and Leeds. Specific attention is paid to how indicators are used to appraise the sustainability of the High-Speed Rail scheme. In addition, the indicators selected to support the scheme are discussed through the describe, forecast, review, diagnose, and decide indicator application areas, highlighting how various frameworks are being used simultaneously to measure performance. The case study also highlights some of the tensions that exist between national efforts to promote sustainable development and transportation infrastructure projects that are conceived in response to other drivers, such as operational priorities.

Chapter 10 moves from the national to state level and takes a detailed look at New York State Department of Transportation's (NYSDOT's) GreenLITES (Leadership In Transportation and Environmental Sustainability) programs. Collectively, the GreenLITES programs represent one of the leading sustainable transportation performance measurement frameworks under development in the USA, which cover project design, maintenance and operations, and regional planning. The case study provides a unique perspective on how an agency's culture plays an essential role in the promotion of a sustainable transportation performance measurement system. Particular attention is paid to how data from the GreenLITES programs are used to support the capital investment decision-making process. An important finding is how sustainability indicators need to compete alongside existing, more traditional, indicators that measure, for example, pavement condition or safety. In this regard, the GreenLITES data have yet to become a leading driving force in transportation investment decisions despite the progress that has been made. The case study concludes by discussing how the indicators and data from the GreenLITES programs are being used in the learn, decide, forecast, and communicate indicator application areas.

Chapter 11, the final case study in Part II, describes the Eco-Model City (EMC) program created by the Japanese Government in 2008, with a specific focus on the regional capital city of Toyama. The objective of the EMC program is to transform cities toward a low-carbon future. Progress toward this objective is measured using a communicative, results-oriented framework involving central and local government, as well as independent experts, where indicators are primarily applied to review, diagnose, and learn. The case study focuses on the indictors used to assess the city of Toyama's EMC program implementation, reduction in greenhouse gas emissions, and spread and extension of best practices.

To close out the book, Chap. 12 provides a concluding commentary and highlights the main lessons and takeaway points from this work.

Like the concepts of sustainable development and sustainable transportation, the preparation of this book can itself be characterized as a *process* of experimentation and learning, driven by a clear purpose to make a contribution to the field of sustainable transportation performance measurement. We hope that our emphasis on the nuanced and context-specific nature of performance measurement frameworks helps advance the general understanding of the subject.

Finally, we are humbled by the complexity of the challenge facing those tasked with realizing more sustainable forms of transportation. Upon reading this text, we invite students, academics, and practitioners to reach out to us with ideas on how the material could be improved for future editions or for additional case studies that could be developed based on the theories, frameworks, and examples introduced.

We have relished the challenge of writing this book and hope this enthusiasm is reflected in the subsequent pages.

## References


African Development Bank (AfDB), Asian Development Bank (ADB), Development Bank of Latin America (CAF), European Bank for Reconstruction and Development (EBRD), European Investment Bank (EIB), Inter-American Development Bank (IADB), Islamic Development Bank (ISDB), & World Bank (WB). (2012). Commitment to sustainable transport. Joint statement to the Rio+20 United Nations conference on sustainable development. Philippines: Asian Development Bank.

African Development Bank (AfDB), Asian Development Bank (ADB), Development Bank of Latin America (CAF), European Bank for Reconstruction and Development (EBRD), European Investment Bank (EIB), Inter-American Development Bank (IADB), Islamic Development Bank (ISDB), & World Bank (WB). (2013). *Progress report (2012–2013) of the MDB Working Group on Sustainable Transport*. Philippines: Asian Development Bank.

Alonso, A., Monzón, A., & Cascajoa, R. (2015). Comparative analysis of passenger transport sustainability in European cities. *Ecological Indicators, 48*, 578–592.

Amekudzi, A. A., Meyer, M. D., Ross, C. L., & Barrella, E. (2011). *Transportation planning and sustainability guidebook*. Washington, DC: U.S. Department of Transportation. Available at http://www.fhwa.dot.gov/hep/climate/sustainability/sustain00.cfm.

Banister, D. (2005). *Unsustainable transport: City transport in the new century*. London: Routledge.

Barrella, E., Amekudzi, A. A., Meyer, M. D., Ross, C. L., & Turchetta, D. (2010). Best practices and common approaches for considering sustainability at U.S. state transportation agencies. *Transportation Research Record, 2174*, 10–18.

Black, W. R. (1996). Sustainable transportation: A U.S. perspective. *Journal of Transport Geography, 4*(3), 151–159.

Black, W. R. (2010). *Sustainable transportation: Problems and solutions*. New York: Guilford Press.

Black, W. R., & Nijkamp, P. (Eds.). (2002). *Social change and sustainable transport*. Bloomington, IN: Indiana University Press.

Bongardt, D., Schmid, D., Huizenga, C., & Litman, T. (2011). *Sustainable transport evaluation: Developing practical tools for evaluation in the context of the CSD process*. Eschborn: The Partnership on Sustainable, Low Carbon Transport (SloCat).

Booz Allen Hamilton. (2014). *Volume 4: Sustainability as an organizing principle for transportation agencies* (NCHRP Report 750). Washington, DC: Transportation Research Board (TRB).

Button, K., & Nijkamp, P. (1997). Social change and sustainable transport. *Journal of Transport Geography, 5*(3), 215–218.

Carr, E. L., Ernst, D. A., Rosenbaum, A., Glass, G., & Hartley, S. (2007). *Analyzing, documenting, and communicating the impacts of mobile source air toxic emissions in the NEPA process* (NCHRP Project 25-25). Fairfax: ICF.

Cox, P. (2010). *Moving people: Sustainable transport development*. London: Zed Books.

Dulca, J. (2013). *Global land transport infrastructure requirements: Estimating road and railway infrastructure capacity and costs to 2050*. Paris: International Energy Agency (IEA).

Faiz, A. (2012). *The promise of rural roads: Review of the role of low-volume roads in rural connectivity, poverty reduction, crisis management, and livability* (Transportation research circular, number E-C167). Washington, DC: Transportation Research Board.

Forman, R. T. T., Sperling, D., Bissonette, J. A., Clevenger, A. P., Cutshall, C. D., Dale, V. H., et al. (2003). *Road ecology: Science and solutions*. Washington, DC: Island Press.

Greene, D. L., & Wegener, M. (1997). Sustainable transport. *Journal of Transport Geography, 5* (3), 177–190.

Gudmundsson, H., & Hojer, M. (1996). Sustainable development principles and their implications for transport. *Ecological Economics, 19*, 269–282.

Hall, R. P. (2006). *Understanding and applying the concept of sustainable development to transportation planning and decision-making in the U.S.* Engineering Systems Division, Massachusetts Institute of Technology, Cambridge. http://esd.mit.edu/students/esdphd/dissertations/hall_ralph.pdf

Holden, M. (2013). Sustainability indicator systems within urban governance: Usability analysis of sustainability indicator systems as boundary objects. *Ecological Indicators, 32*, 89–96.

Holden, E., Linnerud, K., & Banister, D. (2013). Sustainable passenger transport: Back to Brundtland. *Transportation Research Part A, 54*, 67–77.

Hoogma, R., Kemp, R., Schot, J., & Truffer, B. (2002). *Experimenting for sustainable transport: The approach of strategic Niche management*. London: Spon Press.

Intergovernmental Panel on Climate Change (IPCC). (2013). Summary for policymakers. In T. F. Stocker, D. Qin, G.-K. Plattner, M. Tignor, S. K. Allen, J. Boschung, et al. (Eds.), *Climate change 2013: The physical science basis*. Cambridge: Cambridge University Press.

International Energy Agency (IEA). (2013a). *$CO_2$ emissions from fuel combustion* (2013 ed.). Paris: International Energy Agency.

International Energy Agency (IEA). (2013b). *World energy outlook 2013*. Paris: International Energy Agency.

Jeon, C. M., Amekudzi, A. A., & Guenseler, R. (2013). Sustainability assessment at the transportation planning level: Performance measures and indexes. *Transport Policy, 25*, 10–21.

Joumard, R., & Gudmundsson, H. (Eds.). (2010). *Indicators of environmental sustainability in transport*. France: INRETS.

Milojevic, A., Wilkinson, P., Armstrong, B., Bhaskaran, K., Smeeth, L., & Hajat, S. (2014). Short-term effects of air pollution on a range of cardiovascular events in England and Wales: Case-crossover analysis of the MINAP database, hospital admissions and mortality. *Heart, 100*, 1093–1098.

Replogle, M. (1991). Sustainability: A vital concept for transportation planning and development. *Journal of Advanced Transportation, 25*(1), 3–17.

Shrouds, J. M. (2009). *Air quality community of practice mobile source air toxics state-of-the-practice*. Washington, DC: American Association of State Highway and Transportation Officials (AASHTO), Center for Environmental Excellence.

Steg, L., & Gifford, R. (2005). Sustainable transportation and quality of life. *Journal of Transport Geography, 13*, 59–69.

The Royal Academy of Engineering (RAE). (2005). *Transport 2050: The route to sustainable wealth creation*. London: The Royal Academy of Engineering (RAE).

UK Round Table on Sustainable Development (UKRTSD). (1996). *Defining a sustainable transport sector*. London: UK Sustainable Development Commission.

United Nations (UN). (2012). *Report of the United Nations conference on sustainable development* (A/CONF.216/16). New York: United Nations.

van der Ree, R., Jaeger, J. A. G., van der Grift, E. A., & Clevenger, A. P. (2011). Effects of roads and traffic on wildlife populations and landscape function: Road ecology is moving toward larger scales. *Ecology and Society, 16*(1), 48.

Wheeler, S. M. (2013). *Planning for sustainability: Creating livable, equitable and ecological communities*. London: Routledge.

Whitelegg, J. (1997). *Critical mass: Transport, environment and society in the twenty-first century*. London: Pluto Press.

World Health Organisation (WHO). (2004). *World report on road traffic injury prevention*. Geneva: World Health Organisation.

World Health Organisation (WHO). (2013). *Gobal status report on road safety 2013: Supporting a decade of action*. Geneva: World Health Organisation.

Zietsman, J., Ramani, T., Potter, J., Reeder, V., & DeFlorio, J. (2011). *A guidebook for sustainability performance measurement for transportation agencies* (NCHRP Report 708). Washington, DC: Transportation Research Board.

# Part I

# Conceptual Foundations

# Sustainable Development 2

## 2.1 Purpose and Content

Since the late 1980s, sustainable development has garnered much interest from government agencies, businesses, nongovernment organizations, and civic groups, resulting in policy initiatives in both the public and private sector. Yet, people and organizations citing sustainable development as an objective often lack a firm grasp of the origins and true meaning of the concept. Such an understanding is important as it provides a holistic perspective on development against which a sectoral—e.g., transportation specific—focus on sustainability can be considered. This chapter explores the evolution of sustainable development through the perspective of international conferences and publications often referred to in discussions of sustainability. The chapter then introduces the challenges that are frequently confronted when trying to conceptualize sustainable development through different disciplinary lenses. It concludes with a discussion of the need to adopt a holistic and integrative approach to the design of policies and initiatives aimed at achieving more sustainable forms of development.

## 2.2 The Emergence of Sustainable Development

The concept of sustainable development obtained formal international recognition at the 1992 United Nations Conference on Environment and Development (UNCED) in Rio de Janeiro, Brazil. However, it is possible to trace the roots of the concept back to the 1950s/1960s, when developed nations were becoming increasingly aware that the local or regional environment was being stressed by rapid industrialization.[1]

[1] The discussion in this section draws from Hall and Ashford (2012) and Ashford and Hall (2011).



H. Gudmundsson et al., *Sustainable Transportation*, Springer Texts in Business and Economics, DOI 10.1007/978-3-662-46924-8_2

### 2.2.1 The Formation of Environmental Movements

Although events such as the London Smog of 1952–1953 illustrated the dangerous effects of pollution (Bell and Davis 2001; Davis et al. 2002), it was the publication of Rachael Carson's book *Silent Spring* in 1962 that focused public attention on the negative impacts of industrial activities (see Fig. 2.1 for a timeline of sustainability-related events). Carson described the potential dangers of the excessive use of pesticides (such as DDT) and argued that it served the interests of chemical companies, industrial agriculture, the military, and universities to ignore these dangers, promote their use, and continue their development. "*Silent Spring* altered a balance of power in the world. No one since would be able to sell pollution as the necessary underside of progress so easily or uncritically" (Hynes 1989, p. 3).[2]

In parallel with the growing distrust of the government-industry complex, arguments warning the environmental problems associated with the prevailing development model of rapid industrialization and economic growth began to surface. Two classic publications which supported this movement were "The Tragedy of the Commons" (Hardin 1968) and *The Population Bomb* (Ehrlich 1968). Hardin (1968) highlighted the natural tendency of private actors to exploit the public/environmental commons to the point where it can no longer support economic activity. Ehrlich (1968) expressed concern that the appetite of a growing population may not be met by a fixed resource base—a similar argument to that made in *Limits to Growth* (Meadows et al. 1972). The latter report was novel in its use of computer simulations to illustrate potentially disastrous future consequences of the continuation of current production and consumption patterns.

In response to public concern in the USA, Congress passed the National Environmental Policy Act (NEPA) and signed it into law on January 1, 1970. NEPA was designed to ensure that the entire federal bureaucracy considered the environmental impacts of its actions (Blumm 1990). Since its passage, more than 100 countries around the world have adopted similar procedures for environmental impact assessments (Jay et al. 2007). In addition to placing the environment on a more equal footing with development, the act influenced the Brundtland concept of sustainable development that followed some two decades later (see Sect. 2.2.5). NEPA required the federal government to "fulfill the responsibilities of each generation as trustee of the environment for succeeding generations" (Sec. 101, (b), 1 [42 USC § 4331]). Intergenerational considerations now form a central element of the idea of sustainable development.

[2] Outside of the UK and the USA, a similar environmental awareness was emerging in other developed regions. In Japan, problems such as the "Minamata" disease (caused by mercury poisoning in the city of Minamata) starkly revealed the downsides of heavy industrial development.

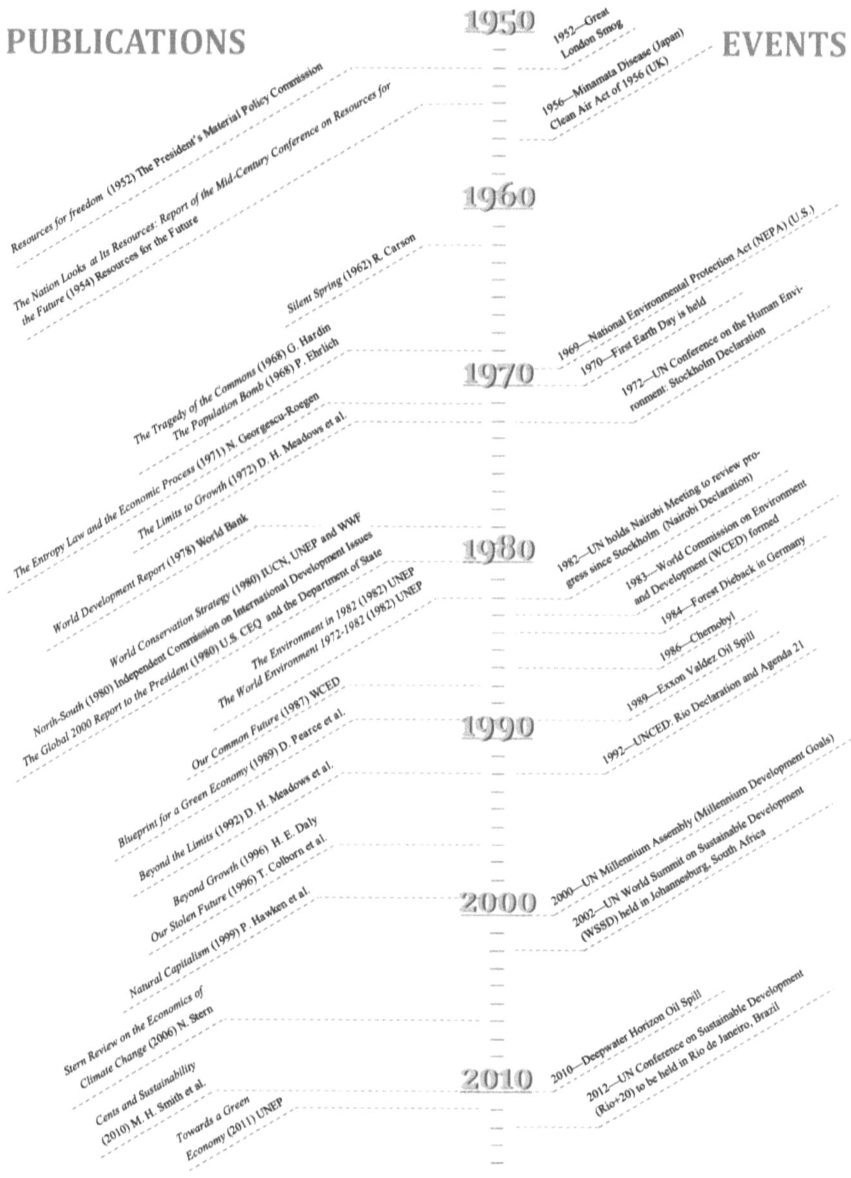


**Fig. 2.1** Timeline of key events and publications

### 2.2.2 The 1972 Stockholm Conference

As a result of growing environmental concerns within industrialized nations and an awareness that these challenges were not confined by national borders, the United Nations held a Conference on the Human Environment in Stockholm in 1972.

The Stockholm Conference brought together the topics of ecosystem integrity, biological diversity, and human health and the issue of ecological and resource limits to growth. The conference discussed the potential problem with toxic substances (in its Action Plan), but this environmental concern remained primarily the focus of national legislation during the 1970s. Toward the end of the 1970s, the international community began to discuss the related concerns of ozone depletion and greenhouse gas emissions. However, it was not until the second half of the 1980s and the 1990s that international action was taken to address ozone depletion and global climate change, respectively.

The Stockholm Conference is considered a defining moment for two reasons (Caldwell and Weiland 1996)—it identified the critical need for all nation states to establish environmental policy at the national level and informed the world community of the vital role that a healthy biosphere plays in sustaining life, placing a concern for the environment on national agendas. The Stockholm Conference also led to the creation of the United Nations Environment Programme (UNEP) to provide the UN with the institutional capacity needed to address and coordinate the recommendations put forward in the Stockholm Action Plan and, more generally, to advocate for the protection and improvement of the environment.

Although the Stockholm Conference and its agreements were influential in advancing concerns for the human environment, many suggest that the conference's major impact came from the intense pre-conference deliberations and from its role as a catalyst for an explosion of literature that raised the world's consciousness about the natural environment (Dernbach 1998; UNEP 1982a, b; Emmelin 1972; Strong 1972; United Nations 1972).

### 2.2.3 The 1980 World Conservation Strategy

One of the foundational texts on sustainable development is the International Union for Conservation of Nature and Natural Resources (IUCN) et al.'s (1980) *World Conservation Strategy* (WCS). The WCS is a synthesis of decades of debate in the international community over the need to protect the environment while continuing the process of development. The WCS used the term "sustainable" to describe development that takes "account of social and ecological factors, as well as economic ones; of the living and non-living resource base; and of the long term as well as short term advantages and disadvantages of alternative actions" (IUCN et al. 1980, p. 18). Acknowledging that "[c]onservation and development have so seldom been combined that they often appear—and are sometimes represented as being—incompatible" (*ibid.*, p. 18), the WCS proceeds to develop its case as to why conservation and economic and social development are mutually supportive endeavors (*ibid.*). "Conservation must ... be combined with measures to meet short term economic needs. The vicious circle by which poverty causes ecological degradation which in turn leads to more poverty can be broken only by development. But if it is not to be self-defeating, it must be sustainable—and conservation helps to make it so" (*ibid.*, p. 19).

The WCS's notion of sustainable development—the idea that economic and social development can occur in unison with the conservation of living resources—presented a different perspective on global problems. While the WCS did not fully integrate development and environmental considerations (Clapp and Dauvergne 2005), its formulation of "sustainable development" informed the World Commission of Environment and Development's (WCED's) report *Our Common Future* (see Sect. 2.2.5) that made the concept a defining and integrating theme of the 1992 UN Conference on Environment and Development (see Sect. 2.2.6) (Caldwell and Weiland 1996).

### 2.2.4 The 1982 Nairobi Meeting

Ten years after Stockholm, the UN convened a meeting in Nairobi to review the progress in implementing the Stockholm Action Plan and make recommendations with respect to prevailing environmental trends for the future actions of the UNEP. The pre-conference reports prepared by UNEP (1982a, c) and the Nairobi Declaration presented a clear message that while nation states had made progress toward environmental protection, their actions were insufficient to reverse the rate of environmental degradation occurring throughout the world. The Nairobi meeting also highlighted the role of economic growth in improving the health and welfare of people and the environment in developing countries (UNEP 1982a, p. 37).

Since the initial concerns for the human environment grew from the negative impacts of industrialization in developed countries, the shift in the international focus toward the environmental problems faced by developing nations is significant. By identifying poverty as a major contributor to environmental degradation, economic growth became more important since it was considered to be the only pragmatic way of alleviating poverty. However, the only way to grow the economy was to follow the path of conventional development. This meant a reliance on technology that was fueled by nonrenewable resources and that generated a significant amount of pollution which would likely damage ecosystems and human health. Thus, developing countries faced a paradox. They needed to develop to not only alleviate poverty but to also protect and improve their environment—upon which their future depended—but in doing so, they would ultimately damage the very environment they wished to safeguard. This contradiction underscored the need for development and environmental protection to advance in unison.

### 2.2.5 The World Commission on Environment and Development: *Our Common Future*

In light of the evidence that environmental conditions around the world were deteriorating (UNEP 1982a; IUNC et al. 1980; Brandt 1980; CEQ 1980) and population and economic growth—two critical factors affecting the environment—were continuing to increase (Strong 2003), the UN General Assembly established a special, independent commission on the environment to create "long-term environmental strategies for achieving sustainable development."[3] As part of its terms of reference, the commission was required to consider the interrelationships between developed and developing nations and between people, resources, the environment, and development. In short, the commission was required to articulate a new vision of development.

Under the chairmanship of former Prime Minister Gro Harlem Brundtland of Norway, the World Commission on Environment and Development (WCED, also known as the Brundtland Commission) was subsequently formed and held its first meeting in Geneva, Switzerland, in October 1984.

Between 1984 and 1987, the Brundtland Commission received advice and support from thousands of individuals, institutions, and organizations from all over the world (WCED 1987, p. 359). The commission also visited each world region to obtain a firsthand view of environment and development issues and to hold deliberative meetings and open public hearings. On December 11, 1987, the commission's "Environmental Perspective to the Year 2000 and Beyond" was adopted by the UN General Assembly.[4] That same year, the Commission's full report was published as *Our Common Future*.

Benefiting from more than a decade of debate over the notion of sustainable development, the Brundtland Commission sought to effectively integrate social and economic development with the need for environmental protection. By combining these elements with a consideration of intergenerational equity, the Commission created what has become the most cited definition of sustainable development.

> Sustainable development is development that meets the needs of the present without compromising the ability of future generations to meet their own needs. It contains within it two key concepts:
>
> - the concept of "needs," in particular, the essential needs of the world's poor, to which overriding priority should be given; and
> - the idea of limitations imposed by the state of technology and social organization on the environment's ability to meet present and future needs.
>
> Thus the goals of economic and social development must be defined in terms of sustainability in all countries—developed or developing, market orientated or centrally planned. Interpretations will vary but must share certain general features and must flow from a consensus on the basic concept of sustainable development and on a broad strategic framework for achieving it (WCED 1987, p. 43).

The latter part of this definition highlights what has since become one of the major issues of contention with sustainable development. The *interpretation* of

[3] Source: UN General Assembly, *Resolution 38/161, Process of preparation of the Environmental Perspective to the Year 2000 and Beyond*, 19 December 1983, Section 8 (a), http://www.un.org/documents/ga/res/38/a38r161.htm (accessed on April 19, 2015).

[4] Source: UN General Assembly, *Resolution 42/186, Environmental Perspective to the Year 2000 and Beyond*, 11 December 1987, 2, http://www.un.org/documents/ga/res/42/a42r186.htm (accessed on April 19, 2015).

sustainable development by one nation might be seen as leading to "unsustainable" development by another.

*Our Common Future* defined the major objective of development as the "satisfaction of human needs and aspirations" (WCED 1987, p. 43). Further, it envisioned sustainable development not as an end state but rather as "*a process of change* in which the exploitation of resources, the direction of investments, the orientation of technological development, and institutional change are made consistent with future as well as present needs" (WCED 1987, p. 9, emphasis added). The Brundtland Commission adopted a highly political agenda by viewing "sustainable development as a policy objective, rather than a methodology. It is an overarching concept. … Such an approach is unapologetically normative, and places both the responsibility for problems, and the political will to overcome them, in the hands of human actors" (WCED 1987, p. 37).

The Brundtland Commission made a convincing argument that environment and development are "inexorably linked" and cannot be treated as separate challenges (WCED 1987, p. 37). It concluded: "[d]evelopment cannot subsist upon a deteriorating environmental resource base; the environment cannot be protected when growth leaves out of account the costs of environmental destruction" (WCED 1987, p. 37). This statement implies that limits need to be placed on environmentally destructive economic activity, and, as stated above, these "limits" should be determined by the "state of technology and social organization," implying that the solutions lie in *better* technology and systems of governance.

*Our Common Future* appeared at a time when the political climate was beginning to become more receptive to the issues raised by the report. Future prospects for economic growth in industrialized nations were beginning to look positive, while global ecosystems were beginning to show signs of distress (Engfeldt 2002). An international audience was eager to learn how to embrace economic growth while reducing pressure on ecosystems. The Commission's insistence that science and technology could be utilized to meet human needs and solve environmental problems was the answer many were looking for. By promoting the role of technological improvements in supporting economic growth, conserving natural resources, and protecting the environment, the Commission gained the support of both developed and developing nations. If science and technological innovation—two mainstays of economic growth in industrial societies—had not been a central theme of sustainable development, national governments (primarily of the north) would most likely have rejected the concept as another radical and politically unrealistic form of environmentalism.

By explicitly bringing science and technology into the development equation, the technologically optimistic Brundtland Commission sought to articulate a new era of economic growth that is decoupled from increasing environmental degradation.

Having articulated a bold new development agenda, the Brundtland Commission highlighted a major problem with the institutional frameworks that would be responsible for implementing the new era of economic and social development (we treat institutional issues and governance in the context of transportation in

Chap. 5). It argued that most governmental environment agencies, especially those in developing nations, "tend to be independent, fragmented, [and] working to relatively narrow mandates with closed decision processes" (WCED 1987, p. 9). It stated the same was true for many international agencies responsible for areas such as development lending, trade regulation, and agricultural development. The Commission believed the solution to these problems lay in ensuring that national and international institutions consider the ecological dimensions of policy at the same time as economic, social, trade, energy, agricultural, and other dimensions. The idea was to develop a more integrated and proactive approach to environmental protection, rather than the more expensive "react and cure" approach that was typical of many government approaches in the post-Stockholm era (Runnalls 2008). In parallel with this, the Commission called for the strengthening of international law and conventions in support of sustainable development and for better implementation of these mechanisms for change.

Box 2.1 presents the broad set of conclusions from *Our Common Future*, which reiterates the above points and presents several additional requirements for the pursuit of sustainable development.

**Box 2.1: Requirements for the Pursuit of Sustainable Development, *Our Common Future* (WCED 1987, p. 65)**

In its broadest sense, the strategy for sustainable development aims to promote harmony among human beings and between humanity and nature. In the specific context of the development and environment crises of the 1980s, which current national and international political and economic institutions have not and perhaps cannot overcome, the pursuit of sustainable development requires:

- A political system that secures effective citizen participation in decision making,
- An economic system that is able to generate surpluses and technical knowledge on a self-reliant and sustained basis,
- A social system that provides for solutions for the tensions arising from disharmonious development,
- A production system that respects the obligation to preserve the ecological base for development,
- A technological system that can search continuously for new solutions,
- An international system that fosters sustainable patterns of trade and finance, and
- An administrative system that is flexible and has the capacity for self-correction.

### 2.2.6 The 1992 United Nations Conference on Environment and Development: The Rio Summit

In response to *Our Common Future*, the UN General Assembly decided to convene the UNCED in Rio de Janeiro, Brazil (also known as the Rio or Earth Summit) in 1992. The UNCED attracted some 178 nation states, including 110 heads of state who attended the final 2-day meeting (UN 1993a–c), an unprecedented global gathering of such leaders.

Two of the official documents from UNCED have since taken a central role in shaping the idea of sustainable development: the Rio Declaration on Environment and Development and Agenda 21. Whereas the Rio Declaration provided a vision of sustainable development, Agenda 21 provided a comprehensive plan of action (a blueprint) that was created to guide and coordinate the work of the UN, governments, and other major groups in their efforts to transition society toward sustainable development. The conference also adopted the Framework Convention on Climate Change, providing the international legal framework for climate policy.

Continuing the Brundtland Commission's conception of sustainable development, the Rio Declaration and Agenda 21 did not supplant previous approaches to development, rather they revised (in fundamental ways) the *conventional development* approach. Prior to the 1990s, the conventional development model (promoted by the international community) incorporated four related concepts: (1) peace and security; (2) economic development; (3) social development; and (4) national governance that secures peace and development (Dernbach 1998, 2004). The Brundtland Commission and UNCED agreements called for environmental concerns to be *integrated* into the conventional development model. Principles 3 and 4 of the Rio Declaration speak directly to this aim.[5]

**Principle 3** The right to development must be fulfilled so as to equitably meet developmental and environmental needs of present and future generations.

**Principle 4** In order to achieve sustainable development, environmental protection shall constitute an integral part of the development process and cannot be considered in isolation from it.

The recognition of the need to protect the environment—upon which the development process depends—can be considered as the *fifth element* of the international notion of development (Dernbach 1998, p. 21). Therefore, **sustainable development** could be crudely considered as: **conventional development + environmental protection/conservation**.

Principles 15 and 16 of the Rio Declaration also articulated the *precautionary* and *polluter pays* principles, respectively, which have since become guiding principles of sustainable development policy and programs.

[5] Source: *UNCED Declaration on Environment and Development*, http://www.un.org/documents/ga/conf151/aconf15126-1annex1.htm (accessed on April 19, 2015).

While the UNCED is considered a watershed event in the formation of the concept of sustainable development, it was not without its critics who pointed to several shortcomings of the meeting and its products (The Ecologist 1993; Korten 1991). Grubb et al. (1993) argue that the principles of the Rio Declaration reveal weaknesses in the compromises that were made to make the Declaration politically palatable. A significant turning point in the negotiations of the Declaration was the success of developing nations in placing their "right to development" at the forefront of considerations (Sachs 2001). The recognition that less-developed nations needed to "develop" meant that the Rio Declaration effectively turned into a "declaration on development, rather than on environment" (Sachs 2001, p. 5). Further, since "development" can be defined in multiple ways, it can be argued that the Rio Declaration supports a business-as-usual approach to development where the environment is more of an afterthought.[6]

Redclift (1996) argues that the UNCED neglected to address important questions relating to population, trade, poverty, the debt crisis (faced by many oil-importing developing nations), and distributional inequality more generally. In addition, he raises an important question about whether the "development" of industrialized nations is what the developing world should be aspiring to achieve.

Criticisms such as these point to the need for careful consideration of *who* the development process is really benefiting and what model of development is being promoted. Critics notwithstanding, *Our Common Future*, the Rio Declaration, and Agenda 21 are typically considered as the building blocks of the notion of sustainable development. Two other notable documents that contribute to an understanding of sustainable development are the *Earth Charter* (prepared by the Earth Council) and the UN *Millennium Declaration*—both published in 2000.

### 2.2.7 The 2002 Johannesburg Summit

In 2002, the Johannesburg Summit was held to review progress since UNCED. During the decade following Rio, the world had experienced a new phase of economic growth that was largely based upon patterns of development, consumption, and lifestyles that had the effect of widening the gap between affluent and poor nations (South Centre 2002).

A new era of economic globalization had changed the approaches necessary to transition the world toward sustainable development. The Johannesburg Declaration stated that "[t]he rapid integration of markets, mobility of capital and significant increases in investment flows around the world have opened new challenges and opportunities for the pursuit of sustainable development." In addition to reaffirming a commitment to sustainable development, the declaration specifically urged developed nations to provide the internationally agreed-upon levels of

[6] Ashford and Hall (2011) argue that a similar situation occurs today with employment, which they view as a critical, but often forgotten, element of sustainable development.

official development assistance (ODA)—set at 0.7 % of GNP in 1969 (Pearson 1969)—to developing nations. Furthermore, the private sector was called upon to recognize its role in achieving sustainable development. The declaration stated it had a "duty to contribute to the evolution of equitable and sustainable communities and societies" and that it should "enforce corporate accountability, which should take place within a transparent and stable regulatory environment." Finally, the declaration stated that the goals of sustainable development would be achieved through "effective, democratic, and accountable international and multilateral institutions," putting multilateralism at the center of sustainable development efforts.

An important recognition at the Johannesburg Summit was the role of voluntary, multi-stakeholder, international-/national-/local-level partnerships for sustainable development (ECOSOC 2002, p. 7). At the time of the summit, over 220 partnerships had been identified with many new partnerships being announced during and after the Summit. However, some caution that NGOs were worried the partnerships might mitigate government obligations, that governments may "lose control" over their sustainable development agendas to the organizations leading the partnerships, and that since the implementation of sustainable development is not a core activity of many organizations, the impacts of the partnerships may be limited (Hens and Nath 2005, p. 33).

Another outcome of the Johannesburg process was the international community's commitment to market mechanisms and capacity building (or capacity development) as critical measures to achieving sustainable development. This transition toward a reliance on the market reflected a continuing ideological shift away from the role of the government as a driving force for development. Indeed, multi-lateralism and the inclusion of a strong business and NGO presence in the delivery architecture for sustainable development means that the governance (steering) of actors is increasingly important. This is not to underestimate the importance of governmental actors but to recognize them as part of a broader constellation.

While the Johannesburg Summit focused on a more comprehensive set of environmental issues than those discussed at the UNCED, in the years following the summit, the international community's attention gravitated toward the challenge of global climate change. The release of Al Gore's documentary *An Inconvenient Truth*, followed by the award of the 2007 Nobel Peace Prize to him and the Intergovernmental Panel on Climate Change "for their efforts to build up and disseminate greater knowledge about man-made climate change,"[7] did much to raise global concern about the issue. Equally important was the publication of the *Stern Review on the Economics of Climate Change* (known as the *Stern Review*) by the UK Treasury on October 30, 2006 (Stern 2007). Although the review was not the first economic analysis of climate change (Cline 1992; Mendelsohn et al. 1998;

[7] Source: Nobel Foundation, The Nobel Peace Prize 2007, http://nobelprize.org/nobel_prizes/peace/laureates/2007/ (accessed on April 19, 2015).

Nordhaus and Boyer 2000), its status as an official government document made it one of the most widely known and debated studies of its kind. The growing dominance of global climate change as *the* environmental concern means that the focus on other important environment and human health concerns is lessened (Ashford and Hall 2011).[8]

### 2.2.8 The 2012 United Nations Conference on Sustainable Development (Rio+20)

From June 20 to 22, 2012, the UN Conference on Sustainable Development (known as Rio+20) was held in Rio de Janeiro, Brazil, 20 years after the first Rio Conference. The primary purpose of the conference was to reinvigorate the international community's efforts to promote sustainable development. While some 50,000 policymakers, environmentalists, and business leaders attended the conference, the inability of delegates to agree on a comprehensive framework with commitments and targets for long-term action left many organizations considering the conference a failure.[9]

The most significant outcome from the Rio+20 Conference was the endorsement of the "green economy" as a flexible mechanism for advancing sustainability. The Rio+20 conference report, entitled *The Future We Want*, provides the following commentary on the green economy in the context of sustainable development and poverty eradication:

> We affirm that there are different approaches, visions, models and tools available to each country, in accordance with its national circumstances and priorities, to achieve sustainable development in its three dimensions which is our overarching goal. In this regard, we consider green economy in the context of sustainable development and poverty eradication as one of the important tools available for achieving sustainable development and that it could provide options for policymaking but should not be a rigid set of rules. We emphasize that it should contribute to eradicating poverty as well as sustain[ing] economic growth, enhancing social inclusion, improving human welfare and creating opportunities for employment and decent work for all, while maintaining the healthy functioning of the Earth's ecosystems (UN 2012, p. 9).

---

[8] The importance of maintaining a holistic approach to development is discussed in Sect. 2.5.

[9] For example, see the Friends of the Earth Rio+20 blog that describes the unwillingness of governments to commit to a new set of principles (source: http://www.foei.org/news/blogs/rio-20/rio20-summit-condemned-as-sell-out-of-people-and-the-planet-2/, accessed on April 19, 2015), and Greenpeace's press statement on Rio+20 that called the conference a "failure of epic proportions" due to its lack of commitments and targets (source: http://www.greenpeace.org/international/en/press/releases/Greenpeace-Press-Statement-Rio20-Earth-Summit-a-failure-of-epic-proportions/, accessed on April 19, 2015).

The emphasis on the green economy continues the modernist development stance established at the 1992 Rio Conference—that is, *green* economic growth (*green growth*) can occur through the deliberate application of science and technology. This stance is reflected in the Rio+20 conference report, which recognizes "the critical role of technology as well as the importance of promoting innovation" to make progress toward sustainable development and reduce poverty (UN 2012, p. 13).

The strategies required to transition to a green *global* economy—such as significantly increasing investment in green technologies combined with more stringent national and international regulations/standards—have revived concerns of emerging economies that such actions may promote green protectionism, conditionality, and subsidies that protect the domestic economies of developed regions (UNCSD and UNCTAD 2011). There is also the concern that only developed nations have the available finance and innovative capacity to create and supply the needed technologies for a green transition—with the possible exception of certain green technology sectors in China (e.g., clean coal technology) and Brazil (e.g., biofuels) (UN 2011). Thus, the technology gap between advanced and emerging economies may increase, placing developing regions at a further disadvantage. Such arguments increase the focus on mechanisms to transfer or share technologies with emerging economies, which raises important questions in areas such as intellectual property.[10]

The Rio+20 Conference provided decision-makers with the opportunity to revisit the message from the 1992 Rio Conference—that economic development (i.e., growth) must be decoupled from environmental harm. The green economy is the mechanism the UN system advanced to achieve this objective (UN 2011, 2012).[11]

Two of the many publications written to inform the preparation of Rio+20[12] were UNEP's (2011) report *Towards a Green Economy: Pathways to Sustainable Development and Poverty Eradication* and Smith et al.'s (2010) book *Cents and Sustainability: Securing Our Common Future by Decoupling Economic Growth from Environmental Pressure*. The message from both publications is that a "green economy" and "decoupling" present new growth opportunities that can help protect the environment, create decent jobs, and help address the challenge of poverty. The

[10] Many of the challenges that will accompany a transition to a green economy are clearly articulated in several preparatory reports for Rio+20 (UN 2011; Ocampo et al. 2011; UNCSD and UNCTAD 2011; UNEP 2011).

[11] The Organisation for Economic Co-operation and Development (OECD) has also published a series of reports that outline its strategy for promoting "green growth"—see OECD (2011a–c).

[12] See, for example, the extensive list of pre-conference publications listed on the website of the United Nations Conference of Sustainable Development, http://www.uncsd2012.org/resources_publications.html (accessed on April 19, 2015).

publications draw on over a decade of experience with decoupling strategies that *Cents and Sustainability*, in particular, presents in an attempt to renew momentum behind the approach.

The principal argument of *Cents and Sustainability* and *Towards a Green Economy* is that economic growth can coexist with environmental protection—reinforcing the message from the 1992 Rio Conference that economic growth and environmental protection can advance in unison while reducing poverty. The publications are based on a premise that a green economy or decoupling agenda presents the most viable pathway toward sustainable development.[13]

### 2.2.9 The Post-2015 Agenda

In 2015, the United Nations (UN) will replace the Millennium Development Goals (MDGs) with a new set of goals and indictors for the 2015–2030 timeframe. The post-2015 agenda represents the next evolution in the concept of sustainable development that is explored below by reviewing the emerging sets of sustainable development goals (SDGs).

The challenge of creating the post-2015 agenda falls primarily on the intergovernmental Open Working Group (OWG) on SDGs that was established following Rio +20. The mandate for the OWG was outlined in the Rio+20 outcome document—*The Future We Want*—which charged the 30-member group to deliver the final and "limited" set of SDGs to the UN General Assembly at its 68th session. While no specific SDGs were provided in the outcome document, it did call for the creation of goals that balanced all three dimensions of sustainability in a coherent and integrated way. The process of developing the post-2015 agenda has led to an unprecedented global dialogue that has involved thematic discussions,[14] national consultations in 88 countries,[15] and the submission of reports and input from the High-level Panel on the Post-2015 Development Agenda

[13] In the aftermath of the 2008 global financial crisis, this position is being challenged by the "new economics" or "degrowth" movement that calls for a fundamental reorganization of social activity, where progress is not measured by economic growth. See, for example, D'Alisa et al. (2014).

[14] These issues cover inequalities, governance, growth and employment, health, education, environmental sustainability, food security and nutrition, conflict and fragility, population dynamics, energy, and water. Source: The World We Want, Thematic Consultations, http://www.worldwewant2015.org/sitemap#thematic (accessed on April 19, 2015).

[15] See the World We Want, National Consultations, http://www.worldwewant2015.org/sitemap#national (accessed on April 19, 2015).

(UN 2013),[16] the Sustainable Development Solutions Network (SDSN 2014),[17] and many other organizations including over 40 entities within the UN system.[18]

With the final SDGs yet to be announced, an analysis of the emerging sets of goals developed by the SDSN, the High-level Panel, and the OWG provides some indication of what factors are being considered. Table 2.1 shows the current MDGs alongside each of the proposed sets of SDGs. The goals have been grouped to enable comparison. The table shows that the original eight MDGs are all covered to a certain extent by each of the proposed sets of SDGs. Beyond the original eight goals, two new goals are present in all three sets of SDGs. These are (1) the need for sustainable economic growth accompanied by the creation of jobs and (2) the need to develop sustainable energy systems that reduce the pressure on the climate.

Two of the sets of SDGs promote the need to establish resilient cities/infrastructure and sustainable/universal access to water and sanitation services, whereas the goal of ensuring sustainable production and consumption is found only in the OWG's set of SDGs.

Interestingly, several of the new SDGs were previously included in the MDGs as targets or indicators. For example, the need for employment for all was a target under first MDG, whereas the need to reduce by half the proportion of people without sustainable access to safe drinking water and basic sanitation was a target under MDG 8. While MDG 7 included an indicator to measure $CO_2$ emissions, there were no specific targets related to reducing climate change emissions. The proposed SDGs related to promoting sustainable energy systems, resilient cities/ infrastructure, and sustainable production and consumption are new. However, with the exception of "resilient" cities/infrastructure, the need to develop sustainable energy and production and consumption systems can be traced back to the 1992 Rio Declaration and Agenda 21. For example, Principle 8 of the Rio Declaration and Chapter 4 for Agenda 21 focus specifically on sustainable patterns of consumption and production, and Chapter 7 of Agenda 21 discusses the need for sustainable energy. Thus, while the post-2015 agenda is likely to restructure the framing of the

[16] The 27 member High-level Panel on the Post-2015 Development Agenda was created in July 2012, by the UN Secretary-General Ban Ki-moon to advise on the global development framework beyond 2015. Information on the activities of the panel can be found on the UN Secretary-General's website: http://www.un.org/sg/management/hlppost2015.shtml (accessed on April 19, 2015).

[17] The Sustainable Development Solutions Network (SDSN) was launched by UN Secretary-General Ban Ki-moon in August 2012, to mobilize "scientific and technical expertise from academia, civil society, and the private sector in support of sustainable development problem solving at local, national, and global scales" (source: SDSN, Vision and Organization, http://unsdsn.org/about-us/vision-and-organization/, accessed on April 19, 2015). The group aims to overcome the compartmentalization of technical and policy work by identifying "integrated" solutions to the environmental, economic, and social challenges confronting the world (see Sect. 2.5 for a discussion of the importance of adopting a holistic and integrative approach to sustainable development).

[18] A detailed list of documents, publications, and statements related to the post-2015 agenda development process can be viewed via the OWG's website: http://sustainabledevelopment.un.org/owg.html (accessed on April 19, 2015).

**Table 2.1** The MDGs and the emerging post-2015 development agenda

| Millennium development goals (MDGs) | SDSN's post-2105 development goals (SDSN 2014) | High-level panel's post-2105 development goals (UN 2013) | OWG's post-2015 development goals (UN 2014) |
|---|---|---|---|
| Goal 1: Eradicate extreme poverty and hunger | Goal 1: End extreme poverty including hunger<br>Goal 6: Improve agriculture systems and raise rural prosperity | Goal 1: End poverty<br>Goal 5: Ensure food security and good nutrition | Goal 1: End poverty in all its forms everywhere<br>Goal 2: End hunger, achieve food security, and improved nutrition and promote sustainable agriculture<br>Goal 10: Reduce inequality within and among countries |
| Goal 2: Achieve universal primary education | Goal 3: Ensure effective learning for all children and youth for life and livelihood | Goal 3: Provide quality education and lifelong learning | Goal 4: Ensure inclusive and equitable quality education and promote lifelong learning |
| Goal 3: Promote gender equality and empower women | Goal 4: Achieve gender equality, social inclusion, and human rights for all | Goal 2: Empower girls and women and achieve gender equality | Goal 5: Achieve gender equality and empower all women and girls |
| Goal 4: Reduce child mortality<br>Goal 5: Improve maternal health<br>Goal 6: Combat HIV/AIDS, malaria, and other diseases | Goal 5: Achieve health and wellbeing at all ages | Goal 4: Ensure healthy lives | Goal 3: Ensure healthy lives and promote well-being for all at all ages |
| Goal 7: Ensure environmental sustainability | Goal 9: Secure biodiversity and ensure good management of water, oceans, forests, and natural resources | Goal 9: Manage natural resource assets sustainably | Goal 14: Conserve and sustainably use the oceans, seas, and marine resources for sustainable development<br>Goal 15: Protect, restore, and promote sustainable use of terrestrial ecosystems, sustainably manage forests, combat desertification, and halt and reverse land degradation and halt biodiversity loss |

(continued)

**Table 2.1** (continued)

| Millennium development goals (MDGs) | SDSN's post-2105 development goals (SDSN 2014) | High-level panel's post-2105 development goals (UN 2013) | OWG's post-2015 development goals (UN 2014) |
|---|---|---|---|
| Goal 8: Develop a global partnership for development | Goal 10: Transform governance and technologies for sustainable development | Goal 10: Ensure good governance and effective institutions<br>Goal 11: Ensure stable and peaceful societies<br>Goal 12: Create a global enabling environment and catalyse long-term finance | Goal 16: Promote peaceful and inclusive societies for sustainable development, provide access to justice for all, and build effective, accountable, and inclusive institutions at all levels<br>Goal 17: Strengthen the means of implementation and revitalize the global partnership for sustainable development |
| – | Goal 2: Promote economic growth and decent jobs within planetary boundaries | Goal 8: Create jobs, sustainable livelihoods, and equitable growth | Goal 8: Promote sustained, inclusive, and sustainable economic growth, full and productive employment, and decent work for all |
| – | Goal 8: Curb human-induced climate change and ensure sustainable energy | Goal 7: Secure sustainable energy | Goal 7: Ensure access to affordable, reliable, sustainable, and modern energy for all<br>Goal 13: Take urgent action to combat climate change and its impacts |
| – | Goal 7: Empower inclusive, productive, and resilient cities | – | Goal 9: Build resilient infrastructure, promote inclusive and sustainable industrialization, and foster innovation<br>Goal 11: Make cities and human settlements inclusive, safe, resilient, and sustainable |
| – | – | Goal 6: Achieve universal access to water and sanitation | Goal 6: Ensure availability and sustainable management of water and sanitation for all |
| – | – | – | Goal 12: Ensure sustainable consumption and production patterns |

critical development concerns, if considered in the broader context of the main sustainable development declarations and texts, it could be argued that the agenda is trying to capture concerns that have been previously well articulated. In this regard, the post-2015 agenda could be viewed as a more comprehensive framing of sustainable development.

As should be evident from the above discussion, the framing of sustainable development is likely to continue to evolve in relation to key events (such as the creation of the post-2015 agenda), new knowledge (such as the emerging interest in resilient cities/infrastructure), and from the actions of nations, regions, and the international community in trying to implement the concept. The review of the emerging post-2015 agenda shows that the future SDGs are likely to build on the "Brundtland-UNCED-Johannesburg-Rio+20" agenda, which can be described as technologically optimistic and market oriented. Whether one agrees or not with this approach to sustainable development is a matter of personal conviction; what is important is that if the "Brundtland-UNCED-Johannesburg-Rio+20-Post-2015" view is adopted, the adopter is aware of the development model being promoted. While the Brundtland formulation of sustainable development is the most widely used and accepted approach, other definitions and formulations exist. The following section highlights two useful perspectives on sustainable development by discussing the "weak" and "strong" forms of sustainability.

**Discussion Topics**

- Given that industrialized nations are primarily responsible for many of the global environmental problems we face today, such as climate change, should these countries be held responsible for remedying these problems—i.e., the *polluter pays* principle is invoked? What actions could be taken? How might these actions impact the development opportunities of emerging economies?
- What kind of development should emerging economies aspire to, and how can it realistically be attained? From an equity standpoint, what right do developed nations have to impose restrictions on developing countries when they have engaged in non-sustainable development for so long? Is there room for compromise?
- What actions can be taken, and by whom, to promote a "green economy"? What challenges and opportunities does the green economy agenda present for developed and developing countries?

## 2.3 Conceptualizing Sustainable Development

Early critiques on the concept of sustainable development revealed a wide range of interpretations and a lack of a sufficiently robust theoretical and analytic framework against which decisions aimed at achieving a more sustainable form of

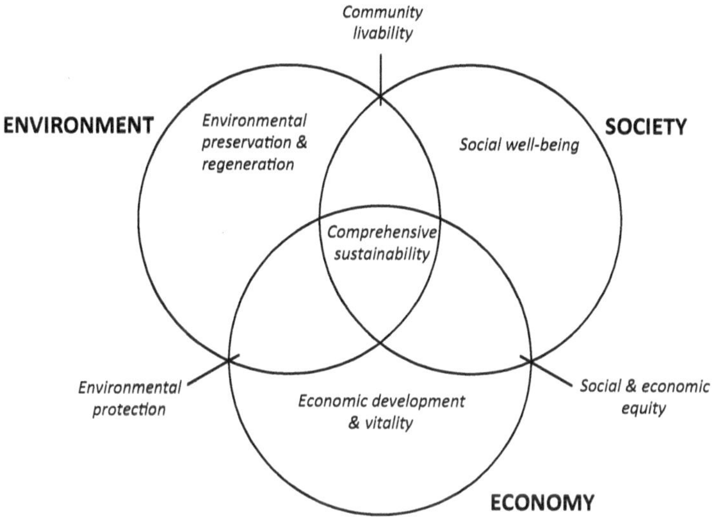


**Fig. 2.2** Comprehensive sustainable development. *Sources*: Adapted from CST (1997, p. 2) and Brodmann and Spillmann (2000, p. 8)

development could be assessed (Holdren et al. 1995; Holmberg and Sandbrook 1992; Shiva 1992; Toman 1992; Lele 1991; Redclift 1991; Dixon and Fallon 1989; Norgaard 1988). These reviews indicate that sustainability should be seen as a broad field of inquiry encompassing issues of cultural integrity, justice, and governance, as well as questions of ecological limits to economic activity, the individual right to a safe and secure livelihood, and the national right to economic development.

The different ways in which sustainable development can be formulated raises challenges to its operationalization and measurement. One is quickly faced with questions such as what is to be sustained, for how long, and who bears the costs? As Richard Norgaard (1988, p. 607) aptly pointed out, "[e]nvironmentalists want environmental systems sustained. Consumers want consumption sustained. Workers want jobs sustained." A further challenge is that the lens or framework (see Chap. 7) through which one views/constructs the problem needing attention can be based on quite different philosophical foundations (Schön and Rein 1994).

Figure 2.2 provides a common visual representation of sustainable development that is often associated with the Brundtland model of development. This comprehensive view implies that progress in all three of the environmental, social, and economic dimensions is necessary for sustainable development. If taken at face value, the diagram indicates that elements of the environmental, social, and economic dimensions can be considered in isolation from each other, which aligns with the "weak" formulation of sustainability discussed below.

A good example of how scientific disciplines can frame the idea of sustainable development quite differently is found in the notions of *substitutability* or *weak sustainability* (Solow 1993) and the *steady-state economy* (SSE) or *strong sustainability* (Daly 1991b, 1996, 2008; Czech and Daly 2004). Both notions view the environment as a special kind of economic asset—called "natural capital."

Solow's (1993) "mainstream" economic lens to sustainability is rooted in the idea that technology can create high degrees of substitutability between one resource and another and, implicitly, that natural and human-made capital are in some sense "fungible." This is what is described as "weak" sustainability, which essentially argues that natural capital can be substituted by human-made capital (Ayres 2007; Beltratti et al. 1995; Neumayer 2003; Hediger 1999). If resources are fungible, it means that society has no obligation to save a resource for future generations as long as an alternative resource is made available. Therefore, Solow (1993, p. 181) defines sustainable development as "an obligation to conduct ourselves so that we leave to the future the option or the capacity to be as well off as we are." The basic model is that under certain conditions, maintaining the "aggregate" capital stock (i.e., manufactured + natural capital + human capital + financial capital) intact provides future generations with the same *opportunity* as the present generation and enables them to choose how they use their endowed capital base. Put differently, by focusing on the aggregate stock of capital, a decline in one form of capital is permitted so long as there is an equivalent increase in another form of capital. While conceptually simple, attempting to combine different forms of capital in this way is a highly complex endeavor, and any effort to do so is likely to be thwarted by theoretical challenges.

In neo-classical economics, technological innovation and reproducible human-made capital are viewed as providing "substitutes" for natural capital (Hartwick 1977, 1978a, b; Solow 1974).[19] Under these assumptions of weak sustainability, consumption can be sustained, environmental externalities can be overcome, and resource scarcity problems can be solved. Neo-classical economists argue that as prices increase due to scarcity, investment in technological innovation creates substitutes to replace the scarce resources, further promoting market-led developments.

In contrast, Daly (1991b) holds a "strong" sustainability position—based on an ecological-economic framework—which states that many of the most fundamental services provided by nature cannot be replaced by services produced by humans or

[19] Ayres (1978) presented a convincing case that the laws of thermodynamics place limits on the ability of human-made resources to replace or substitute natural capital. The basic argument is that human-made capital is built and maintained using natural capital. Thus, both forms of capital are complementary and cannot be substituted for one another. It follows that the maintenance of natural capital stock is, therefore, *essential* for the economic process. A reduction in the availability of natural capital will reduce the productivity of human-made capital that depends upon ecosystem goods and services. The same argument is also made by Georgescu-Roegen (1993). Similarly, Ayres (1997) argues that the neo-classical view of externalities as exceptional occurrences in a larger economic context is incorrect. He considers environmental externalities to be *pervasive*, since the real economy depends upon extracting, processing, and converting materials (and energy), which creates waste residuals that can have negative environmental and economic consequences. Since these consequences are not priced in the real economy, the environment is treated as a free good and medium for disposal.

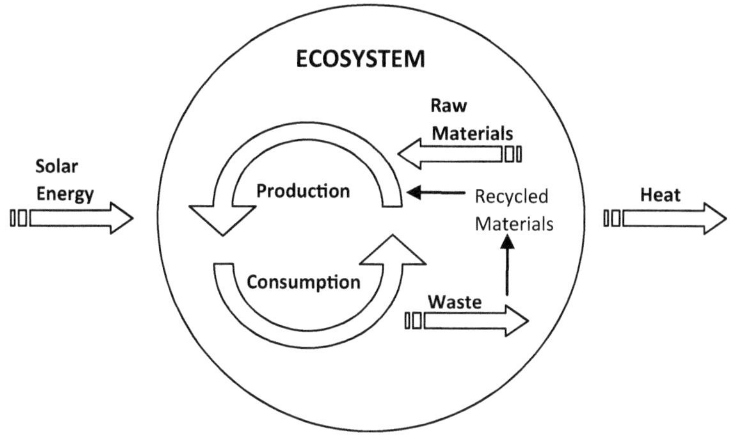


**Fig. 2.3** Steady-state economics view of production and consumption cycles in equilibrium with the macroecosystem. *Source*: Adapted from Daly (1991b, p. 181)

man-made capital. Daly (1991b) provides what is probably the most well-developed vision of an economy which functions within macroecological limits—see also Brown et al. (2012) and Burger et al. (2012) for a discussion of the energetic limits to growth and the macroecology of sustainability, respectively.[20] Arguing from the first principle of thermodynamics, Daly describes a SSE as one in which births replace deaths and production replaces depreciation. The objective of the SSE is to keep the throughput of raw materials (low entropy) and waste (high entropy) to levels within the regenerative and assimilative capacity of the macroecosystem. Whereas neo-classical economics views the growth economy as a continual expansion of production and consumption, the SSE considers these cycles to be in equilibrium with the macroecosystem (Fig. 2.3).

Within the SSE, technology, knowledge, the distribution of income, and the allocation of resources are fluid.[21] Since a fixed amount of resources will yield constant flows of goods and services (all else being equal), technological progress is one way in which more (or more highly valued) goods and services can be produced

[20] The macroecology of sustainability is based on the principles that "1) physical conservation laws govern the flows of energy and materials between human systems and the environment, 2) smaller systems are connected by these flows to larger systems in which they are embedded, and 3) global constraints ultimately limit flows at smaller scales" (Burger et al. 2012, p. 1). Thus, the macroecological perspective requires that all systems and their interrelations must be considered within the context of the global system. Developing a decision-support framework in which such an analysis can occur is perhaps the most important challenge for sustainability science. See Holden et al. (2013) for a commentary on the need to link sustainable passenger transportation to ecological sustainability at a global level.

[21] In general, ecological economists, especially those focusing on steady-state economics, are concerned with the size of the economy relative to the ecosystem. The efficient allocation of resources is a concern, but it is not the primary focus as in neoclassical economics.

**Fig. 2.4** A "strong" model of sustainable development. *Source*: IUCN (2004, p. 10)

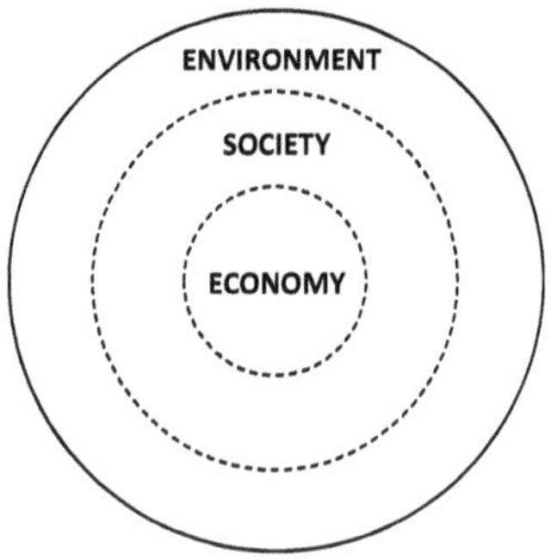


(Czech and Daly 2004; Czech 2003).[22] However, given the laws of thermodynamics there are limits to what is technologically feasible. Thus, there is a theoretical maximum size (an ecological carrying capacity) at which a SSE may exist.

The core principles by which human activities will be kept within the earth's carrying capacity are most clearly articulated by Herman Daly:

1. The main principle is to limit the human scale (throughput) to a level which, if not optimal, is at least within carrying capacity and therefore sustainable. . . .The following principles aim at translating this general macro level constraint to micro level rules.
2. Technological progress for sustainable development should be efficiency increasing rather than throughput increasing.[23] . . .
3. Renewable resources, in both their source and sink functions, should be exploited on a profit-maximizing sustained yield basis and in general not driven to extinction (regardless of the dictates of present value maximization), since they will become ever more important as nonrenewables run out . . . Specifically this means that: (a) harvesting rates should not exceed regeneration rates; and (b) waste emissions should not exceed the renewable assimilative capacity of the environment.
4. Nonrenewable resources should be exploited, but at a rate equal to the creation of renewable substitutes (Daly 1991a, pp. 44–45).

Costanza and Daly (1992) later added the principle that the use of *replenishable* (i.e., nonliving) forms of natural capital (e.g., groundwater and the ozone layer) should not exceed their rates of replenishment or recharge. While Daly's (1991a) second principle highlights technological innovation as an important factor in reducing humanity's ecological impact, social, institutional, and organizational innovation are equally important considerations (Ashford and Hall 2011). Indeed, a more balanced (systems) approach that integrates and co-optimizes technological, social, institutional, and organizational innovation is likely to be more effective at satisfying basic needs while making our resources go further.

Figure 2.4 provides a visual representation of the "strong" form for sustainable development. The figure implies that the economy exists within society (or is a

[22] To help describe the SSE, Daly (1991b) compares it to a steady-state library, where the addition of a new book would mean the removal of an old book. Thus, while the quantitative physical scale remains constant, the library would continue to improve in a qualitative sense.

[23] This principle relates to the rebound effect, whereby efficiency gains can result in additional consumption due to lower costs that undermine or eclipse the environmental gains.

product of social interaction) and that both society and the economy depend upon the environment. Therefore, if human activity exceeds the carrying capacity of the environment, this outcome must affect social well-being and the economy. In Fig. 2.2 (shown previously), one could get the impression that the environment only affects certain aspects of society or that the economy can operate separately from the environment.

A useful critique of both "strong" and "weak" sustainability proponents is provided by Ayres (2007). In his view, while the mathematics of Solow's argument are "impeccable," the underlying assumptions, or what Ayres calls "the physics," are not. Ayres (2007) believes that the proponents of "strong" sustainability are right to point out the relevance of entropy law, the second law of thermodynamics, and the impossibility of perpetual motion machines; however, they are wrong to assert that human civilization is totally dependent on a finite stock of high quality (low entropy) resources stored in the earth's crust. "The fact that much of our industrial base currently utilizes fossil fuels and high-quality metal ores is merely due to the ready availability of these resources at low cost. It does not follow from the entropy law that there are not substitutes" (Ayres 2007, p. 116). Nonetheless, Ayres (2007, p. 126) concludes by saying that: "I have to reiterate that, while there is plenty of room for substitution and some possibility of major breakthroughs (e.g., in manufacturing room temperature super-conductors or carbon nanotubes) the pessimists—those who espouse the notion of "strong sustainability" appear to be closer to the truth than the optimists who believe in more or less unlimited substitution possibilities."

In summary, the basic distinction between "weak" and "strong" sustainability has important implications as to whether environmental systems and resources should be kept intact by themselves or if the environment can decline as long as the overall value of society's economic capital is kept intact. The choice of a "weak" or "strong" perspective could have considerable consequences for how environmental sustainability is defined, measured, and verified. From a decision-making perspective, the adoption of a "weak" or "strong" approach will have important implications for the type of tools that can be used to support decisions. For example, tools such as cost-benefit analysis that are compatible with "weak" sustainability may be rejected from a "strong" sustainability perspective on the grounds that the environmental costs being accounted for run against the principle of maintaining the stock of natural capital (Marsden et al. 2010).

While the possible frameworks through which sustainable development can be considered present a range of formulations, the general principles that inform these frameworks remain the same. We identify the following principles—adapted from Zietsman et al. (2011)—that have emerged from the sustainability literature and reflect the international perspective of sustainable development discussed in the previous section.

*Sustainability entails meeting human needs for the present and future, while:*

- *Preserving environmental and ecological systems*
- *Improving quality of life*

- *Promoting economic development that includes the creation of meaningful and well-paid jobs*
- *Ensuring equity between and among population groups and over generations*

What is not listed in these principles is governance that promotes peace and development, since it is assumed that this is a necessary condition for society to be able to seriously address sustainable development.

**Discussion Topics**

- Create a list of the main principles of sustainable development discussed in this chapter. Do you think proponents of "weak" sustainability would place more value on certain principles? How might supporters of "strong" sustainability value the principles?
- Think carefully about your own perspective on the purpose of development. Does your perspective align with either the "weak" or "strong" form of sustainability? Why have you adopted this position—e.g., is your position influenced by your education, professional experience, etc.? If so, what are the implications of this for advancing a weak or strong sustainability agenda?

## 2.4 Measuring Sustainable Development

The term *development* implies a continual process of change. *Sustainable* development, therefore, describes a process of change that promotes the principles of sustainability (described previously). The only way of knowing whether progress is being made toward sustainable development is to measure how we are doing based on existing and prior performance and to use this information to consider what change is likely, under different development scenarios/strategies, in the future. This action requires the use of indicators to quantify the key parameters that define sustainable development. Thus, indicators and performance measures (or targets) are paramount to any attempt to implement a sustainable development agenda.[24]

> We need many indicators because we have many different purposes—but there may be over-arching purposes that transcend nations and cultures, and therefore there may be overarching indicators.
>
> We need many indicators because we have many worldviews—but indicators may help narrow the differences between worldviews (Meadows 1998, p. viii).

At a basic level, the problem of sustainable development can be measured using indicators that capture rates/flows, stocks/conditions, and feedback (Sterman 2000). Such information can inform a society/government of how its actions might be

[24] The concepts of an indicator and performance measure are discussed in detail in Chap. 6.

beneficial and/or harmful/unsustainable, enabling adjustments to be made to avoid serious problems and maintain overall societal well-being.

The discourse on indicators of sustainable development is fueled by the fact that different knowledge domains (such as economics, ecology, sociology, and psychology) view sustainable development and its indicators differently (Simon 2003). Similarly, different societies and cultures place different values on what is deemed acceptable in an environmental, social, and economic sense. Further, uncertainty relating to causal chain mechanisms and gaps in information, and differences between how the public and experts perceive information, all combine to make the task of defining, measuring, and responding to perceived problems highly complex (Reiner 2002). It therefore seems unlikely that there will be one golden set of sustainability indicators that are applicable, or acceptable, to all nations and communities. One way to address this problem is to develop overarching frameworks that can guide indicator development using a "fitness-for-purpose" approach—i.e., "using different indicator sets for different purposes. Although, … *different* does not mean *unconnected* or *inconsistent*" (Levett 1998, p. 291). This general approach to the selection and application of indicators is adopted in this book and is expanded on in Chap. 6.

The creation of indicators of sustainable development can be placed at the end of a long history of indicator development that emerged during the twentieth century (Innes 1990; Hodge et al. 1999; Hodge 1995, 1997). With the emergence of sustainable development during the 1970s/1980s came the need for more holistic indicators that were capable of measuring progress at a system—rather than a domain/sector—level (Hodge et al. 1999; Hodge 1997). *Our Common Future* laid the foundation for these indicators by arguing that economic measures alone are an inadequate measure of social well-being (WCED 1987). It called for the creation of an overarching framework to integrate economic, environmental, and social concerns relating to human development. This call was later reinforced at the 1992 Rio Conference by "Agenda 21" (Chapter 40), at the 2002 Johannesburg Summit by the "Johannesburg Plan of Implementation" (Chapter X), and at the 2012 Rio+20 Conference where goals, targets, and indicators were seen as essential to "measuring and accelerating progress" (UN 2012, p. 21). The intent of Agenda 21 was to encourage governments, as well as international governmental and non-governmental organizations, to develop a series of indicators for sustainable development that would form the building blocks for decision-making at all levels. Emphasis was placed on *harmonizing* the indicators across geographic levels and on creating a set of indicators at the international level that would be made widely available and kept up to date.

From these initial calls for better ways to measure progress toward sustainable development, a wide variety of indexes/indicator frameworks have emerged. One of the most comprehensive lists of indicator initiatives relating to sustainable development can be found in the International Institute for Sustainable

Development's (IISD's) *Compendium of Sustainable Development Indicator Initiatives*.[25] In 2015, the compendium listed some 895 initiatives that range in scale from international to community-focused indicator projects.

At the international level, the MDGs and their supporting indicators provide a framework against which to assess progress toward sustainable development. In 2015, the original set of eight goals is set to be expanded (see Table 2.1 discussed previously) with the launch of the post-2015 development agenda. The final set of goals and indicators selected are likely to "frame" international and national development efforts for the coming decade. Thus, any national, regional, or local effort to measure progress toward sustainable development is likely to be directly or indirectly influenced by the post-2015 development agenda.

In this book, we explore the critical features of indicators (Chap. 6) and indicator frameworks (Chap. 7) and how they can be used to support decision-making for sustainable transportation (Chaps. 8–11). The purpose of Chaps. 6 and 7 is to provide basic/foundational knowledge that can be applied in the development of indicator frameworks. The case study chapters (Chaps. 8–11) then shed some light on how indicators are used in practice. While macro indicator systems such as the post-2015 agenda provide useful national indicators of progress toward sustainable development, at the state and local level, the selection and use of indicators are likely to be driven by agency priorities and the need to cater to organizational, political, and geographic/system realities. Thus, having the knowledge to develop indicator frameworks that can respond to contextual factors while attempting to make connections with theory (e.g., weak vs. strong sustainability) and overarching frameworks (such as the post-2015 agenda) is likely to be more important than having access to lists of indicators. Put differently, the real challenge is to ensure that the indicators selected align with an indicator application area and are representative, practical, and context-specific—i.e., they are *embedded* within an organization culture (see Chap. 6 and the case study in Chap. 10).

## 2.5 The Importance of a Holistic and Integrative Approach to Sustainable Development

Sustainable development requires a holistic and integrative approach to the design of policies and initiatives for its achievement in order to capture the broad array of environmental, social, and economic development issues that need to be considered. These issues tend to resist easy classification and cut across areas of government and economic activity, promoting the need for an integrative approach to addressing them.

[25] See the International Institute for Sustainable Development (IISD), *Compendium of Sustainable Development Indicator Initiatives*, http://www.iisd.org/measure/compendium/ (accessed on April 19, 2015).

Dernbach (2003, p. 250) goes as far as saying that "integrated decision-making is the foundational principle of sustainable development. . . . Thus, sustainable development requires that fragmentation in decision-making be eliminated." He identifies four areas where integrated decision-making could occur (Dernbach 2003, pp. 259–282), which include:

1. **Selection of the decision-making process**—e.g., a procedural or substantive mechanism could be used to either consider or achieve desired objectives, respectively;
2. **Scope of the decision-making process**—e.g., decisions could be integrated around a resource, issue, activity, or geographic place;
3. **Time horizon**—e.g., the decision-making process could integrate both short- and long-term objectives; and
4. **Selection of an implementation method**—e.g., legal and policy tools could be integrated to achieve a desired outcome and decision-makers could take action to overcome horizontal/vertical integration barriers to decision-making processes.

An integrative, trans-disciplinary approach is also required to overcome the fragmentation and inadequacy of the knowledge base that leads to the creation of single purpose or narrowly fashioned solutions to complex problems.

Figure 2.5 attempts to provide a visual representation of how existing government structures (or activity areas—the rings) can independently focus on the main sustainable development challenges (the wedges). Thus, a transportation agency could focus on climate change independently from agencies that are, for example, responsible for the environment, energy, or agriculture. The intent of the diagram is to reveal the need for a holistic and integrative approach to sustainable development. The diagram is illustrative of the general activity areas of government and the sustainability challenges facing society and can be adjusted to be more relevant to a specific level of government or geographic region.

Four critical environmental concerns related to sustainable development are highlighted by Fig. 2.5 (Ashford and Hall 2011).[26] In addition, it captures important

[26] Over the past 40 years, the environmental factors that underlie the concern for sustainable development incorporated—to varying degrees and at different times—what can now be identified as four different environmental concerns (Ashford and Hall 2011). First is the disruption of ecosystems and loss of biological diversity and the indirect effects these have on human health and well-being. The second concern relates to the world's finite resources and energy supplies, and asks the question of whether there are sufficient resources to fuel the economy in its current form. A corollary concern is what will the environmental impact be of using a significant proportion of the existing resources? The third concern is that toxic pollution directly affects human health and the health of other species. The final concern is that greenhouse gases from anthropocentric (human-driven) sources are leading to a disruption of the global climate. The first, third, and fourth environmental concerns are connected with the unintended effects of human development/growth, while the second deals with increasing shortages of resources needed to fuel development/growth.

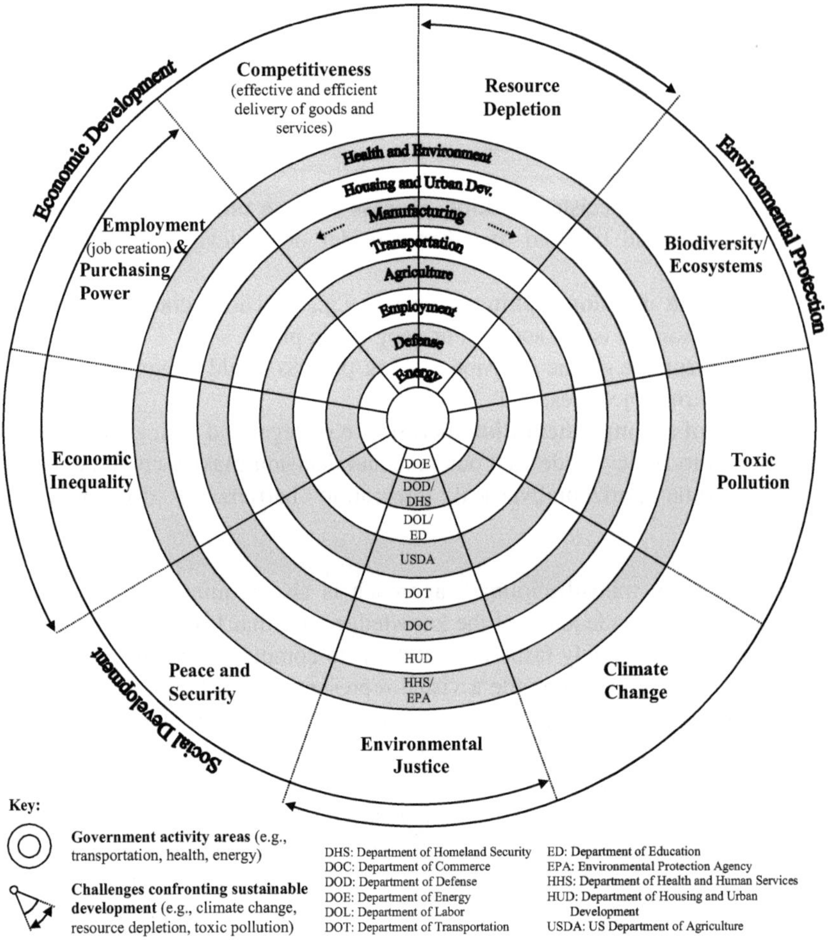


**Fig. 2.5** Government activity areas and challenges confronting sustainable development. *Source*: Adapted from Ashford and Hall (2011)

social concerns such as the need for peace, security, and equality, both in terms of environmental justice and income. Employment is also placed alongside these concerns given its critical role in raising purchasing power and providing sufficient income to make essential goods and services accessible to all. A "competitiveness" wedge is included to account for the economic challenge of delivering effective and efficient goods and services. The rationale is that competitiveness is a critical factor of economic growth and one that is closely related to technological innovation. Further, focusing on the competitive delivery of goods and services is also more likely to lead to long-term economic benefits than a focus on short-term economic growth.

The three arrows that follow the circumference of the outer circle in Fig. 2.5 identify which challenges relate to environmental protection, social development, and economic development. The rings in the figure represent several government activity areas—that is, those areas where government provides basic goods and services. There is no hierarchy to the activity areas shown.

Figure 2.5 shows that focusing on, for instance, climate change as *the* major challenge confronting sustainable development ignores the importance of other environmental, social, and economic challenges. In addition, single-purpose policies designed to confront climate change may inadvertently worsen problems in other areas. For example, increasing the percentage of ethanol in gasoline to reduce $CO_2$ emissions might lead to the production of additional toxic air pollutants and to an increased use of pesticides, worsening the toxics problem, as well as raising the cost of food and actually increasing greenhouse emissions through additional land use (Searchinger et al. 2008). Thus, a major advance in confronting sustainable development would be the integration of government decision-making to address environmental, social, and economic problems that are not constrained by instuitional missions or the fragmentation of activities within government agencies (Hall 2006). Such an endeavor is undoubtedly very complex to deliver. One of the key aims of this book is to show how it is possible to achieve more integrated decisions and to demonstrate why information, indicators, and the decision-making frameworks that they are used in are the critical glue which make more integrated decisions possible.

**Discussion Topics**

- Select one of the sustainability challenges shown in Fig. 2.5. How does each government activity area address, or not, this challenge? What policy connections exist between the government agencies addressing the challenge? If evidence of connections can be found, are the government agencies *coordinating* their independent activities or attempting to *integrate* their activities by working closely together toward a common goal/objective?
- How similar/consistent are the various policy responses to each of the sustainability challenges? For example, are the policy approaches focused on increasing employment and earning capacity consistent with policies for climate change or biodiversity/ecosystem health?

## 2.6 Conclusions

This chapter provides an overview of the concept of sustainable development from both a historical and conceptual perspective. The historical perspective reveals the compromises that were made when crafting key foundational texts on sustainable development such as *Our Common Future* and the Rio Declaration. It also highlights what form of development is being endorsed when the Brundtland-UNCED-Johannesburg-Rio+20-Post-2015 formulation of sustainable development is invoked—i.e., one of technological optimism and market liberalization. In contrast, the conceptual perspective removes the historical and political dimensions and presents two different ways in which progress toward sustainability could be achieved. While the "weak" and "strong" forms of sustainability are somewhat academic, they help frame a continuum between an environmentally oriented business-as-usual approach to development and a radical reformulation of economic activity to keep it within macroecological limits. Regardless of where one stands on this continuum, policies and initiatives designed to promote sustainable development need to stem from a holistic and integrative process in an effort to overcome shortfalls that occur in the creation of single-purposed or narrowly fashioned solutions to complex problems.

## References

Ashford, N. A., & Hall, R. P. (2011). *Technology, globalization, and sustainable development: Transforming the industrial state*. New Haven, CT: Yale University Press.

Ayres, R. U. (1978). Application of physical principles to economics. In R. U. Ayres (Ed.), *Resources, environment and economics: Applications of the materials/energy balance principle* (pp. 37–71). New York: Wiley.

Ayres, R. U. (1997). Environmental market failures: Are there any local market-based corrective mechanisms for global problems? *Mitigation and Adaptation Strategies for Global Change, 1*, 289–309.

Ayres, R. U. (2007). On the practical limits to substitution. *Ecological Economics, 61*, 115–128.

Bell, M. L., & Davis, D. L. (2001). Reassessment of the lethal London fog of 1952: Novel indicators of acute and chronic consequences of acute exposure to air pollution. *Environmental Health Perspectives, 109*, 389–394.

Beltratti, A., Chichilnisky, G., & Heal, G. (1995). Sustainable growth and the green golden rule. In I. Goldin & L. A. Winters (Eds.), *The economics of sustainable development* (pp. 147–172). Cambridge: Cambridge University Press.

Blumm, M. C. (1990). Symposium on NEPA at twenty: The past, present and future of the National Environmental Policy Act: Introduction: The National Environmental Policy Act at twenty: A preface. *Environmental Law, 20*(3), 447–483.

Brandt, W. (1980). *North-South: A programme for survival, report by the Independent Commission on International Development Issues*. Cambridge: MIT Press.

Brodmann, U., & Spillmann, W. (2000). *Verkehr—Umwelt—Nachhaltigkeit: Standortbestimmung und Perspektiven. Teilsynthese des NFP 41 aus Sicht der Umweltpolitik mit Schwerpunkt Modul C*. Berne: Nationales Forschungsprogramm.

Brown, J. H., Burnside, W. R., Davidson, A. D., DeLong, J. P., Dunn, W. C., Hamilton, M. J., et al. (2012). Energetic limits to economic growth. *Bioscience, 61*, 19–26.

Burger, J. R., Allen, C. D., Brown, J. H., Burnside, W. R., Davidson, A. D., Fristoe, T. S., et al. (2012). The macroecology of sustainability. *PLoS Biology, 10*, 1–7.

Caldwell, L. K., & Weiland, P. S. (1996). *International environmental policy: From the twentieth to the twenty-first century* (3rd ed.). Durham: Duke University Press.

Centre for Sustainable Transportation (CST). (1997). *Definition and vision of sustainable transportation*. Ontario: The Centre for Sustainable Transportation.

Clapp, J., & Dauvergne, P. (2005). *Paths to a green world: The political economy of the global environment*. Cambridge: MIT Press.

Cline, W. R. (1992). *Optimal carbon emissions over time: Experiments with the Nordhaus DICE model*. Washington, DC: Institute for International Economics.

Costanza, R., & Daly, H. E. (1992). Natural capital and sustainable development. *Conservation Biology, 6*, 37–45.

Council on Environmental Quality (CEQ). (1980). *The global 2000 report to the President: Entering the twenty-first century* (Vol. 1). Washington, DC: U.S. Government Printing Office.

Czech, B. (2003). Technological progress and biodiversity conservation: A dollar spent a dollar burned. *Conservation Biology, 17*, 1455–1457.

Czech, B., & Daly, H. E. (2004). In my opinion: The steady state economy—What it is, entails, and connotes. *Wildlife Society Bulletin, 32*, 598–605.

D'Alisa, G., Demaria, F., & Kallis, G. (Eds.). (2014). *Degrowth: A vocabulary for a new era*. New York: Routledge.

Daly, H. E. (1991a). Elements of environmental macroeconomics. In R. Costanza & L. Wainger (Eds.), Ecological economics: The science and management of sustainability (pp. 32–46). New York: Columbia University Press.

Daly, H. E. (1991b). *Steady-state economics*. Washington, DC: Island Press.

Daly, H. E. (1996). *Beyond growth: The economics of sustainable development*. Boston: Beacon Press.

Daly, H. E. (2008, April 24). A steady-state economy. A failed growth economy and a steady-state economy are not the same thing; they are the very different alternatives we face. Paper written for the Sustainable Development Commission, UK.

Davis, D. L., Bell, M. L., & Fletcher, T. (2002). A look back at the London smog of 1952 and the half century since. *Environmental Health Perspectives, 110*, a734–a735.

Dernbach, J. C. (1998). Sustainable development as a framework for national governance. *Case Western Reserve Law Review, 49*, 1–103.

Dernbach, J. C. (2003). Achieving sustainable development: The centrality and multiple facets of integrated decisionmaking. *Indiana Journal of Global Legal Studies, 10*, 247–285.

Dernbach, J. C. (2004). Making sustainable development happen: From Johannesburg to Albany. *Albany Law Environmental Outlook, 8*, 173–186.

Dixon, J. A., & Fallon, L. A. (1989). The concept of sustainability: Origins, extensions, and usefulness for policy. *Society and Natural Resources, 2*, 73–84.

Ehrlich, P. R. (1968). *The population bomb*. New York: Ballantine Books.

Emmelin, L. (1972). The Stockholm conferences. *Ambio, 1*, 135–140.

Engfeldt, L. G. (2002). The road from Stockholm to Johannesburg. *United Nations Chronicle, 39* (3), 14–17.

Georgescu-Roegen, N. (1993). Energy and economic myths. In D. E. Herman & T. N. Kenneth (Eds.), *Valuing the earth: Economics, ecology, ethics* (pp. 89–112). Cambridge: MIT Press.

Grubb, M., Koch, M., Munson, A., Sullivan, F., & Thomson, K. (1993). *The "Earth Summit" agreements: A guide and assessment: An analysis of the Rio '92 UN conference on environment and development*. London: Earthscan.

Hall, R. P. (2006). *Understanding and applying the concept of sustainable development to transportation planning and decision-making in the U.S.* Engineering Systems Division, Massachusetts Institute of Technology, Cambridge. http://esd.mit.edu/students/esdphd/dissertations/hall_ralph.pdf

Hall, R. P., & Ashford, N. A. (2012). Primer on the Emergence and Evolution of Sustainable Development (1951 to 2012). Virginia Tech, VA. Available at: https://ralphphall.files.wordpress.com/2012/11/primer_sustdev_2012-11-26.pdf

Hardin, G. (1968). The tragedy of the commons. *Science, 162*, 1243–1248.

Hartwick, J. M. (1977). Intergenerational equity and the investing of rents from exhaustible resources. *American Economic Review, 67*, 972–974.

Hartwick, J. M. (1978a). Investing returns from depleting renewable resource stock and intergenerational equity. *Economics Letters, 1*, 85–88.

Hartwick, J. M. (1978b). Substitution among exhaustible resources and intergenerational equity. *Review of Economic Studies, 45*, 347–354.

Hediger, W. (1999). Reconciling "weak" and "strong" sustainability. *International Journal of Social Economics, 26*, 1120–1143.

Hens, L., & Nath, B. (2005). The Johannesburg conference. In L. Hens & B. Nath (Eds.), *The world summit on sustainable development: The Johannesburg conference* (pp. 1–39). Dordrecht: Springer.

Hodge, R. A. (1995). *Assessing progress toward sustainability: Development of a systemic framework and reporting structure*. School of Urban Planning, McGill University, Montreal.

Hodge, T. (1997). Toward a conceptual framework for assessing progress toward sustainability. *Social Indicators Research, 40*, 5–98.

Hodge, T., Hardi, P., & Bell, D. V. J. (1999, May 6–9). Seeing change through the lens of sustainability. In *Beyond delusion: Science and policy dialogue on designing effective indicators of sustainable development*, Costa Rica. The International Institute for Sustainable Development.

Holden, E., Linnerud, K., & Banister, D. (2013). Sustainable passenger transport: Back to Brundtland. *Transportation Research Part A, 54*, 67–77.

Holdren, J. P., Daily, C., & Ehrlich, P. R. (1995). The meaning of sustainability: Biogeophysical aspects. In M. Munasinghe & W. Shearer (Eds.), *Defining and measuring sustainability: The biogeophysical foundations*. Washington, DC: World Bank. Distributed for the United Nations University.

Holmberg, J., & Sandbrook, R. (1992). Sustainable development: What is to be done? In J. Holmberg (Ed.), *Making sustainable development* (pp. 19–38). Washington, DC: Island Press.

Hynes, P. H. (1989). *The recurring silent spring*. New York: Pergamon Press.

Innes, J. E. (1990). *Knowledge and public policy: The search for meaningful indicators*. New Brunswick, NJ: Transaction.

International Union for Conservation and Nature and Natural Resources (IUCN). (2004). The IUCN Programme 2005-2008. Many voices, one earth. International Union for Conservation and Nature and Natural Resources (IUCN)—The World Conservation Union, Gland. http://www.iucn.org/ourwork/ppet/index.htm

International Union for Conservation and Nature and Natural Resources (IUCN), United Nations Environment Programme (UNEP) & World Wildlife Fund (WWF). (1980). *World conservation strategy*. Gland: IUCN.

Jay, S., Jones, C., Slinn, P., & Wood, C. (2007). Environmental impact assessment: Retrospect and prospect. *Environmental Impact Assessment Review, 27*, 287–300.

Korten, D. C. (1991, April 15). The UN conference on environment and development: Unasked questions. PCD Forum, Column #12.

Lele, S. M. (1991). Sustainable development: A critical review. *World Development, 19*, 607–621.

Levett, R. (1998). Sustainability indicators—Integrating quality of life and environmental protection. *Journal of the Royal Statistical Society, 162*, 291–302.

Marsden, G., Kimble, M., Nellthorp, J., & Kelly, C. (2010). Sustainability assessment: The definition deficit. *International Journal of Sustainable Transportation, 4*, 189–211.

Meadows, D. H. (1998). *Indicators and information systems for sustainable development. A report to the Balaton Group*. Hartland: The Sustainability Institute.

Meadows, D. H., Meadows, D. L., Randers, J., & Behrens, W. W. (1972). *The limits to growth*. New York: Potomac.

Mendelsohn, R. O., Morrison, W., Schlesinger, M. E., & Andronova, N. A. (1998). Country-specific market impacts of climate change. *Climatic Change, 54*, 553–569.

Neumayer, E. (2003). *Weak versus strong sustainability: Exploring the limits of two opposing paradigms*. Cheltenham: Edward Elgar.

Nordhaus, W. D., & Boyer, J. G. (2000). *Warming the world: Economic models of global warming*. Cambridge: MIT Press.

Norgaard, R. B. (1988). Sustainable development: A co-evolutionary view. *Futures, 20*, 606–620.

Ocampo, J. A., Cosbey, A., & Khor, M. (2011). *Report by Panel of Experts on the Transition to a Green Economy: Benefits, challenges, and risks from a sustainable development perspective*. New York: United Nations Department of Economic and Social Affairs (UNDESA), United Nations Environment Programme (UNEP), and United Nations Conference on Trade and Development (UNCTAD).

Organisation for Economic Cooperation and Development (OECD). (2011a). *Tools for delivering on green growth*. Paris: Organisation for Economic Cooperation and Development.

Organisation for Economic Cooperation and Development (OECD). (2011b). *Towards green growth*. Paris: Organisation for Economic Cooperation and Development.

Organisation for Economic Cooperation and Development (OECD). (2011c). *Towards green growth: Monitoring progress—OECD indicators*. Paris: Organisation for Economic Cooperation and Development.

Pearson, L. B. (1969). *Partners in development*. New York: Praeger.

Redclift, M. (1991). The multiple dimensions of sustainable development. *Geography, 76*, 36–42.

Redclift, M. (1996). *Wasted: Counting the costs of global consumption*. London: Earthscan.

Reiner, D. M. (2002). *Causal reasoning and goal setting: A comparative study of air pollution, antitrust and climate change policies*. Department of Political Science, Massachusetts Institute of Technology, Cambridge, 440.

Runnalls, D. (2008). *Our common inaction: Meeting the call for institutional change environment magazine, 50*, 18–28.

Sachs, W. (2001). *Rio+10 and the North-South divide*. Berlin: Heinrich Boll Foundation.

Schön, D. A., & Rein, M. (1994). *Frame reflection: Toward the resolution of intractable policy controversies*. New York, NY: Basic Books.

Searchinger, T., Heimlich, R., Houghton, R. A., Dong, F., Elobeid, A., Fabiosa, J., et al. (2008). Use of U.S. croplands for fuels increases greenhouse gases through emissions from land-use change. *Science, 319*, 1238–1240.

Shiva, V. (1992). Recovering the real meaning of sustainability. In D. E. Cooper & J. A. Palmer (Eds.), *The environmental question* (pp. 187–193). New York: Routledge.

Simon, S. (2003). Sustainability indicators. In E. Neumeyer (Ed.), *Web encyclopedia of ecological economics*, http://www.ecoeco.org/publica/encyc_entries/SustIndicator.pdf

Smith, M. H., Hargroves, K. C., & Cheryl, D. (2010). *Cents and sustainability: Securing our common future by decoupling economic growth from environmental pressure*. London: Earthscan.

Solow, R. M. (1974). Intergenerational equity and exhaustible resources. *Review of Economic Studies, 41*, 29–46.

Solow, R. M. (1993). Sustainability: An economist's perspective. In R. Dorfman & N. S. Dorfman (Eds.), *Economics of the environment: Selected readings* (pp. 179–187). New York: W. W. Norton.

South Centre. (2002). The South, the North and sustainable development: The continuity of basic issues. In South Centre (Ed.), *The South and sustainable development conundrum: From Stockholm 1972 to Rio 1992 to Johannesburg 2002* (pp. 4–30). Geneva: South Centre.

Sterman, J. D. (2000). *Business dynamics: Systems thinking and modeling for a complex world*. Boston: Irwin McGraw-Hill.

Stern, N. (2007). *Stern review on the economics of climate change*. Cambridge: Cambridge University Press.

Strong, M. F. (1972). The Stockholm conference—Where science and politics meet. *Ambio, 1*, 73–78.

Strong, M. F. (2003). Global sustainable development. In S. Vertovec & D. A. Posey (Eds.), *Globalization, globalism, environments, and environmentalism: Consciousness of connections* (pp. 103–122). Oxford: Oxford University Press.

Sustainable Development Solutions Network (SDSN). (2014, April 11). *Proposed sustainable development goals (SDGs) and targets*. New York: Sustainable Development Solutions Network.

The Ecologist. (1993). *Whose common future? Reclaiming the commons*. London: Earthscan.

Toman, M. A. (1992). The difficulty in defining sustainability. *Resources, 106*, 3–6.

United Nations. (1972). *Development and environment: Report and working papers of a panel of experts convened by the Secretary-General of the United Nations conference on the human environment (Founex, Switzerland, June 4-12, 1971)*. Paris: Mouton.

United Nations (UN). (1993a). *Report of the United Nations conference on environment and development, Volume I, Resolutions adopted by the conference*. New York: United Nations.

United Nations (UN). (1993b). *Report of the United Nations conference on environment and development, Volume II, Proceedings of the conference*. New York: United Nations.

United Nations (UN). (1993c). *Report of the United Nations conference on environment and development, Volume III, Statements made by heads of state or government at the summit segment of the conference*. New York: United Nations.

United Nations (UN). (2011). *Objective and themes of the United Nations conference on sustainable development. Report of the Secretary-General. 22 December 2010* (A/CONF.216/PC/7). New York: United Nations.

United Nations (UN). (2012). *The future we want. A/CONF.216/L.1*. New York: United Nations.

United Nations (UN). (2013). *A new global partnership: Eradicate poverty and transform economies through sustainable development. The report of the high-level panel of eminent persons on the post-2015 development agenda*. New York: United Nations.

United Nations (UN). (2014). *Open working group proposal for sustainable development goals. Full report of the open working group of the general assembly on sustainable development goals is issued as document A/68/970*. New York: United Nations.

United Nations Commission on Sustainable Development (UNCSD) & United Nations Conference on Trade and Development (UNCTAD). (2011). Trade and green economy. In *Rio+20 issues briefs*. New York: UNCSD Secretariat and UNCTAD.

United Nations Economic and Social Council (ECOSOC). (2002). *Follow-up to the World summit on sustainable development and the future role of the Commission on Sustainable Development: The implementation track, Report of the Secretary-General*. New York: United Nations.

United Nations Environment Programme (UNEP). (1982a). *The environment in 1982: Retrospect and prospect*. Nairobi: United Nations Environment Programme (UNEP).

United Nations Environment Programme (UNEP). (1982b). *The state of the environment 1972-1982*. Nairobi: United Nations Environment Programme.

United Nations Environment Programme (UNEP). (1982c). *The world environment 1972-1982: A report by the United Nations Environment Programme*. Dublin: Tycooly.

United Nations Environment Programme (UNEP). (2011). Towards a green economy: Pathways to sustainable development and poverty eradication. United Nations Environment Programme. www.unep.org/greeneconomy

World Commission on Environment and Development (WCED). (1987). *Our common future*. Oxford: Oxford University Press.

Zietsman, J., Ramani, T., Potter, J., DeFlorio, J., & Reeder, V. (2011). *A guidebook for sustainability performance measurement for transportation agencies* (NCHRP report 708). Washington, DC: Transportation Research Board of the National Academies.

## Additional Readings

Ayres, R. U. (2006). Turning point: The end of exponential growth? *Technological Forecasting and Social Change, 73*, 1188–1203.
Dryzek, J. S. (1997). *The politics of the earth: Environmental discourses*. New York: Oxford University Press.
McCormick, J. (1995). *The global environmental movement* (2nd ed.). New York: Wiley.
UNEP. (2011). Towards a green economy: Pathways to sustainable development and poverty eradication. United Nations Environment Programme (UNEP). http://www.unep.org/greeneconomy/

# Planning for Transportation

# 3

## 3.1 Purpose and Content

Transportation is fundamental to the development of society. It provides opportunities to interact with others, moves the goods we need, and supports a vibrant economy. This chapter introduces the demand for transportation and explores key trends and growth forecasts. These underline the ongoing importance of transportation to social progress and the significant challenge that lies ahead in planning transportation in the face of growing population, rising incomes, and technological change.

The chapter reviews several conceptualizations of the transportation system in order to fully understand how it intersects with discussions centered around sustainable development. This enables us to identify the boundaries around what we mean by "transportation." These conceptualizations inform a broad definition of a transportation system which:

1. Is a critical component of a broader economic system which supports business and social development;
2. Is an open system, which requires natural and man-made inputs and produces outputs which impact on the environment;
3. Is part of a social system that shapes and is shaped by that social system, including other policy areas;
4. Comprises a series of physical subsystems each of which has a range of physical and operational components and which are organized through formal and informal conventions; and
5. Is typically a fragmented series of partly connected yet partly competing subsystems with complex and varying governance arrangements.

As well as being fundamental to economic and social progress, the use of the transportation system gives rise to major negative environmental, social, and economic impacts such as road accidents, toxic air pollution, local environmental


H. Gudmundsson et al., *Sustainable Transportation*, Springer Texts in Business and Economics, DOI 10.1007/978-3-662-46924-8_3

disturbance, land-take, and congestion. These key impacts are reviewed and the chapter concludes by identifying the synergy between the positive and negative impacts of transportation and the core notions of sustainable development. These concepts are then explored further in Chap. 4.

## 3.2 Mobility and Travel Demand

The movement of people, goods, ideas, and information has been fundamental to the development of society and remains so today. The development of transportation technologies over the centuries has revolutionized the scale over which these movements have occurred (Geels 2002). Shipping, canals, bicycles, railways, motor vehicles, and aircraft have each in turn had a significant impact on economic and social progress (Cowie 2010). These changes in the means and speed of movement have also been driven by and have driven changes to the systems of production and consumption that we see today. Headicar (2009, p. 70), for example, identifies the development of the railways as being critical to an expansion of suburbanization in Britain, but it also generated changes in social conventions such as the adoption of Greenwich Mean Time as the standard time for the whole country, the growth of national daily newspapers (which could be distributed from London), and mass leisure excursions to the seaside.

One definition of the purpose of the transportation system is to enable "people or freight to move between origins and destinations" (Hall 2006, p. 448). The demand for passenger transportation is generally considered to be derived from people's desires to take part in activities (SACTRA 1999). These activities can be diverse (as shown from example statistics from the UK and the USA in Table 3.1). Over time, mass motorization and the expansion of the road system have afforded more opportunities to travel faster and further and to take part in more activities (Metz 2010).

In general, transportation is considered to be an intermediate good that exists to serve the needs/desires of the user/customer. The conventional view in transportation economics is to regard the journey as a cost which travelers incur (by means of travel time and out of pocket expenses). Journeys are undertaken when the value of participating in the activity exceeds the cost to the traveler of the journey (Cowie 2010). The main drivers of demand are identified as:

- Disposable income;
- The price of travel (and the availability of alternatives);
- Journey times; and
- Population (Mallard and Glaister 2008).

More recently, the "Mobilities" literature has extended the discussion on the extent to which all types of journey can be reduced to a time-based economic cost (Cresswell 2006). Watts and Urry (2008) argue that activities or tasks are embedded in journeys, and these form a part of the positive function of journeys. They cite, for

**Table 3.1** Journey purposes and proportions in the UK and the USA

| Journey purpose | % UK (2009) | % US (2009) |
|---|---|---|
| Commuting, business, and education—a reflection of the role of transportation in supporting access to training, participation in the labor market, and interaction between businesses | 29 | 28 |
| Shopping and personal business—consumption-related trips related to fulfilling basic needs such as access to food and health as well as to purchase goods | 40 | 43 |
| Visiting friends and relatives—arising from some form of human need for social interaction and to remain connected | 16 | 27 |
| Other leisure—such as tourism and getting out and about | 15 | 2 |

There is not a direct mapping of activities in the UK and the USA. The categories listed are those used in the UK. The main variations are that in the US education includes church-related trips, personal business includes personal and family errands, visiting friends and relatives is social and recreational, with the final category listed as "other."
*Sources*: DfT, Statistics, http://www.dft.gov.uk/statistics/ and FHWA, National Household Travel Survey, 2009. http://nhts.ornl.gov/2009/pub/stt.pdf (accessed on August 8, 2012)

example, the benefit that some people derive from having a separation between home and work and the time that people spend on rail journeys writing e-mails, talking on mobile phones, and reading. Virtual mobility, now afforded through high speed internet access and portable mobile internet devices, makes it common place in many parts of the world to order goods from another country, talk to business colleagues or relatives via an internet video link, and to network with friends and communities in ways which were prohibitively difficult even 20 years ago (Lyons 2009). The movement of ideas does not require physical mobility. While the nature and meaning of travel is constantly evolving, it is an undeniably important component in the bringing together of people and activities.

As nations develop economically, citizens have increased disposable income. At the same time, technological progress has significantly lowered the costs of motorized travel. Schafer and Victor (2000) forecast that by 2050, the amount that the average world citizen would travel per year would more than double from around 4400 km to around 10,500 km/year, as far as the average West European traveled in 1990. When these factors are taken together with the significant growth in world population, this presents a powerful cocktail for growth in the demand for travel. Dargay et al. (2007) and Sperling and Gordon (2009) both estimate that the total world vehicle stock will have increased from around 150 million in 1960 to over two billion cars by 2030 or before.[1] The largest areas for growth are non-OECD countries which will comprise more than half of the total world vehicle stock. Brazil, China, and India will form the largest proportions of this growth, fueled by increases in income levels and personal vehicle ownership. Figure 3.1 shows one set of projections of personal transportation activity based on work by

[1] This includes motorcycles, scooters, and powered three wheelers.

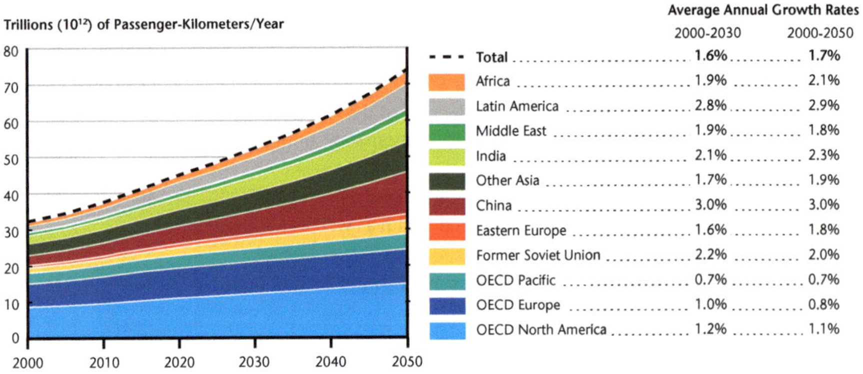


**Fig. 3.1** WBCSD forecast personal transportation activity by world region. *Source*: WBCSD (2004, p. 30)

the World Business Council for Sustainable Development (WBCSD 2004). The implications of such growth, even if tempered by the global financial crisis, are significant for congestion, energy use, and other impacts of transportation.

Zhao (2014, p. 55) reviews the growth of traffic and congestion in 11 Chinese cities finding that "in Beijing, the total number of daily trips by private cars increased from 5.2 to 10.7 million during the period 2000–2010. In Shanghai, since 1995, the number of daily trips by private cars has increased on average by 29 % annually, reaching 9.8 million per day in 2009. Trips by private cars have also increased rapidly in other cities. For example, in Tianjin and Qingdao, the number of daily trips by private cars saw a fourfold increase in the 2000s." Singh (2005) suggests that in India, even allowing for technological progress, the total energy demand for personal transportation will quadruple to 4545 PJ per year over 20 years to 2021 with the average energy consumption per capita over the same period more than trebling to 3419 MJ through a shift to increased use of motorized modes, particularly the car.

While the drivers for growth in developing countries seem clear, the continuation of existing trends in developed countries is more contested. Millard-Ball and Schipper (2011) suggest that developed countries may reach a saturation of travel demand earlier than previous forecasts have estimated which may be a result of congestion, sustained high oil prices, and the diminishing marginal utility of travel beyond a certain level.[2] This conforms with Schafer et al.'s (2009) analysis which suggests that car travel is peaking in developed countries with growth occurring in higher speed modes, primarily aviation. In the UK, for example, air passenger numbers were forecast to almost double between 2007 and 2030 to levels four times higher than in 1990 (DfT 2009). Global demand stands at 2681 million passengers per year, up 50 % over the last decade (IATA 2011). Although the economic

[2] While this analysis extends well beyond the USA, the levels of car ownership per license holder in the USA today are already at 1.15 according to Sperling and Gordon (2009).

downturn has led to declines in passenger numbers from a peak in 2007 in some European markets, there remains significant potential for growth across all markets. For example, in a developed market such as the UK, the average person takes less than 1.5 international flights per year, while in the wealthier South East of England, the number is over three flights per year (DfT 2003). This difference implies that flights per year will increase with rising income.

The demand for freight transportation is driven by consumption and patterns of consumer demand and the structure of global industries and supply chains. Goods have to be moved from their place of production to the point of sale, and inputs have to be brought together at the place of production (SACTRA 1999). As highlighted in Chap. 2, one element of the sustainable development debate is the growth in consumption which accompanies rising prosperity. As consumption increases, so do freight movements. Hesse and Rodrigue (2004, p. 171) suggest that the growth in freight flows over time has been "a fundamental component of contemporary changes in economic systems at the global, regional and local scales." This reflects, in part, structural changes to the economy with new manufacturing processes and the changing supplies of cheap labor and raw materials. Rodrigue (2006, p. 510) identifies trends in "the spatial and functional fragmentation of manufacturing," which combined with reducing inventories "have led to smaller, more frequent and synchronized shipments, transforming the logistics industry, but placing intense pressures on transportation systems to support these flows." It also reflects the way logistics systems have adapted to the availability of improved road and air links which have changed feasible geographies of supply and distribution. Tian et al. (2014) explore the implications of both the growth in domestic consumption but also the shift in global production patterns on greenhouse gas emissions from freight in China. They conclude that "the total amount of GHG emissions caused by the Chinese freight transport sector reached 978 million tons in 2011, indicating an average annual growth of 74 million tons $CO_2e$ for the last decade" (*ibid.*, p. 43).

Gillingwater et al. (2003) note that globally connected manufacturing systems are increasingly reliant on just-in-time delivery, with limited inventories of materials being held at factories and the efficient processing and movement of products into and out of sites being critical to their efficient functioning. Global air freight increased from 32.3 million tons in 2001 to 45.8 million tons in 2010 (IATA 2011), despite drops resulting from the global economic downturn, and can be expected to continue to expand. Levels of maritime container traffic are also growing significantly, with volumes almost trebling between 1995 and 2008 and forecast to continue their sharp growth (GIA 2010).

Despite the clear importance of freight to society and to the transportation system, the availability of data on freight demand and elasticities of demand is patchy (McKinnon 2008). The WBCSD (2004) estimated that freight growth rates for surface transportation alone will average around 2.5 % per annum globally, with emerging economies far higher (e.g., 4.2 % for India) as shown in Fig. 3.2.

From a sustainable development perspective, transportation-enabled trade can improve the environmental and economic performance of cities, regions, and nations, through the import of goods and services which are not available locally

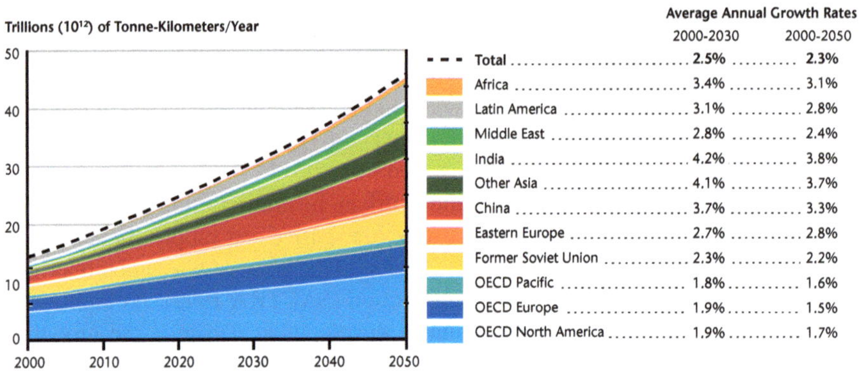


**Fig. 3.2** WBCSD forecast freight transportation activity by world region (excluding air, pipeline and waterborne). *Source*: WBCSD (2004, p. 32)

(e.g., bio-mass plants). However, the increased flows of trade also raise an important issue relating to the balance of responsibility for pollution as the burdens are typically borne in the regions in which they are created rather than where they are consumed. Thus, a macroecological perspective is needed when considering the environmental impacts of a society (Burger et al. 2012).

In summary, mobility of people and goods has been central to the development of society. Rising incomes, changing technologies, and growing populations all provide strong driving forces that suggest significant increases in mobility in the coming years. The growth will be larger in developing countries, where current per capita mobility levels are far lower than in developed countries. In the following section, we examine the transportation system and the different ways of viewing its role in providing the expected levels of future mobility, before reflecting on the evidence of the impacts, both positive and negative, of this mobility.

## Discussion Topics

- Review Millard-Ball and Schipper (2011) and look at current national transportation statistics for a developed country. To what extent is the demand for car travel saturating? If it is, then what might the broader implications of this trend be?
- List five everyday items such as food you eat and clothing you are wearing and explore where they are sourced from. What does this say about freight movements?

## 3.3 The Characteristics of Transportation Systems

There is a need to define what a transportation system is, and what it consists of, in order to explore how transportation interacts with the notions of sustainable development. The definition of a system varies according to the disciplinary lens used to view it, with fields such as planning, engineering, ecology, sociology, and economics each having something different to add to the picture (Leleur 2008). An overarching system definition is provided by Haines (2010, p. 2) who defines it as "a set of elements or components that work together in relationships for the overall objectives/vision of the whole."

Two different ways to define a transportation system can generally be found in transportation studies literature (e.g., see Van Acker and Witlox 2005): one that refers to the physical and organizational elements that produce transportation, or the "supply" of transportation, and a broader one that incorporates the interaction between demand and supply of transportation.

A classical definition of the former type is given by Marvin Manheim (1979, p. 11), who sees the transportation system as consisting of "the persons and things being transported; the vehicles in which they are conveyed; and the network of facilities through which the vehicles, passengers, and cargoes move, including terminals and transfer points as well as line—haul facilities."

Sussman (2000) sets out some of the components that are internal to the transportation system (Fig. 3.3). The specification of the internal components is particularly useful in breaking down the physical components and infrastructure that are typically seen as making up the system (the roads, railway carriages, bicycles, and traffic lights) from the functions and processes that make the system work. The operators are engaged with the day-to-day running of services and equipment, their maintenance, and their long-range planning. These are all critical functions of the transportation planning task. The operating plans are the tactical

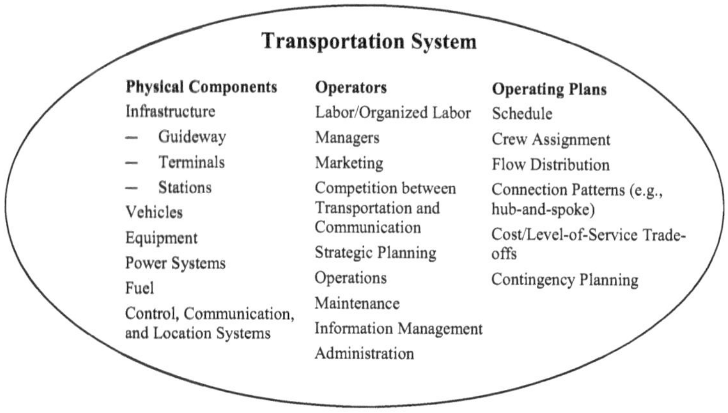


**Fig. 3.3** Internal components of the transportation system. *Source*: Adapted from Sussman (2000, pp. 11–25)

decisions which underpin the configuration of networks and the operation of services.

The second type of transportation system definitions focuses on supply and demand. A good example is provided by Cascetta et al. (2007, p. 339): "A Transportation system can be defined as the combination of elements and their interactions which produce the demand for travel within a given area and the supply of transportation services to satisfy this demand."

Section 3.2 introduced the key drivers of demand in the transportation sector. The supply of transportation can be seen as the physical infrastructure, management systems, and operational practices shown in Fig. 3.4, combined with a key aspect of supply, the cost of using the transportation system (SACTRA 1999). It is commonplace in transportation planning to consider cost as a combination of the financial costs (fares, parking fees, fuel) along with the time costs of conducting a journey (Börjesson et al. 2010, p. 41), referred to as generalized cost. SACTRA (1999, p. 41) suggests that it "can be expected that the demand for transport will be inversely related to its costs as perceived by the users."

The supply of transportation can be altered by improving existing assets (e.g., electrification of an existing rail line), renewing existing assets (e.g., road resurfacing), investing in new assets (e.g., an additional runway), removing or reallocating existing assets (e.g., pedestrianization or introducing bus lanes in place of existing road space), making better use of existing infrastructure (e.g., real-time parking information), and changing prices (e.g., congestion charging or high occupancy toll lanes). Of course, such supply changes may also lead to changes in demand as new or improved transportation options become available.

Managing the transportation system requires an understanding of the drivers of demand for travel and how they relate to the supply of transportation. Where demand exceeds supply, there will be congestion which can be observed through traffic jams, queues, and overcrowding on public transportation and delays at airports.

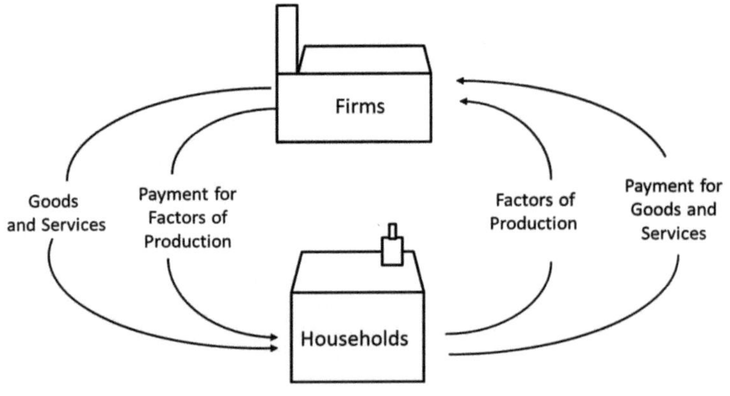


**Fig. 3.4** Simplified version of local economy (adapted from Cowie 2010)

While these definitions are undoubtedly useful, they also typically afford the transportation system an analytical position of primacy. However, given that most travel is seen as a derived demand from the need to take part in other activities, it is essential to take a broader view of what societal functions the transportation system exists to support. In addition, the transportation system is constructed and operated with natural resources and human capital, and it has impacts on humans and the natural environment. One of the purposes of this book is to explore how the principles of sustainable development can be integrated with transportation to ensure that the approaches taken are consistent with sustainable futures. To do this, a range of additional conceptualizations of the transportation system are considered below.

### 3.3.1 Transportation as Part of an Economic System

Transportation has a number of important roles as part of the broader economy. Figure 3.4 shows a simplified picture of the local economy. In this simplified representation, the system is a closed circular system, whereby households provide factors of production to firms by means of labor, capital, and raw materials and in return, they receive payment which they can use for rent and the purchase of goods which the firms produce. Gross Domestic Product can be calculated by adding up the total household income, all household expenditure, or the output of firms (Cowie 2010). The diagram can be expanded to consider a more open economy where there are imports and where loans are provided by external agencies and firms purchasing directly from each other. Further, investments in firms can promote the growth of the system, whereby more people are hired to support the expansion of economic activity, who then spend their new income on goods and services.

Transportation can be recognized as a factor in the production costs of industry. "A traditional theoretical view suggests that a transportation improvement which reduces transportation costs (through shorter journey times and lower vehicle operating costs) enables firms to sell their products more cheaply. This stimulates greater demand, so that as firms enjoy enhanced scale economies, a virtuous circle of further cost reductions and sales growth is set in motion" (SACTRA 1999, p. 30). It is estimated that transportation typically forms 5–10 % of production costs (*ibid.*).

Transportation also plays a major role with respect to the degree to which it makes markets more accessible. Improved transportation links can widen the markets which can be served by businesses and the pool of labor that can access employment. Economic theory suggests that the more flexible the local labor market, the more likely the most highly skilled person will fill a particular vacancy and therefore the more productive the labor (Cowie 2010). Businesses have also sought to reorganize to take advantage of improved transportation which serves to underline that the relationships between transportation investment and the economy are complex and bidirectional (McKinnon 2008). Causality has proven difficult to establish in many cases (Banister and Berechman 2001).

Transportation also has a direct and indirect impact on employment. Mallard and Glaister (2008) estimate that households from the EU25 spent more than 800 billion euros on transportation in 2006. Investments in new transportation systems and infrastructure offer a source of temporary employment, while the on-going operation of systems provides more permanent jobs. It is estimated that more than eight million people are employed in providing transportation services (including freight, public transportation, and travel agency services) in the EU25 (*ibid.*).

There is considerable debate about the extent to which new investment in transportation systems can stimulate economic growth. In developed economies, it is argued that the mature nature of the networks and existing high levels of accessibility mean that the marginal improvements of new road or public transportation improvements are relatively small (Eddington 2006). However, congestion acts to increase the costs of moving goods and reduces accessibility of labor markets. Policies that tackle congestion through demand or supply side interventions may increase the potential for further interaction. In developing economies, the step change in accessibility that new facilities can provide can still be transformative to accessibility.

### 3.3.2 Transportation as an Open System with Inputs and Outputs

A transportation system can be described as an *open system*—i.e., it interacts with its environment. The *environment* within which the transportation system operates can be defined in terms of people (i.e., society—where the government, stakeholders, and users/customers play a critical role), physical components, the economy, and the natural environment. The transportation system is *large* in that it spreads across nations and provides access to almost every corner of the world.

In both natural and social systems, "[m]ovement takes place within a system of origins and destinations" (Gudmundsson and Hojer 1996, p. 274). In the transportation system, the focus is primarily on movement within social systems. It is possible to identify three general types of movement that are associated with system *inputs* and *outputs* (Fig. 3.5). First is the movement of *people* and *freight* (including oil and gas transported via pipelines), which enter the system at an origin and leave the system at a final destination.[3]

Second is the movement of *energy* and *matter*, which enter the system as transportation-related fuel, construction material, and products (such as vehicles, equipment, etc.) and leave the system as emissions, waste, or material that is recycled/down-cycled for other products or purposes. The quantity and type of

[3] These movements exist to fulfil the broader socio-economic trends which drive travel demand and that reflect transportation supply described above.

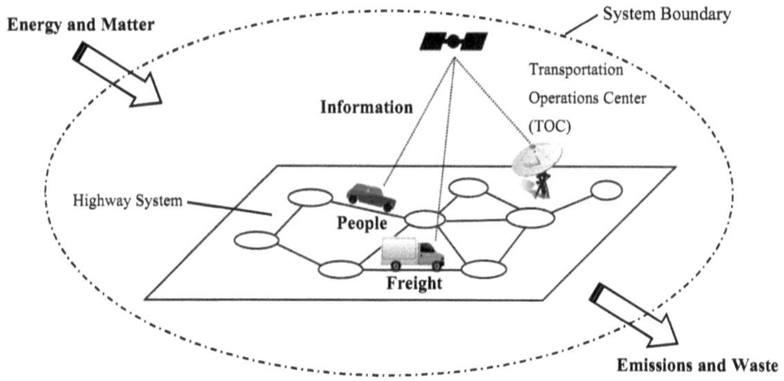


**Fig. 3.5** Input and outputs to and from the transportation system. *Note*: Icons were obtained from https://www.iconfinder.com/

energy and matter that is supplied to the system is *derived* from the demand for passenger and freight transportation services.

Finally there is the movement of *information* that can enhance the management, operation, and performance of the transportation system (e.g., through intelligent transportation systems, ITS) or alleviate the need for people to travel (e.g., through information and communication technology, ICT, which enables activities such as teleworking). It is not yet commonplace to treat the movement of information and the ability to travel virtually as equally important in transportation policy despite its growing importance (Lyons 2009).

All three types of movement are highly interconnected and can affect human well-being in different ways. For example, while the mobility and accessibility provided by the automobile might satisfy our psychological need for connectedness, the associated emissions (from the combustion of fuel/energy) can lower our physiological health. Further, the impacts of these emissions not only affect the driver (Peters et al. 2004) but also the community at large (Gorham 2002), which means that the distributional impacts of the transportation system must be considered with care.

This conceptualization of the transportation system makes the crucial point that the system is a social system and that it requires natural and manmade inputs that produce outputs which, in turn, impact the environment. It does not capture the interaction with other types of systems (e.g., energy or healthcare) directly, although these all intersect with the transportation system in some way.

### 3.3.3 Transportation as a Socio-technical System

Geels (2005) views the transportation system as a socio-technical system and does not make an obvious distinction between physical and social elements. The socio-technical systems approach considers that the systems we have in place are created and refined over time as a result of the actions of a range of actors including

**Table 3.2** Elements of a socio-technical system

| Element | Transportation examples |
|---|---|
| Infrastructure | Road network, runways, bike hire bays with bikes, rail stations |
| Maintenance and distribution infrastructure | Car Garages, rail maintenance depots, airport maintenance hubs |
| Fuel infrastructure | Airport fuel depots, electrification system for underground, petrol station network |
| Production system | Car manufacturer and supplier configuration |
| Policies and regulations | Regulation of rail fares, speeding laws, vehicle taxation, commuter parking policies |
| Socio-cultural norms | Attitudes to drink-driving, culture of cycling, road user hierarchy, culture of time keeping |
| Markets and user practices | Mobility patterns, driver preferences, yield management techniques |

*Source*: Adapted from Geels (2002)

government, companies, interest groups, and the public as well as changing technology. The "functioning" of the system is dependent on a series of interconnected elements (Geels 2005) (see Table 3.2).

Analysis of the transportation system through the socio-technical systems lens allows one to understand the full range of factors that explain the different ways in which systems operate and develop within and across countries. So, while the infrastructure, fueling, maintenance, and production systems for public transport are largely similar across Europe, the approach to regulation and policy makes a substantial difference to the nature and operation of the system. The stronger free-market philosophy within the UK, for example, has shaped the nature of the more open rail market relative to Germany (Lodge 2003). Similarly, the UK, Sweden, and the Netherlands have outperformed much of mainland Europe on road safety. This can be explained through different approaches to policy and infrastructure management and a different view of what is normal and acceptable with respect to driving practices (Wegman et al. 2005).

While the transportation system is indeed large-scale, complex, and designed by humans, its use has evolved in ways which the system designers did not imagine, producing significant unintended consequences (Fischer 1992; Hubers et al. 2011). This includes the facilitation of large-scale urban sprawl and the growing market for international weekend leisure breaks. As well as generating direct travel opportunities, it also acts as part of a bigger system of consumption and is subject to demands that arise out of changes to non-transport technologies. For example, the freezer and the supermarket trolley changed the volumes of food which can be easily purchased and stored and therefore the nature of food shopping (*ibid.*). The impacts of the closure of European airspace in 2010 as a result of an ash cloud from the Icelandic volcano Eyjafjallajökull showed how globalized and interconnected travel patterns are today and how fragile some aspects of our lives are to the impacts of disruptive events which impact on transportation (Guiver and Jain 2011).

It is, we argue, important to see the transportation system not just as a system of physical components and operational plans but as part of a much broader social system. The transportation system shapes and is shaped by that system. This perspective can be expanded further by locating the transportation system within the natural environment in which it operates. Given the global reach of the transportation system, a macroecological perspective (Burger et al. 2012) may be required to fully account for the environmental impacts from transportation (see Sect. 4.4).

### 3.3.4 Transportation as an Integrated System

To view the transportation system as "one system" would be a gross simplification of what has been the result of the emergence and reconfiguration of several transportation systems over a long period of time (Geels 2002, 2005). The entire transportation system is made up of subsystems that can be defined by infrastructure type, transportation mode, or by the subsystem's purpose (e.g., movement of passengers, freight, etc.). It is common, for example, for policies in the rail and road sectors to be developed by different agencies. These subsystems (or subnetworks) interact to make an intermodal and integrated system. For example, a city that develops a cycle network will typically develop the network using a mix of segregated and shared facilities and will seek to integrate the network with public transportation (e.g., through safe parking at public transportation hubs). The intermodal nature of the transportation system makes it flexible—i.e., there are typically many options for the same trip. This flexibility—which is an attribute of system "resilience"—means that the system is able to provide mobility following a shock to the system or even the temporary loss of a subsystem (such as the closure of air space or rail lines due to strike action, terrorism, or natural disaster). Surface transportation systems (including underground systems) are related to each other in the sense that they occupy geographical space and they connect together at key interchanges (some of which are intermodal). The air transportation system connects to the surface transportation system at airports (or nodes); otherwise, this system operates above the earth's surface.

In theory, a transportation system *should* operate as an integrated network of highways, roads, railways, walkways, bike paths, canals and rivers, and air corridors with coordinated transportation services which facilitate the movement of different modes of transportation (including bicycles, electric vehicles, motorbikes, automobiles, trucks, trains, boats, and aircraft). Travelers and logistics providers seek to make journeys or move goods to places they wish to go using multiple parts of the system and expect an integrated experience when they do so. The reality is that transportation infrastructure and services are typically owned and operated by a large range of public, private, and pseudo state owned enterprises each operating with a different set of objectives. Chapter 5 reviews the overarching rationales for governing transportation systems and the pros and cons of market competition within the transportation system. For now, it is necessary to conclude that while the transportation system is integrated, this is incomplete and sometimes constrained by the interests of the providers of the transportation infrastructure and services.

### 3.3.5 Transportation in a Systems Hierarchy

Another way to view the transportation system is to use a hierarchy (Fig. 3.6). "A system hierarchy … provides order and function to the operation of the individual components [of a system] in the context of more global system goals. How this system hierarchy is defined affects how one views problems and conducts planning" (Meyer and Miller 2001, p. 91).

The benefit of using a hierarchy to describe a system is that it enables the analyst/decision-maker to consider how changes in one system might affect other systems. When considering sustainable development, understanding how policies aimed at the transportation system might impact land use and the livability of communities, for instance, is important. Conversely, understanding how changes to other systems (such as the energy, power, and/or agricultural system) might impact the transportation system is of equal interest. For example, in considering whether electrification

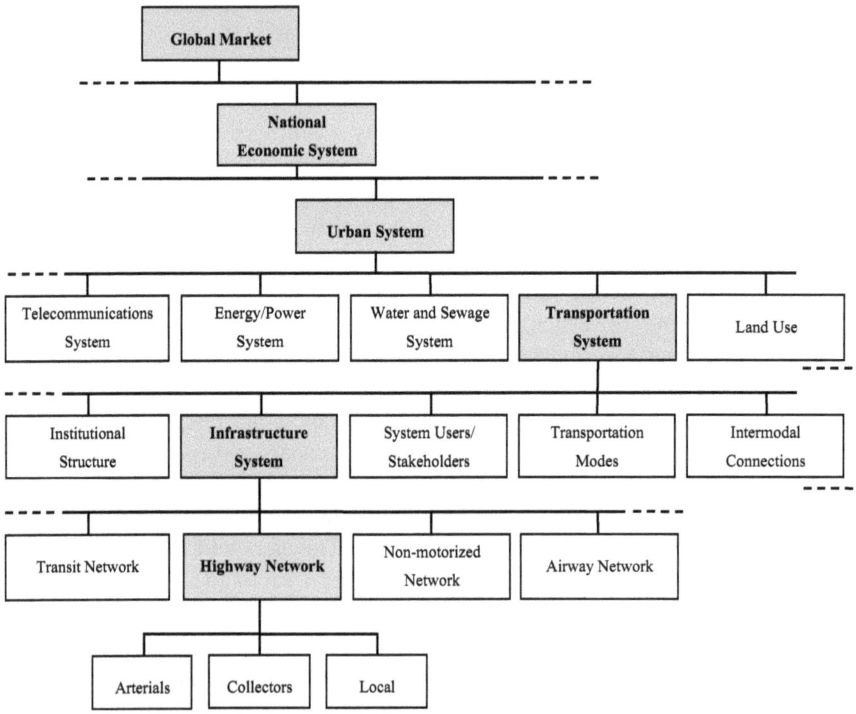


**Fig. 3.6** Transportation in a systems hierarchy. *Source*: Adapted from Meyer and Miller (2001, p. 91)

of the vehicle fleet could reduce dependence on fossil fuels, it is essential to understand the extent to which the electricity system will provide low-carbon electricity.

### 3.3.6 Systems Discussion

It is clear that each type of system representation provides certain advantages and disadvantages. For example, a detailed focus on physical subsystems could enhance our understanding of mobility patterns but provides limited information on the political context in which the systems operate and how transportation interacts with other sectors such as energy. Likewise, a system hierarchy presents a useful way to identify the relationship of critical components; however, one might question whether it is possible to make such a clear distinction between the levels shown in the hierarchy.

The characterizations discussed previously may be helpful in determining the boundaries of any particular system of study. For the purpose of understanding how the transportation planning process interacts with notions of sustainable development, we draw out the following key issues from within the range of definitions. The transportation system:

1. Is a critical component of a broader economic system which supports business and social development;
2. Is an open system, which requires natural and man-made inputs and produces outputs which impact on the environment;
3. Is part of a social system and shapes and is shaped by that social system, including other policy areas;
4. Comprises a series of physical subsystems each of which has a range of physical and operational components and which are organized through formal and informal conventions; and
5. Is typically a fragmented series of partly connected yet partly competing subsystems with complex and varying governance arrangements.

This conceptualization of the transportation system is intended to provide clarity about why we can expect transportation to be central to discussions about sustainable development. It explains why looking just at the siloed operation of the modes of transportation will only give a partial account of the benefits and impacts of transportation. One of the key challenges for transportation planners is to develop policies that create positive system-wide impacts which can successfully integrate potentially competing elements of the system.

**Discussion Topics**

- Select a local transportation system in your area. Using one or more of the frameworks presented in this section, define your system paying close attention to the system boundaries. How easy/difficult was this task? What are the implications of the system boundaries that you drew from a management/operations, political, and financial perspective?
- How is the transportation system integrated with the energy system? How might this change if the vehicle fleet is electrified? What would some of the benefits and risks be to such a change?
- Why is it difficult to establish clear evidence on the relationships between transportation and the wider economy?

## 3.4 Impacts of Transportation

Section 3.2 highlighted the importance of transportation to the economy and for people's participation in employment. It also demonstrated that transportation is used more for leisure, shopping, and visiting friends and relatives than it is for work or work-related journeys. The sheer volume and range of activities which transportation supports serve to underline the positive role it can play in our society. However, transportation can also have negative impacts on the environment, society, and ultimately the economy (Reardon et al. 2011). It is the presence of these negative impacts and the inability of the current systems of provision to control them that provides the case for state intervention in the planning and operation of transportation. This section summarizes the main negative impacts of transportation through a number of detailed examples.[4] Their relationship with notions of sustainable development is reviewed in Chap. 4 before the role of transportation governance in tackling these issues and maximizing the positive impacts of transportation is discussed in Chap. 5.

There is a significant literature around the negative impacts of transportation that has led to the identification of a range of problems shown in Table 3.3.[5] While some of these impacts have emerged as a part of the sustainable development debate (e.g., global climate change and ecosystem damage), others such as accidents, local air pollution, and congestion pre-date these debates.

[4] This section draws extensively on material prepared for May and Marsden (2010).

[5] Mallard and Glaister (2008, p. 151) define an externality as "an unconsidered cost or benefit experienced by a third party due to an economic decision made by others." A common example is congestion, whereby drivers only take account of the delay they expect to experience and not the delay that their journey will cause others.

**Table 3.3** Negative impacts associated with transportation

| Environmental | Social | Economic |
|---|---|---|
| Air pollution | Accidents | Costs of transportation to customers/consumers |
| Consumption of land/urban sprawl | Declining community livability/community partitioning | Costs relating to accidents |
| Depletion of the ozone layer | Human (psychological and physiological) health impacts | Depletion of nonrenewable resources and energy supplies (*also an environmental and intergenerational equity concern*) |
| Disruption of ecosystems and habitats | Inequalities associated with negative environmental and health impacts | Traffic congestion |
| Global climate change | Mobility barriers/inequalities for the disadvantaged | Transportation facility costs |
| Hydrologic impacts | Time wastage | Transportation-related health costs |
| Introduction of exotic species | Visual pollution | |
| Light pollution | | |
| Noise pollution | | |
| Release of toxic/hazardous substances | | |
| Solid wastes | | |
| Vibration pollution | | |
| Visual intrusion and aesthetics | | |
| Water pollution | | |

*Sources*: Black (2005), Button (1993), Maddison et al. (1996), Rothengatter (2003), Spellerberg (2002), TRB (1997), Wachs (2005), Whitelegg (1993, 1997), Whitelegg and Haq (2003), VTPI (2005), and Zietsman and Rilett (2002)

## 3.4.1 Environmental

Transportation has a number of direct and indirect impacts on the environment. Direct impacts include land-take for new infrastructure, visual intrusion, severance, and noise and vibration from construction and use of infrastructure. Indirect impacts include local air pollution, acidification, water pollution, and climate change. Comprehensive methods for identifying and mitigating significant environmental impacts from new construction exist through techniques such as Environmental Impact Assessment (EIA) in the USA and Europe (Fry and Therivel 2009) and Strategic Environmental Assessment (Dalal-Clayton and Sadler 2005) in Europe. Even in countries with strong environmental regulation, a number of significant environmental concerns remain.

The key material inputs to the transportation sector are energy, materials for construction, and vehicles and land. The transportation sector is currently heavily dependent on nonrenewable fossil fuel sources as the major source of propulsion energy. In 2008, transportation accounted for 27 % of world energy use, and 95 % of this was generated by fossil fuel-burning internal combustion engines (EIA 2011). The transportation system relies on nonrenewable resources and energy supplies to build/maintain infrastructure and manufacture transportation vehicles/equipment (Hille 1997; Geiser 2001; Rodrigue et al. 2009). Land is also an important input to the transportation system, and the negative environmental impacts of land-take can be an important factor in decisions about whether or not to proceed with new infrastructure developments (Fahrig and Rytwinski 2009; Nellthorp and Mackie 2000).

Some key environmental outputs (or emissions and waste) include toxic waste from vehicle disposal, noise, local air pollution, and climate change emissions. While significant improvements have been made to the proportions of vehicle waste that can be recycled at the end of the vehicles' life (Vermeulen et al. 2011), it is estimated that around 25 % of a vehicle remains nonrecyclable and that much of this could be considered as hazardous (Kanari et al. 2003). Whereas in developed countries, there is a regulated process for managing the disposal of waste at the end of a vehicle's life, this is not the case globally.

Traffic noise can disturb sleep patterns, affect cognitive functioning, and aggravate some cardiovascular problems (den Boer and Schroten 2007). As noise is related to the amount of activities being conducted and the noise intensity of those activities, it is not surprising that it is largely an urban problem. Den Boer and Schroten (2007) estimated the social cost of road traffic noise in the EU22 at 38 (30–46) billion euros per year, while for rail, the estimates were about 2.4 (2.3–2.5) billion euros (in total around 0.4 % of GDP). This is forecast to rise over the next decade.

As knowledge on the relationship between toxic emissions and health outcomes has grown, the emphasis given to local air quality in transportation planning has increased. Several pollutants have significant adverse health impacts, particularly on those who, perhaps through preexisting respiratory conditions, are more vulnerable to them. The main pollutants of concern from transportation sources are particulate matter ($PM_{10}$ and $PM_{2.5}$), oxides of nitrogen, and hydrocarbons. The exact nature of the problem can vary significantly with the climate and topography of the cities. The World Health Organisation (WHO) suggests that poor air quality brings forward around two million deaths per year annually (WHO 2009) and that the disease burden "falls most heavily on developing countries, particularly those in Asia" (Krzyzanowski and Cohen 2008, p. 7). Most countries have adopted standards of environmental protection which aim to limit the negative health impacts of local air quality on the population, including specific provisions to target transportation sources. A major source of air quality improvements has been the adoption of improved vehicle technologies with lower emission rates per mile driven. This can be particularly significant in developing countries where many older vehicles remain in use. However, there are limits as to the extent to which vehicle and fuel technologies can tackle this issue—Nitrogen dioxide is, for example, produced even in cases of complete and efficient combustion, and it is proving difficult to achieve the stricter air quality standards being adopted in Europe.

The combustion of fossil fuels gives rise to carbon dioxide emissions, a major contributor to man-made climate change. The Intergovernmental Panel on Climate Change estimates that "Unless there is a major shift away from current patterns of energy use, projections foresee a continued growth in world transportation energy use of 2 % per year, with energy use and carbon emissions about 80 % above 2002 levels by 2030. In developing countries, transportation energy use is rising faster (3–5 % per year) and is projected to grow from 31 % in 2002 to 43 % of world transportation energy use by 2025" (Metz et al. 2007, p. 48). Of course, countries with high motorization have far higher per capita emissions than developing countries (Short et al. 2009). For example, the average $CO_2$ per capita in London is 1.3 tons, while in Delhi it is 0.4 tons (Hickman and Banister 2009).

In contrast to toxic air pollutants that have a local and regional impact, the origin of greenhouse gas emissions is largely irrelevant to its impacts on the global environment. While the need to reduce greenhouse gas emissions poses significant challenges to urban transportation planning in all countries, many cities will also require significant adaptation to climate change. The impacts of climate change can include more intensive and more frequent rain storms and hurricanes and more sustained periods of higher temperatures. Whereas the exact nature and severity of the change is not wholly clear and will be context specific, the vulnerability of communities to these effects and the costs which they impose are becoming more apparent. Hurricane Katrina, for example, was estimated to have created $100 billion worth of damage in the USA alone.[6]

### 3.4.2 Social

Table 3.3 highlights a broad range of potential social impacts. Some of these impacts are direct (e.g., time wastage), while some operate in complex ways in interaction with broader societal trends (e.g., spatial environmental inequalities). This section reviews one direct (safety) and one indirect (social exclusion) impact as exemplars.

In 2004, the WHO estimated that almost 1.2 million people were killed in road traffic accidents, while as many as 50 million people were estimated to be injured, representing "the combined population of five of the world's large cities" (WHO 2004, p. 3). It is the main cause of death in those aged 40 and under, and the direct economic costs globally have been estimated at US$518 billion (*ibid.*).

It is estimated that road traffic accidents cost typically between 1 and 2 % of GDP for both developed and developing countries (WHO 2004, 2013). The total global cost of accidents was estimated in 2000 to be almost US$518 billion per year. There are important differences in global road safety records which suggest that there is much to be learnt in some areas (see Table 3.4).

Even within Europe, there is variation from the best (Sweden and UK) to the worst (Portugal) (Eurostat 2013). While developed countries are forecast to cut road traffic fatalities by an average of 28 % over the period 2000–2020, developing countries are expected to see this rate rise (South Asia by 144 % and Sub-Saharan Africa by 80 %) (Commission for Global Road Safety 2007).

Social exclusion was identified by the European Commission as a policy issue in 1989. There have been several definitions, but the clearest is probably that it is "a

[6] Source: Infoplease, Billion Dollar U.S. Weather Disasters, 1980–2010. http://www.infoplease.com/ipa/A0882823.html (accessed on April 19, 2015).

**Table 3.4** Predicted road traffic fatalities by region

| Region[a] | Number of countries | 1990 | 2000 | 2010 | 2020 | Change (%) 2000–2020 | Fatality rate (deaths/100,000 persons) | |
|---|---|---|---|---|---|---|---|---|
| | | | | | | | 2000 | 2020 |
| East Asia and Pacific | 15 | 112 | 188 | 278 | 337 | 79 | 10.9 | 16.8 |
| East Europe and Central Asia | 9 | 30 | 32 | 36 | 38 | 19 | 19 | 21.2 |
| Latin America and Caribbean | 31 | 90 | 122 | 154 | 108 | 48 | 26.1 | 31 |
| Middle East and North Africa | 13 | 41 | 56 | 73 | 94 | 68 | 19.2 | 22.3 |
| South Asia | 7 | 87 | 135 | 212 | 330 | 144 | 10.2 | 18.9 |
| Sub-Saharan Africa | 46 | 59 | 80 | 109 | 144 | 80 | 12.3 | 14.9 |
| Sub-total | 121 | 419 | 613 | 862 | 1,124 | 83 | 13.3 | 19 |
| High-income countries | 35 | 123 | 110 | 95 | 80 | −27 | 11.8 | 7.8 |
| Total | 156 | 542 | 723 | 957 | 1,204 | 67 | 13 | 17.4 |

*Source*: Adapted from WHO (2004, p. 39)
[a]Data are displayed according to the regional classifications of the World Bank

shorthand label for what can happen when individuals or areas suffer from a combination of linked problems such as unemployment, poor skills, low incomes, poor housing, high crime environments, bad health and family breakdown" (quoted in DETR 2000; SEU 2003).

Poor transportation access is not identified directly in the definition but is clearly linked to employment opportunities, and probably health as well. The UK Social Exclusion Unit report on the role of transportation in social exclusion describes a number of ways that people can be excluded from the activities they wish to undertake (SEU 2003):

- Spatially, because they cannot get there at all;
- Temporally, because they cannot get there at the appropriate time;
- Financially, because they cannot afford to get there; and
- Personally, because they are physically or mentally unable to use the transport available.

In the UK, almost three quarters of households now have access to a car compared to 91 % in the USA. Car ownership allows people to travel further, within the same travel time budget, to access a wider range of services and facilities. The commercial sector has reacted to these opportunities with the provision of large out-of-town shopping centers and a boom in edge of town supermarkets, designed principally for those with vehicles. This in turn has brought a reduction in the viability and attractiveness of local shops as well as a decrease in their overall number. All of this acts to reduce the options available locally to those without access to a car, creating a vicious circle. The ability to travel further to work has seen a lengthening of commuting trips by 16 % in the last decade (DfT 2010). This trend, along with cheaper land on the edges of cities, has continued to promote the decentralization of cities. In turn, the viability of public transportation services is undermined, which further limits opportunities for those dependent on them.

There is likely to be growing pressure to make the transportation system more accessible to those with mobility restrictions, both as a result of changing attitudes to disability and the ageing of the population. Around one in six (ten million) people have a limiting long-term illness, impairment, or disability in Great Britain, a figure typical of many other European countries (ODI 2011). Although age is not a perfect predictor of mobility problems, on average, mobility problems increase with age as a result of declining levels of physical function (Frye 2005).

The impacts of ageing and poor accessibility can be quite different within cities, across regions, as well as between countries. Factors such as the availability of a broader social security system and the nature of family support structures matter. Nonetheless, there is increasing awareness of the impacts of poor access and loss of independent mobility on economic potential, health, and well-being (Lucas 2004). There are clearly very different issues at play between developed and developing countries. This relates both to the general position on accessibility, where car ownership levels are much lower and dependence on public transport far higher, but also to the treatment of social inequality relating to gender and disability (UN 2006). So, while the general bullet points identified above might have relevance in a range of countries, the relative importance of them and their connection to other transportation issues will look quite different.

### 3.4.3 Economic

As well as bringing substantial economic benefit, the transportation system also generates economic costs. As described above, road traffic accidents and air quality both lead to the loss of life or the diminution in quality of life which have direct costs to the economy in terms of lost productivity and lower citizen welfare. There are also costs to the government of providing the transportation system and associated subsidies for travelers which need to be funded through either taxation or the fare box. In the UK, in 2010/2011, the government expenditure on transportation was £12.3 billion (US$18.4 billion). While a significant outlay, it forms only 1.8 % of the £691.7 billion (US$1033 billion) spent across the whole of government, with pensions, welfare, education, and health typically forming at least half of total expenditure (Rogers 2011). It is also considerably less than income from fuel tax and the annual vehicle ownership duty which was just over £30 billion in the UK in 2010 (ONS 2011). Global estimates of the required infrastructure expenditure on land transportation are $45 trillion by 2050 (Dulac 2013).

The depletion of nonrenewable energy sources was raised in Sect. 3.4.1. As well as reducing the options available to future generations, it is anticipated that the growing global demand for oil, of which transportation is a key element, will mean that demand will ultimately outstrip available supply capacity, with the markets correcting for this through increased prices. As Fig. 3.7 shows, there has already been a major increase in fuel price over the past decade. It is suggested by some commentators (e.g., Hirsch 2005) that we have already passed "Peak Oil," a phenomenon defined as "the point where the high practicable rate of global production has been achieved and from which future levels of production will either plateau or begin to diminish" (Industry Taskforce on Peak Oil and Energy Security 2010, p. 9). Whether or not this is yet the case, it is clear that slower rates of new discovery and more difficult extraction conditions, combined with high demand, are putting pressure on prices, and this trend is expected to continue (Almeida and Silva 2009). However, a decline in the availability of crude oil may be compensated by coal liquefaction that would enable the mass production of oil, using current technology, at prices of US$60–80 per barrel, even with coal prices at US$100 per short ton (von Weizsäcker et al. 2010). With the geological reach of coal spanning some 200 years, petroleum-fueled vehicles may prove difficult to displace without specific strategies to promote alternative fuels or vehicle power systems. The example above is illustrative of the potential for technological

**Fig. 3.7** Pump prices for a liter of fuel in the UK (April 2001 = 100). *Source*: ONS (2011)

innovation to advance a "weak" form of sustainable development. In contrast, a "strong" sustainability approach would simply see this as a transfer of the diminution of nonrenewable resources.

Congestion imposes a significant cost on the economies of all countries. It is a highly visible indication of the mismatch between supply and demand (as is overcrowding on public transport). Sperling and Gordon (2009, p. 5) point out that "Some traffic congestion is desirable—the absence of congestion would indicate a depressed economy, a somnolent society, or an overinvestment in infrastructure. But congestion levels in most large cities of the world are severe enough to harm economic and social activity."

While congestion can occur on inter-urban routes, the majority of congestion costs occur in urban areas. Even in the UK, where inter-urban commuting trips are commonplace, 89 % of lost time on roads is in urban areas (Eddington 2006). The full calculation of the costs of congestion on the economy requires a good understanding of both the nature of the congestion and the value of time lost by the affected road users (Grant-Muller and Laird 2006). Global estimates of congestion cost are therefore difficult to establish; however, estimates for the USA are around $87 billion per annum (Schrank and Lomax 2009). It is clear that the average traffic speeds in the largest and most rapidly growing cities of the middle income and developing countries are also very low. In India, for example, average speeds are below 20 km/h in the largest cities of Hyderabad, Chennai, Bangalore, Delhi, and Mumbai (Alam and Ahmed 2013), while in China speeds in major arterials in Beijing and Shanghai are reported to have speeds below 10 km/h (Peng 2011). Of great concern for the future is the potential for congestion to increase rapidly. As more and more routes become oversaturated, the delay costs can spiral. In the USA, Schrank and Lomax (2009) estimate that in the past 25 years, urban congestion costs have increased fourfold, while the UK estimates that under a business-as-usual scenario, the costs to the economy could increase by as much as one-third year on year by 2025 (Eddington 2006). Although the levels of congestion in central areas tend to stabilize once very high levels are reached, there is an inevitable spread of the peak hours over longer periods and wider areas that are less well served by public transportation. Goodwin et al. (1991) concluded that there is no prospect of building our way out of congestion so the solution must be to manage the demand for transportation.

**Discussion Topics**

- How interconnected are the impacts of transportation? Take an impact such as congestion or accidents and consider how they may in turn have impacts on the economy, environment, and society.
- Select an impact area from the economy, environment, and society and discuss the potential for the impacts to fall to different groups in society in different ways, generating inequalities.

– Look back at the definitions of transportation systems. Which of the system definitions seem best matched to capturing the inputs, operation, and outputs of the transportation system set out in Sect. 3.4?

## 3.5 Conclusions

The movement of people, goods, and information has been, and remains, a fundamental part of our social progress. Transportation and the transportation system have been integral to this, both being shaped by and shaping major social trends such as suburbanization and globalization. Major changes continue to challenge our ingenuity to plan ahead such as revolutions in mobile telecommunications and internet capability and ever-increasing networks of low-cost air travel. These trends will play out differently in different parts of the world. Developed countries are seeking to manage already crowded infrastructure to ensure that the potential for economic growth is not choked off. In developing countries, the dual impacts of growth in disposable income and population, combined with rapid urbanization, present challenges of urban transportation planning on a scale not seen before.

This chapter has shown how the transportation system needs to be conceptualized in the broadest sense. Traditionally, it was common to concentrate on how the components of the different transportation modes and networks fit together and are organized. This indeed remains a challenge which we explore further in Chap. 5. However, to do only this overlooks the broader societal role that the system plays in accommodating (or not) the demand for travel. As an illustration, it is worth reflecting on why UK citizens in Edinburgh and Manchester rejected the proposed urban congestion charging schemes when the schemes, and their associated public transportation investment, would have served the economic good of the area. By contrast, Norway and Sweden have established a tradition of using tolling or congestion charging to fund new infrastructure. The way the transportation system develops is shaped by a range of interests including those of the public, manufacturing interests, and lobby groups as well as the more typical constraints of historical investments, limited finance, and capacity to deliver.

This chapter has also demonstrated that the transportation system is an important part of the broader physical environment. It is highly dependent on nonrenewable resources for construction of infrastructure and vehicles, and particularly for powering vehicles. The use of the transportation system also produces major negative environmental, social, and economic impacts such as road accidents, toxic air pollution, local environmental disturbance, land-take, and congestion. It is difficult to disentangle the benefits from the negative impacts and to understand the contribution that each of the many hundreds of decisions about the composition and operation of the transportation system makes to these impacts. An integrated approach to analyzing the problem and to developing solutions is required. As Sperling and Gordon (2009, p. 5) point out "the solutions to oil security and climate change also can resolve local air pollution, traffic congestion, and urban livability. . . .

The need to address traffic congestion and escalating infrastructure costs could engage the transportation community in reducing oil use and greenhouse gas emissions."

To make a difference through transportation policies requires a rounded understanding of the role the transportation system plays in society and its broader connection to other debates such as individual rights, poverty, energy security, and climate change. Chapter 4 takes forward the discussion of the integration of transportation impacts and benefits with the core notions of sustainable development. Chapter 4 also addresses the important question of whether we should seek a sustainable transportation system (drawing a tight boundary around the system) or if working toward a transportation system that supports sustainable development (a broader interpretation) is a more desirable approach.

This chapter has identified and discussed the scale of some of the impacts of transportation. The review is not comprehensive nor could it be given the scale of global diversity and the very different levels of information that is collected (or not) in different ways across the world. The context is different between cities within a country and so the differences between countries become even more significant. In this book, the benefits of standardizing information collection for learning are recognized (see Chap. 6), but this is a long-term endeavor and will never fully overcome the important resource and contextual differences that might require or produce different data on the connections between transportation and sustainability in different places. As noted in Chap. 1, the production of a blueprint or one global standard set of indicators which can be taken from one place and applied in another is, in our view, the wrong approach. This book therefore focuses its attention on how to use and organize information to make more informed decisions which support sustainable development. A framework approach (Chap. 7) allows a diversity of priorities to be respected and accommodated, while still providing some clear tools as to how those priorities can be used to work to achieve better transportation decisions.

## References


Alam, M. A., & Ahmed, F. (2013). Urban transport systems and congestion: A case study of Indian Cities, Transport and Communications Bulletin for Asia and the Pacific, Number 83. Accessed November 14, 2014, from www.unescap.org/sites/default/files/b82_Chapter3.pdf.

Almeida, P., & Silva, P. (2009). The peak of oil production—Timings and market recognition. *Energy Policy, 37*(4), 1267–1276.

Banister, D., & Berechman, Y. (2001). Transport investment and the promotion of economic growth. *Journal of Transport Geography, 9*, 209–218.

Black, W. R. (2005). Sustainable transport: definitions and responses. In *TRB conference on integrating sustainability into the transportation planning process*, 11–13 July 2004, Baltimore, TRB conference proceedings 37.

Börjesson, M., Fosgerau, M., & Algers, S. (2010). Identifying the value of travel time distribution—Evidence from the Swedish value of time study 2008. In *Proceedings of the European Transport Conference 2010.*

Burger, J. R., Allen, C. D., Brown, J. H., Burnside, W. R., Davidson, A. D., Fristoe, T. S., et al. (2012). The macroecology of sustainability. *PLoS Biology, 10*(6), 1–7.
Button, K. (1993). *Transport, the environment and economic policy*. Aldershot: Edward Elgar.
Cascetta, E., Pagliara, F., & Papola, A. (2007). Governance of urban mobility: Complex systems and integrated policies. *Advances in Complex Systems, 10*(2), 339–354.
Commission for Global Road Safety. (2007). Make roads safe: A decade of action for road safety. Accessed August 4, 2012, from http://www.makeroadssafe.org/publications/Documents/decade_of_action_report_lr.pdf.
Cowie, J. (2010). *The economics of transport: A theoretical and applied perspective*. Oxon: Routledge.
Cresswell, T. (2006). *On the move: Mobility in the modern Western world*. London: Routledge.
Dalal-Clayton, B., & Sadler, B. (2005). *Strategic environmental assessment: A sourcebook and reference guide to international experience*. London: Earthscan.
Dargay, J., Gately, D., & Sommer, M. (2007). Vehicle ownership and income growth, worldwide: 1960-2030. *The Energy Journal, 28*(4), 143–169.
Den Boer, L. C., & Schroten, A. (2007). *Traffic noise reduction in Europe Health effects, social costs and technical and policy options to reduce road and rail traffic noise*. Delft: CE Delft.
DETR. (2000). *Social exclusion and the provision and availability of public transport*. London: Department of Environment, Transport and the Region.
DfT. (2003). The future of air transport white paper. Department for Transport, London. Accessed August 4, 2012, from http://collections.europarchive.org/tna/20080728103028/http:/www.dft.gov.uk/about/strategy/whitepapers/air/.
DfT. (2009). *UK air passenger demand and* $CO_2$ *forecasts*. London: Department for Transport.
DfT. (2010). *Transport statistics Great Britain* (36th ed.). London: Department for Transport.
Dulac, J. (2013). Global land transport infrastructure requirements: Estimating road and railway infrastructure capacity and costs to 2050. Information paper for the International Energy Agency. Accessed November 7, 2014, from http://www.iea.org/publications.
Eddington, R. (2006). The case for action: Sir Rod Eddington's advice to government. HM Treasury. Accessed August 4, 2012, from http://www.hm-treasury.gov.uk.
EIA. (2011). International Energy Outlook. US Energy Information Administration. Accessed August 4, 2012, from http://205.254.135.7/forecasts/ieo/world.cfm.
Eurostat. (2013). Energy transport and environment indicators. Eurostat Pocketbooks, ISSN 1725-4566.
Fahrig, L., & Rytwinski, T. (2009). Effects of roads on animal abundance: An empirical review and synthesis. *Ecology and Society, 14*(1), 1–20.
Fischer, C. S. (1992). *America calling: A social history of the telephone to 1940*. Berkeley: University of California Press.
Fry, C., & Therivel, R. (2009). Transport. In P. Morris & R. Therivel (Eds.), *Methods of environmental impact assessment* (3rd ed., pp. 94–119). Oxon: Routledge.
Frye, A. (2005). Mobility in old age—A European policy perspective. In *Proceedings of the SIZE conference*, 19th May. Accessed August 4, 2012, from http://size-project.at/conference05.htm.
Geels, F. W. (2002). Technological transitions as evolutionary reconfiguration processes: A multi-level perspective and a case-study. *Research Policy, 31*, 1257–1274.
Geels, F. W. (2005). The dynamics of transitions in socio-technical systems: A multi-level analysis of the transition pathway from horse-drawn carriages to the automobile (1860–1930). *Technology Analysis and Strategic Management, 17*(4), 445–476.
Geiser, K. (2001). *Materials matter: Toward a sustainable materials policy*. Cambridge: MIT Press.
GIA. (2010). Maritime containerization: Market research report. Global Industry Analysts Inc. Accessed August 4, 2012, from http://www.strategyr.com/Maritime_Containerization_Market_Report.asp.

Gillingwater, D., Humphreys, I., & Watson, R. (2003). Air freight and global supply chains: The environmental dimension. In P. Upham (Ed.), *Towards sustainable aviation* (pp. 149–162). London: Earthscan. Chapter 9.

Global Commission for Road Safety. (2007). Make roads safe: A new priority for sustainable development. Accessed August 4, 2012, from www.roadsafe.org.

Goodwin, P., Hallett, S., Kenny, F., & Stokes, G. (1991). Transport: The new realism. Report to the Rees Jeffreys Road Fund, TSU Oxford.

Gorham, R. (2002). Air pollution from ground transportation. An assessment of causes, strategies and tactics, and proposed actions for the international community. Division for Sustainable Development, Department of Economic and Social Affairs, United Nations, New York.

Grant-Muller, S., & Laird, J. (2006). Costs of congestion: Literature based review of methodologies and analytical approaches. Transport Research Findings No.1/2006, Scottish Government, Edinburgh.

Gudmundsson, H., & Hojer, M. (1996). Sustainable development principles and their implications for transport. *Ecological Economics, 19*, 269–282.

Guiver, J., & Jain, J. (2011). Grounded: Impacts of and insights from the volcanic ash cloud disruption. *Mobilities, 6*(1), 41–45.

Haines, S. (2010). Systems thinking research rediscovered: Ludwig von Bertalanffy and the Society for General Systems Research's relevance in the 21st century. In *Proceedings of the 54th annual meeting of the ISSS.*

Hall, R. P. (2006). *Sustainable transportation.* PhD. Thesis, Massachusetts Institute of Technology.

Headicar, P. (2009). *Transport policy and planning in Great Britain.* Oxon: Routledge.

Hesse, M., & Rodrigue, J.-P. (2004). The transport geography of logistics and freight distribution. *Journal of Transport Geography, 12*, 171–184.

Hickman, R., & Banister, D. (2009, September 28–30). Backcasting for low carbon transport—London and Delhi. In *University of Oxford, 4 Degrees and beyond conference*, St Anne's College.

Hille, J. (1997). The concept of environmental space—Implications for policies, environmental reporting and assessments. European Environment Agency (EEA), Copenhagen.

Hirsch, R. (2005). The inevitable peaking of world oil production. The Atlantic Council of the US. Accessed August 4, 2012, from http://www.hilltoplancers.org/stories/hirsch0502.pdf.

Hubers, C., Lyons, G., & Birtchnell, T. (2011, January 5–7). The unusual suspects: The impacts of non-transport technologies on social practices and travel demand. In: *43rd Universities Transport Study Group conference*, Milton Keynes, UK.

IATA. (2011, June). IATA industry fact sheets. Accessed August 4, 2012, from http://www.iata.org/pressroom/facts_figures/fact_sheets/Pages/index.aspx.

Industry Taskforce on Peak Oil and Energy Security. (2010). The oil crunch: A wake up call for the UK economy. Industry Taskforce on Peak Oil and Energy Security. Accessed August 4, 2012, from http://peakoiltaskforce.net/wp-content/uploads/2010/02/final-report-uk-itpoes_report_the-oil-crunch_feb20101.pdf.

Kanari, N., Pineau, J.-L., & Shallari, S. (2003). End-of-life vehicle recycling in the European Union. *Journal of the Minerals Metals and Materials Society, 55*, 15–19.

Krzyzanowski, M., & Cohen, A. (2008). Update of WHO air quality guidelines. *Air Quality and Atmospheric Health, 1*, 7–13. doi:10.1007/s11869-008-0008-9.

Leleur, S. (2008). Systems science and complexity: Some proposals for future development. *Systems Research and Behavioral Science, 25*, 67–79.

Lodge, M. (2003). Institutional choice and policy transfer: Reforming British and German Railway Regulation. *Governance, 16*(2), 159–178.

Lucas, K. (Ed.) (2004). Transport and social exclusion: A G7 comparison. FIA Foundation.

Lyons, G. (2009). The reshaping of activities and mobility through new technologies. *Journal of Transport Geography, 17*, 81–82.

Maddison, D., Pearce, D., Johansson, O., Calthrop, E., Litman, T., & Verhoef, E. (1996). *The true cost of road transport*. London: Earthscan.

Mallard, G., & Glaister, S. (2008). *Transport economics: Theory, application and policy*. Basingstoke: Macmillan.

Manheim, M. L. (1979). *Fundamentals of transportation systems analysis* (Vol. 1). Cambridge, MA: MIT Press.

May, A. D., & Marsden, G. (2010). Urban Transport and Mobility, Forum paper 5, International transport forum 2010, Transport and innovation: Unleashing the potential, Leipzig. http://www.internationaltransportforum.org/Pub/pdf/10FP05.pdf

McKinnon, A. (2008, May 28–30). *The potential of economic incentives to reduce $CO_2$emissions from goods transport*. Paper prepared for the 1st International Transport Forum on 'Transport and Energy: The Challenge of Climate Change', Leipzig. Accessed August 4, 2012, from http://www.internationaltransportforum.org/Topics/Workshops/WS3McKinnon.pdf.

Metz, D. (2010). Saturation of demand for daily travel. *Transport Reviews, 30*(5), 659–674.

Metz, B., Davidson, O. R., Bosch, P. R., Dave, R., & Meyer, L. A. (2007). *Contribution of Working Group III to the Fourth Assessment Report of the Intergovernmental Panel on Climate Change, 2007*. Cambridge, UK: Cambridge University Press. Accessed August 4, 2012, from http://www.ipcc.ch/publications_and_data/ar4/wg3/en/contents.html.

Meyer, M. D., & Miller, E. J. (2001). *Urban transportation planning* (2nd ed.). Singapore: McGraw-Hill. ISBN 0-07-120000-2.

Millard-Ball, A., & Schipper, L. (2011). Are we reaching peak travel? Trends in passenger transport in eight industrialized countries. *Transport Reviews, 31*(3), 357–378.

Nellthorp, J., & Mackie, P. J. (2000). The UK roads review: A hedonic model of decision-making. *Transport Policy, 7*(2), 127–138.

ODI. (2011). Disability facts and figures. Office for Disability Issues. Accessed August 4, 2012, from http://odi.dwp.gov.uk/disability-statistics-and-research/disability-facts-and-figures.php#gd.

ONS. (2011). Government revenues from environmental taxes 1993-2010. Office for National Statistics. Accessed August 4, 2012, from http://www.ons.gov.uk/ons/publications/re-reference-tables.html?edition=tcm%3A77-224120.

Peng, Z. (2011). Urban transportation strategies in Chinese cities and their impacts on the urban poor. Wilson Center. Accessed November 14, 2014, from http://www.wilsoncenter.org/publication/urban-transportation-strategies-chinese-cities-and-their-impacts-the-urban-poor.

Peters, A., von Klot, S., Heier, M., Ines Trentinaglia, I., Hörmann, A., Wichmann, E., et al. (2004). Exposure to traffic and the onset of myocardial infarction. *The New England Journal of Medicine, 351*(17), 1721–1731.

Reardon, L., Abdallah, S., & Seaford, C. (2011). The road to well-being: The relationships between transport and well-being: A report on the existing literature. New Economics Foundation. Accessed August 4, 2012, from www.neweconomics.org.

Rodrigue, J.-P. (2006). Transportation and the geographical and functional integration of global production networks. *Growth and Change, 37*(4), 510–525.

Rodrigue, J.-P., Comtois, C., & Slack, B. (2009). *The geography of transport systems*. New York: Routledge.

Rogers, S. (2011). Government spending by department, 2010-11: Get the data. *The Guardian* on-line. Accessed August 4, 2012, from http://www.guardian.co.uk/news/datablog/2011/oct/26/government-spending-department-2010-11.

Rothengatter, W. (2003). Environmental concepts—Physical and economic. In D. A. Hensher & K. J. Button (Eds.), *Handbook of transport and the environment* (pp. 9–35). London: Elsevier.

Schafer, A., Heywood, J., Jacoby, H. D., & Waitz, I. (2009). *Transportation in a climate constrained world*. Cambridge, MA: MIT Press.

Schafer, A., & Victor, D. G. (2000). The future mobility of the world population. *Transportation Research Part A, 24*, 171–205.

Schrank, D., & Lomax, T. (2009). 2009 Urban mobility report. Texas Transportation Institute. Accessed August 4, 2012, from http://mobility.tamu.edu.

SEU. (2003). Making the connections: Final report on transport and social exclusion. Social Exclusion Unit London.

Short, J., Van Dender, K., & Crist, P. (2009). Transport policy and climate change. In D. Sperling & J. Cannon (Eds.), *Reducing climate impacts in the transportation sector* (pp. 35–48). New York: Springer. ISBN 978-1-4020-6978-9.

Singh, S. K. (2005). Future mobility in India: Implications for energy demand and $CO_2$ emission. *Transport Policy, 13*, 398–412.

Spellerberg, I. F. (2002). *Ecological effects of roads*. Enfield: Science.

Sperling, D., & Gordon, D. (2009, November–December). Two billion cars: Transforming a culture. Transportation Research News 259. Transportation Research Board.

Standing Advisory Committee on Trunk Road Appraisal (SACTRA). (1999). Transport and the Economy: Full report (SACTRA). Department for Transport (DfT), London.

Sussman, J. M. (2000). *Introduction to transportation systems*. Boston: Artech House.

Tian, Y., Zhu, Q., Lai, K.-H., & Venus Lun, Y. H. (2014). Analysis of greenhouse gas emissions of freight transport sector in China. *Journal of Transport Geography, 40*, 43–52.

Transportation Research Board (TRB). (1997). *Toward a sustainable future: Addressing the long-term effects of motor vehicle transportation on climate and ecology*. Washington, DC: National Academy Press.

United Nations. (2006). Some facts about persons with disabilities. Accessed November 14, 2014, from http://www.un.org/disabilities/convention/pdfs/factsheet.pdf.

Van Acker, V., & Witlox, F. (2005). Exploring the relationships between land-use system and travel behaviour concepts: Some first findings. *Solstice: An Electronic Journal of Geography and Mathematics, XVI*(2), 1–14.

Vermeulen, I., Van Caneghem, J., Block, C., Baeyens, J., & Vandecasteele, C. (2011). Automotive shredder residue (ASR): Reviewing its production from end-of-life vehicles (ELVs) and its recycling, energy or chemicals' valorization. *Journal of Hazardous Materials, 190*(1–3), 8–27.

Victoria Transport Policy Institute (VTPI). (2005). Sustainable transportation and TDM. Planning that balances economic, social and ecological objectives. TDM Encyclopedia.

von Weizsäcker, E. U., Hargroves, K. C., Smith, M. H., Desha, C., & Stasinopoulos, P. (2010). *Factor five: Transforming the global economy through 80% improvements in resource productivity*. London: Earthscan.

Wachs, M. (2005). What are the challenges to creating sustainable transportation? In *TRB conference on integrating sustainability into the transportation planning process*, 11–13 July 2004, Baltimore, TRB Conference Proceedings 37.

Watts, L., & Urry, J. (2008). Moving methods, travelling times. *Environment and Planning Part D: Society and Space, 26*, 860–874.

WBCSD. (2004). Mobility 2030: Meeting the challenges to sustainability. World Business Council for Sustainable Development, The Sustainable Mobility Project: Full Report, Geneva. Accessed August 4, 2012, from http://www.wbcsd.org.

Wegman, F., Eksler, V., Hayes, S., Lynam, D., Morsink, P., & Oppe, S. (Eds.). (2005). *SUNflower +6: A comparative study of the development of road safety in the SUNflower+6 countries; Final report*. Leidschendam: SWOV.

Whitelegg, J. (1993). *Transport for a sustainable future: The case for Europe*. London: Belhaven Press.

Whitelegg, J. (1997). *Critical mass: Transport, environment and society in the twenty-first century*. London: Pluto Press.

Whitelegg, J., & Haq, G. (2003). The global transport problem: Same issues but different place. In J. Whitelegg & G. Haq (Eds.), *World transport: Policy and practice* (pp. 3–25). London: Earthscan.

WHO. (2004). *World report on road traffic injury prevention*. Geneva: World Health Organization. ISBN 92 415 6260 9.

WHO. (2009). Air quality and health: Fact sheet no 313. World Health Organization. Accessed August 4, 2012, from http://www.who.int/mediacentre/factsheets/fs313/en/index.html.
WHO. (2013). *Gobal status report on road safety 2013: Supporting a decade of action*. Geneva: World Health Organisation.
Zhao, P. (2014). Private motorised urban mobility in China's large cities: The social causes of change and an agenda for future research. *Journal of Transport Geography, 40*, 53–63.
Zietsman, J., & Rilett, L. R. (2002). *Sustainable transportation: Conceptualization and performance measures* (p. 149). Austin, TX: University of Texas.

## Additional Readings: Planning

Banister, D. (2002). *Transport planning*. London: Routledge.
Cowie, J. (2010). *The economics of transport: A theoretical and applied perspective*, Oxon: Routledge.
Mallard, G., & Glaister, S. (2008). *Transport economics: Theory, application and policy*, Basingstoke: Macmillan.
Sperling, D., & Gordon, D. (2009). *Two billion cars: Driving toward sustainability*. Oxford: Oxford University Press.

# 4 Transportation and Sustainability

## 4.1 Purpose and Content

As discussed in Chap. 3, the transportation system is often envisioned as the engine of development. It is seen as the backbone of the twentieth century's economic and social progress and is the means by which humans access goods and services and connect to communities. Yet, it is also a major contributor to environmental degradation and community disruption, which is often inequitably distributed. The falling costs and increasing efficiency of the transportation system have enabled the emergence of the throughput society (see Sect. 2.3). The ease with which materials and goods can be moved within and between nations has transformed the structure of national economies, enabling connectivity across the world. This process is further enhanced by the global emergence of information and communication technology (ICT). The complexity unleashed by the integration of regional and national economies means that tracing who or what is responsible for negative externalities is not a simple question to ask or answer. With the possibility of two billion vehicles on the horizon (Sperling and Gordon 2010) and growing mobility trends around the world, the challenges presented by transportation are likely to command public attention for the foreseeable future.

In this chapter, we explore the emerging field of sustainable transportation. We look at how the concept has been informed by the key events and documents that describe sustainable development. Having conceptualized transportation in Chap. 3, we revisit some of the material introduced in Chap. 2 through the lens of transportation to track how core sustainability concepts have permeated the field of sustainable transportation. The discussion follows the emergence of definitions and principles of sustainable transportation since the early 1990s. A take-away from this discussion is that sustainable transportation can be framed in two ways. It can be considered as a subject in its own right where transportation is the center of attention *or* the transportation system can be viewed based on its contribution to sustainable development. The former perspective is what we refer to as the transportation-centered view, whereas the latter is called the holistic view based



H. Gudmundsson et al., *Sustainable Transportation*, Springer Texts in Business and Economics, DOI 10.1007/978-3-662-46924-8_4

on the need to adopt a multi-sector approach. As will be argued, these two ways to frame sustainable transportation have important implications for the scope of issues that the transportation profession might address. The chapter concludes by considering how the transportation-centered and holistic views can be used to take advantage of the strengths of both approaches and what this means for sustainable transportation practitioners.

## 4.2 The Emergence of Sustainable Transportation

> There can be no sustainable development without sustainable transportation. It is an essential component not only because transportation is a prerequisite to development in general but also because transportation, especially our use of motorized vehicles, contributes substantially to a wide range of environmental problems, including energy waste, global warming, degradation of air and water, noise, ecosystem loss and fragmentation, and desecration of the landscape. Our nation's environmental quality will be sustainable only if we pursue transportation in a sustainable way (Benfield and Replogle 2002, p. 647).

The concept of sustainable transportation can be described as "an expression of sustainable development in the transportation sector" (Zietsman and Rilett 2002, p. 10). This expression is informed by the declarations and texts from the Stockholm (1972) and Rio (1992 and 2012) conferences, from key publications such as *Our Common Future*, from government policies and international treaties focused on reducing the negative impacts from transportation, and more recently from international discussions on the post-2015 agenda. The following text takes a close look at how the concept of sustainable transportation has emerged over the past four decades, paying attention to the different ways in which transportation is framed—i.e., as an integral part of a larger global system or as a somewhat independent sector that can make substantive changes by itself.

### 4.2.1 From Stockholm (1972) to Rio (1992)

The 1972 Stockholm conference was the first international conference to call the prevailing industrialization processes into question (see Sect. 2.2). While transportation was not directly addressed at the conference, the Stockholm Declaration did articulate principles that would later become key tenets of the notion of "sustainable transportation." Principle 15 stated that "[p]lanning must be applied to human settlements and urbanization with a view to avoiding adverse effects on the environment and obtaining maximum social, economic and environmental benefits for all."[1]

[1] Source: The United Nations Declaration of the United Nations Conference on the Human Environment, http://www.unep.org/Documents/Default.asp?DocumentID=97&ArticleID=1503 (accessed on April 19, 2015).

Principle 13 stated that "States should adopt an integrated and coordinated approach to their development planning so as to ensure that development is compatible with the need to protect and improve [the] environment for the benefit of their population." The importance of adopting a holistic and integrated approach to planning and decision-making for sustainable development is discussed in Sect. 2.5 and later in Sect. 4.3.

A year after the Stockholm Conference, the 1973 International Convention for the Prevention of Pollution from Ships (MARPOL) was established in the wake of several major oil spills, creating the first treaty to directly target the impacts of transportation on the environment. The services that transportation provides also indirectly link the sector to a number of multilateral environmental agreements (MEAs). For example, the 1973 Convention on International Trade in Endangered Species (CITES) and the 1989 Basel Convention on the Control of Transboundary Movements of Hazardous Wastes and Their Disposal both relate to problems enabled by transportation. In the urban context, the role of transportation (or mobility) in creating socially and environmentally sustainable towns and cities has been a primary focus of the UN-Habitat since its formulation in 1978.[2]

Some 15 years after the Stockholm Conference, the Brundtland Commission's report, published as *Our Common Future*, provided the first comprehensive formulation of sustainable development (see Sect. 2.2.5). The report adopted a holistic (cross-sectoral) approach to development in an effort to capture the global and interconnected nature of the problems faced, which tended to fall outside the purview of traditional sectors: "Until recently, the planet was a large world in which human activities and their effects were neatly compartmentalized within nations, within sectors (energy, agriculture, [transportation,] trade), and within broad areas of concern (environment, economics, social). These compartments have begun to dissolve. This applies in particular to the various global 'crises' that have seized public concern, particularly over the past decade. These are not separate crises: an environmental crises, a development crisis, and energy crisis. They are all one" (WCED 1987, p. 4).

Since the report avoided adopting a "sectoral" approach, transportation is only briefly considered in the context of nonrenewable energy, urban mobile source emissions, and the need for public transportation systems (WCED 1987, pp. 198–199). It is also discussed in connection with the challenges facing urban settlements. Given the central role that transportation plays in enabling economic activity, it is surprising that it did not receive greater attention. Notwithstanding this observation, the report has become a foundational text to virtually all efforts to advance sustainability.

In 1992, the concept of sustainable transportation began to take shape at the first Rio conference. Agenda 21—the action plan of the Rio Declaration—included two chapters that address transportation. Chapter 7, which covered sustainable human

[2] See, for example, the 2013 report by the UN-Habitat on "Planning and Design for Sustainable Urban Mobility: Global Report on Human Settlements."

settlements, called for a comprehensive approach to urban transportation planning that focused on ways to promote "efficient and environmentally sound urban transport systems in all countries" (UN 1993, § 7.52). More specifically, each nation was asked to:

(a) Integrate land use and transportation planning to encourage development patterns that reduce transport demand;
(b) Adopt urban-transport programmes favouring high-occupancy public transport in countries, as appropriate;
(c) Encourage nonmotorized modes of transport by providing safe cycleways and footways in urban and suburban centres in countries, as appropriate;
(d) Devote particular attention to effective traffic management, efficient operation of public transport, and maintenance of transport infrastructure;
(e) Promote the exchange of information among countries and representatives of local and metropolitan areas;
(f) Reevaluate the present consumption and production patterns in order to reduce the use of energy and national resources (UN 1993, § 7.52).[3]

The second transportation-related chapter, Chapter 9, focused on the protection of the atmosphere.[4] The primary concern with transportation lay with its contribution to atmospheric emissions (UN 1993, § 9.13). To address this problem, a program area was developed to encourage nations "to develop and promote cost-effective policies or programmes, as appropriate, to limit, reduce or control … harmful emissions into the atmosphere and other adverse environmental effects of the transport sector, taking into account development priorities as well as the specific local and national circumstances and safety aspects" (UN 1993, § 9.14).

In addition, signatory governments to Agenda 21 were asked to "[d]evelop and promote … cost-effective, more efficient, less polluting and safer transport systems," as well as integrate transportation planning in rural and urban areas (UN 1993, § 9.15.a).

While Agenda 21's text on transportation is somewhat limited, there are several broad principles within the Rio Declaration[5] that are particularly relevant to transportation. First, the Declaration states that environmental protection must be integrated into the development process (Principle 4). To help achieve this objective, the Declaration endorses three different approaches: (1) to act with *precaution* where future outcomes are uncertain (Principle 15); (2) to develop economic instruments that *internalize* the costs of negative externalities (Principle 16); and

[3] These objectives were supported by a call for *public awareness campaigns* and *human resource development* to highlight and support the need for change (UN 1993, § 7.53).

[4] Chapters 7 and 9 of Agenda 21 had an influential role in shaping the President's Council on Sustainable Development's (PCSD's) approach to transportation discussed in Sect. 4.2.3. Specifically, the PCSD (1996b, 1999) considered transportation in the context of sustainable communities and global climate change.

[5] Source: The United Nations Rio Declaration on Environment and Development, http://www.unep.org/Documents/Default.asp?DocumentID=78&ArticleID=1163 (accessed on April 19, 2015).

(3) to undertake an *environmental impact assessment* when a proposed activity is likely to have a significant adverse impact on the environment (Principle 17). All three of these approaches are directly applicable to transportation, and many of them are currently used in transportation planning and decision-making throughout the world.

Second, the Rio Declaration states that stakeholder participation is essential when addressing environmental issues (Principle 10). The provision of information to stakeholders and their involvement in decision-making are directly relevant to addressing concerns for "environmental justice" and problems such as social exclusion discussed in Sect. 3.4.2.

Finally, the Rio Declaration states that national activities should not lead to environmental damage in other states (Principle 2). This principle is directly relevant to the emission of greenhouse gases and other air pollutants from transportation motor vehicles and equipment and applies to situations where transportation-related regional air pollutants cross national jurisdictions.

### 4.2.2 The Commission of the European Communities' 1992 Green Paper

At the time of the Rio Conference, national governments and international agencies began to address what the new sustainable development agenda meant for transportation policy. A notable example is the Commission of the European Communities' Green Paper on "The Impact of Transport on the Environment" (EC 1992). This paper is one of the first policy texts to focus on sustainable transportation/mobility, and provides an early example of the transportation-centered perspective.

The Green Paper defined a sustainable mobility framework as one that:

- Contains "the impact of transport on the environment[;]"
- Allows "transport to continue to fulfil its economic and social functions[;]"
- Contributes "to social and economic cohesion ... and to the creation of new opportunities for the peripheral regions[;]"
- Safeguards "the freedom of choice for the user[;]" and
- Identifies, "in accordance with the principle of subsidiarity, the responsibility which the different actors will have to assume in order to achieve the objective of the strategy" (EC 1992, p. 55).

This framework was criticized for being "a typical political compromise containing something for everyone but with no clear indications of new policy directions" (Short 1995, p. 9). While valid, this criticism does not acknowledge the significance of having sustainable transportation as an agenda item for the European Community. The momentum behind much of the work in Europe on sustainable transportation over the past decade has been made possible by early policy documents such as this.[6] In this regard, the Green Paper should be seen as a milestone in Europe's formulation of, and debate on, the concept of sustainable transportation. It had the effect of moving global environmental issues from the periphery to the center of transportation policy and placed sustainable mobility at the top of analysts' agendas (Gudmundsson and Hojer 1996).[7] See Chap. 8 for a discussion of the 2011 European Union White Paper that provides a comprehensive framework for an integrated sustainable transportation strategy in Europe.

### 4.2.3 The President's Council on Sustainable Development (1993–1999)

The USA followed a different pathway. Since the federal government does not have a national sustainable development strategy, efforts to promote the concept have tended to focus on specific issues—such as linking air quality and transportation legislation (Lyons 2000; Weiner 1997; Hall 2006; Pollard 2009) or focusing on livable/sustainable communities and climate change (PCSD 1996b, 1999). The most significant federal initiative that looked specifically at creating a national strategy on sustainable development—the President's Council on Sustainable Development (PCSD) established under the Clinton Administration—did not consider transportation in a comprehensive manner. As Benfield and Replogle (2002, p. 650) comment, "the work of the PCSD related to transportation planning and management was somewhat scattered and arbitrary, varying in emphasis and structure from one report to another." Notwithstanding this criticism, the PCSD did manage to capture many of the fundamental elements of sustainable transportation within the themes of sustainable communities and climate change.

In its first report,[8] the PCSD (1996b) included transportation in its section on "strengthening communities." The Council identified four steps that could be taken to move the transportation system toward sustainability and presented four indicators to measure progress toward this objective (Box 4.1). In addition, with regard to community growth and management, the Council recommended that the federal government "encourage shifts in transportation spending toward transit, highway maintenance and repair, and expansion of transit options rather than new highway or beltway construction" (PCSD 1996b, p. 99). Further, the principle of an

---

[6] For a summary of the latest European work on sustainable transportation, see the European Commission, Sustainable Transport, http://ec.europa.eu/transport/themes/sustainable/index_en.htm (accessed on April 19, 2015).

[7] It would be almost a decade until the European Council (2001) endorsed its internationally-accepted definition of sustainable transportation that is discussed in Sect. 4.2.4.

[8] During its 6-year existence (1993–1999), the PCSD prepared three reports (PCSD 1996a, b, 1999) that are often referred to as a basis for a national strategy on sustainable development (Dernbach 2002; Dernbach and Bernstein 2003; Spyke 2005).

accessible transportation system was included in the sixth goal of the PCSD, which focused on sustainable communities.[9]

**Box 4.1: PCSD's (1996b, p. 54) Recommendations and Indicators Relating to Transportation and Sustainable Development**

[The PCSD outlined four] ... steps that can be taken by government at all levels, communities, businesses, and residents to address the challenge of a sustainable transportation system.

- Improve community design to contain sprawl better, expand transit options, and make efficient use of land within a community to locate homes for people of all incomes, places of work, schools, businesses, shops, and transit in close proximity and in harmony with civic spaces.
- Shift tax policies and reform subsidies to improve economic and environmental performance and equity in the transportation sector significantly.
- Make greater use of market incentives in addition to changes in tax and subsidy policies to achieve environmental objectives.
- Accelerate technology developments and encourage public–private collaboration to move industrial sectors closer to economic, environmental, and equity goals.

Progress in the transportation sector could be measured using the following indicators:

- **Congestion**: Decrease in congestion in metropolitan areas.
- **National Security**: Increase in economic and national security through reduced dependency on oil imports.
- **Transportation Efficiency**: Decrease in the rates of freight and personal transportation emissions of greenhouse gases and other pollutants, including carbon monoxide, lead, nitrogen oxides, small particulate matter, sulfur dioxide, and volatile organic compounds.
- **Transportation Patterns**: Progress toward stabilizing the number of vehicle-miles traveled per person, while increasing the share of trips made using alternative transportation modes.

[9] The ten national goals for promoting sustainable development were grouped under the following headings: (1) Health and the Environment; (2) Economic Prosperity; (3) Equity; (4) Conservation of Nature; (5) Stewardship; (6) Sustainable Communities; (7) Civic Engagement; (8) Population; (9) International Responsibility; and (10) Education (PCSD 1996b).

In 1993, President Clinton and Vice President Gore released their "Climate Change Action Plan" (Clinton and Gore 1993)—which included a specific set of actions directed at transportation.[10] While many initiatives in the Climate Change Action Plan proved to be unsuccessful (Hahn et al. 2003; Brunner and Klein 1999), President Clinton continued to show support for the issue by revising the PCSD's charter in 1997 and requesting the Council to advise him on the domestic implementation of policy options to reduce greenhouse gas emissions.

As a result of this new mandate, the PCSD devoted the first substantive chapter of its final report, *Towards a Sustainable America*, to address climate change. Within this chapter, the Council put forward three recommendations to reduce the transportation sector's impact on climate change. These were to:

1. Reduce greenhouse gas emissions from vehicles.
2. Accelerate development and use of cleaner fuels and engines.
3. Reduce vehicle-miles traveled (PCSD 1999, p. 21).

To support the above recommendations, the Council developed ten action items that focused on a wide array of transportation-related initiatives (Box 4.2). Many of the items in Box 4.2 are central to achieving a more sustainable transportation system. In addition, these action items incorporated the three transportation initiatives put forward in the 1993 Clinton–Gore Climate Change Action Plan. The PCSD's (1999) final report continued the Council's focus on the concept of sustainable communities; however, this time, less emphasis was given to transportation.

**Box 4.2: PCSD's (1999, p. 22) Ten Transportation Action Items to Address Climate Change**

**Action 1**: Government and businesses should accelerate efforts to procure clean fuel/engine fleet vehicles and fuel them in ways that result in real reductions of greenhouse gas emissions.

**Action 2**: Establish consumer tax incentives for purchase of efficient, advanced technology vehicles.

**Action 3**: Establish new programs and strengthen existing policies that foster alternative transportation choices and provide an incentive to drive fewer miles including:

(continued)

[10] Three recommendations put forward by the Climate Change Action Plan focused on transportation. These were to (1) provide workers with the option to cash-in the value of their employer-paid parking spaces to pay for commuting alternatives to the automobile; (2) reduce VMT; and (3) create a tire labeling program to help consumers identify tires that have low rolling resistance (Clinton and Gore 1993).

**Box 4.2** (continued)

(a) Policies that encourage the use of mass transit such as tax benefits for employer-subsidized transit pass and parking cash-out programs.
(b) Credits or incentives for compact development.
(c) Policies that promote car-sharing programs such as those already established in Europe and the United States, which offer the potential to reduce greenhouse gas emissions by lowering the total number of vehicle trips and vehicle-miles traveled within major cities.
(d) Public education and outreach efforts to identify and promote the benefits of efficient vehicles and other transportation choices to stimulate demand for these technologies.
(e) Research on the impact of telecommuting, information technologies, and Internet commerce on reducing greenhouse gas emissions.

**Action 4**: Improve infrastructure for intermodal transportation (i.e., bike racks, bus shelters, train stations).

**Action 5**: States and localities should establish appropriate road pricing policies that reduce congestion, mitigate greenhouse gases, and mitigate any impact on low-income commuters.

**Action 6**: In cases where greenhouse gas reductions can be quantified and verified against credible benchmarks, give communities the opportunity to receive credit when they use community design to lower traffic by adopting zoning codes and other changes that encourage more efficient land use patterns to reduce pollution from motor vehicles.

**Action 7**: Increase and redirect existing support for research, development, and deployment and production of advanced vehicle components toward technologies that enable greater efficiency including hybrid electric systems, lightweight materials, clean engines, energy storage systems, and fuels.

**Action 8**: Support research to determine the potential of intelligent transportation systems (a group of technologies that could improve the flow of traffic through urban areas) to reduce greenhouse gas emissions.

**Action 9**: Prioritize and accelerate efforts to develop infrastructure for alternative-fueled vehicles that reduce greenhouse gas emissions.

**Action 10**: Perform additional research on how to reflect the number of vehicle-miles traveled as a variable cost of insurance so that drivers better understand the price associated with the number of miles they drive.

While transportation was not the primary focus of the PCSD, the council's work helped lay the foundation for current federal transportation activities promoting sustainable communities[11] and climate change.[12] Yet, the lack of a national sustainable development strategy under which transportation policies and programs can be developed leaves state and local governments to formulate their own sustainable transportation policies and programs. For example, see the case study of New York State Department of Transportation's (NYSDOT's) GreenLITES rating systems in Chap. 10. While this flexibility has its advantages in terms of testing innovative approaches, it also presents challenges in terms of aligning approaches across states and regions.

### 4.2.4 International Sustainable Transportation Initiatives (1995–2001)

Following the 1992 Rio Conference, one of the earliest international efforts to develop the concept of sustainable transportation was the OECD project on Environmentally Sustainable Transport (EST) initiated in 1995.[13] During the first phase of the EST project, the project team developed a qualitative definition of sustainable transportation that was very much inspired by Daly's principles of ecological carrying capacity (see Sect. 2.3). The team initially defined an EST system as one in which "[t]ransportation … does not endanger public health or ecosystems and meets mobility needs consistent with (a) use of renewable resources at below their rates of regeneration and (b) use of non-renewable resources at below the rates of development of renewable substitutes" (OECD 1996, p. 54). This aligns well with the input–output transportation system model set out in Chap. 3 (Sect. 3.3.2).

In later work, the EST definition was revised by expanding on its basic principles and relating them to quantified international environmental and health criteria and targets. The revised definition is presented below:

> [A] sustainable transport system is one that throughout its full life-cycle operation:

[11] See, for example, the 1999 Clinton–Gore Livable Communities Initiative, http://clinton3.nara.gov/CEQ/livability.html (accessed on April 19, 2015) and the 2009 HUD-DOT-EPA Partnership for Sustainable Communities, http://www.epa.gov/smartgrowth/partnership/index.html (accessed on April 19, 2015), and the FHWA's Livability Initiative, http://www.fhwa.dot.gov/livability/ (accessed on April 19, 2015).

[12] See the US DOT Transportation and Climate Change Clearinghouse, http://climate.dot.gov/ (accessed on April 19, 2015).

[13] The EST project consisted of six teams of experts from nine countries. Each team focused on one of the following geographic regions—Sweden, the Netherlands, Germany, the Quebec-Windsor corridor in Canada, the greater Oslo region in Norway, and the Alpine region that consisted of parts of Austria, France, Italy, and Switzerland.

- allows generally accepted objectives for health and environmental quality to be met, for example, those concerning air pollutants and noise proposed by the World Health Organization (WHO);
- is consistent with ecosystem integrity, for example, it does not contribute to exceedances of critical loads and levels as defined by WHO for acidification, eutrophication, and ground-level ozone; and
- does not result in worsening of adverse global phenomena such as climate change and stratospheric ozone depletion (OECD 2000, p. 35).

This revised EST definition—while more comprehensive than the original one—still does not include other important social and economic criteria of sustainable development.[14]

In parallel with the early stages of the EST project, the OECD held an important conference in Vancouver (24–27 March, 1996)—*Towards Sustainable Transportation*—that pulled together some 400 transportation stakeholders from 25 nations to develop a vision and chart a course for sustainable transportation (OECD 1997). As one of its key outcomes, the conference endorsed the so-called "Vancouver principles of sustainable transportation" that covered a range of environmental, social, and economic issues (Box 4.3). The principles also highlighted the importance of adopting an open and inclusive decision-making process. We dedicate Chap. 5 to the *governance* of the transportation system and why this is critical for making progress toward sustainable transportation.

**Box 4.3: The Vancouver Principles of Sustainable Transportation**

**Access:** Improve access to people, goods, and services but reduce demand for the physical movement of people and things.

**Decision-making:** Make transportation decisions in an open and inclusive manner that considers all impacts and reasonable options.

**Urban planning:** Limit sprawl, ensure local mixes of land uses, fortify public transport, facilitate walking and bicycling, protect ecosystems, heritage, and recreational facilities, and rationalise goods movement.

**Environmental protection:** Minimise emissions and reduce waste from transport activity, reduce noise and use of nonrenewable resources, particularly fossil fuels, and ensure adequate capacity to respond to spills and other accidents.

**Economic viability:** Internalise all external costs of transport including subsidies but respect equity concerns, promote appropriate research and

(continued)

[14] In a paper reviewing the main results of the OECD EST project, Caid et al. (2002, p. 220) present a slightly revised EST definition that includes a fourth component: "provides for safe, economically viable, and socially acceptable access to people, places, goods, and services" (p. 220). Caid et al.'s (2002) addition to the EST definition and the European Council's revision of the CST definition (discussed below) provide good examples of how definitions of sustainable transportation are seldom fixed and are continually evolving.

**Box 4.3** (continued)
development, consider the economic benefits including increased employment that might result from restructuring transportation, and form partnerships involving developed and developing countries for the purpose of creating and implementing new approaches to sustainable transportation.
Source: OECD (1997, p. 36).

The Vancouver principles should be regarded as a first step toward a comprehensive understanding of sustainable transportation. Indeed, the OECD conference report concludes that "[e]very effort should be made to encourage and invite further work on the development and wider dissemination of this set of principles" (OECD 1997, p. 68). While many national bodies such as the UK Round Table on Sustainable Development (UKRTSD 1996) and the U.S. Transportation Research Board (TRB 1997) were engaged in advancing the concept, Canada's work proved to be influential in establishing an internationally accepted definition of sustainable transportation.

In response to the progress made at the Vancouver conference and the call for further work, Environment Canada and Transport Canada—two agencies of the Canadian government—created the Centre for Sustainable Transportation (CST) (Yevdokimov 2004). The Centre developed one of the first definitions that explicitly covered the three dimensions of sustainable development (see Sects. 2.3 and 4.2.6). This definition formed the basis for what has since become the most cited political definition of sustainable transportation worldwide. The definition's visibility grew significantly when the European Union used the CST's work to develop a notion of a "sustainable transport system" that was subsequently adopted by the European Council of Ministers of Transport and Telecommunication in Luxembourg in 2001 (European Council 2001). Both the CST and European Council definitions are included in Table 4.1 for comparison. While the definitions are similar, the European Council's version adopts Daly's (1991a) terminology to describe the use of renewable and nonrenewable resources. The definition also highlights the importance of balanced regional development—which refers to the European agenda to integrate newer and poorer member states into the north–west dominated economic center and market—and expands the CST's focus on individuals and societies to include companies.

**Table 4.1** CST and European Council definitions of sustainable transportation

| Centre for Sustainable Transportation (CST) | European Council |
|---|---|
| "A sustainable transportation system is one that:<br>– allows the basic access needs of individuals and societies to be met safely and in a manner consistent with human and ecosystem health and with equity within and between generations.<br>– is affordable, operates efficiently, offers choice of transport mode, and supports a vibrant economy.<br>– limits emissions and waste within the planet's ability to absorb them, minimizes consumption of non-renewable resources, limits consumption of renewable resources to the sustainable yield level, reuses and recycles its components, and minimizes the use of land and the production of noise" (CST 1997, p. 1). | "THE COUNCIL ... RECOGNISES, that there is a need for further action in order to attain a sustainable transport system defined as one that<br>– allows the basic access and development needs of individuals, *companies* and societies to be met safely and in a manner consistent with human and ecosystem health, and promotes equity within and between successive generations;<br>– is affordable, operates fairly and efficiently, offers choice of transport mode, and supports a *competitive economy, as well as balanced regional development*;<br>– limits emissions and waste within the planet's ability to absorb them, *uses renewable resources at or below their rates of generation, and, uses non-renewable resources at or below the rates of development of renewable substitutes* while minimising the impact on the use of land and the generation of noise" (European Council 2001, pp. 15–16). |

### 4.2.5 Rio+20 and the Post-2015 Agenda

The 2012 Rio+20 Conference was instrumental in reaffirming the importance of transportation and mobility in achieving sustainable development. The following excerpt from the report on the outcome of the conference, entitled *The Future We Want*, provides the first global (UN General Assembly) statement on sustainable transportation and builds on the ideas raised at the first Rio conference in 1992.

> **Sustainable transport**
>
> 132. We note that transportation and mobility are central to sustainable development. Sustainable transportation can enhance economic growth and improve accessibility. Sustainable transport achieves better integration of the economy while respecting the environment. We recognize the importance of the efficient movement of people and goods and access to environmentally sound, safe and affordable transportation as a means to improve social equity, health, resilience of cities, urban–rural linkages and productivity of rural areas. In this regard, we take into account road safety as part of our efforts to achieve sustainable development.
>
> 133. We support the development of sustainable transport systems, including energy efficient multimodal transport systems, notably public mass transportation systems, clean fuels and vehicles, as well as improved transportation systems in rural areas. We recognize the need to promote an integrated approach to policymaking at the national, regional, and local levels for transport services and systems to promote sustainable development. We also recognize that the special development needs of landlocked and transit developing countries need to be taken into account while establishing sustainable transit transport systems. We acknowledge the need for international support to developing countries in this regard (UN 2012, p. 25).

**Table 4.2** Transportation-related targets in the emerging SDGs

| SDSN's post-2105 development goals (SDSN 2014) | High-level panel's post-2105 development goals (UN 2013) | OWG's post-2015 development goals (UN 2014) |
|---|---|---|
| **Goal 6: Improve Agriculture Systems and Raise Rural Prosperity**<br>6c. Ensure universal access in rural areas to basic resources and infrastructure services (land, water, sanitation, modern energy, transport, mobile and broadband communication, agricultural inputs, and advisory services)<br>**GOAL 7: Empower Inclusive, Productive, and Resilient Cities**<br>7b. Ensure universal access to a secure and affordable built environment and basic urban services including housing; water, sanitation and waste management; low-carbon energy and transport; and mobile and broadband communication | **Goal 7: Secure Sustainable Energy**<br>7c. Double the global rate of improvement in energy efficiency in buildings, industry, agriculture, and transport<br>**8. Create Jobs, Sustainable Livelihoods, and Equitable Growth**<br>8c. Strengthen productive capacity by providing universal access to financial services and infrastructure such as transportation and ICT | **Goal 11: Make Cities and Human Settlements Inclusive, Safe, Resilient, and Sustainable**<br>11.2 By 2030, provide access to safe, affordable, accessible and sustainable transport systems for all, improving road safety, notably by expanding public transport, with special attention to the needs of those in vulnerable situations, women, children, persons with disabilities and older persons |

In Chap. 2, we argue that the post-2015 agenda is likely to provide a more comprehensive articulation of concerns relating to sustainable development. While transportation was not included within the original eight Millennium Development Goals (MDGs) or any of the related targets, this situation is set to change with the post-2015 agenda. A read of the three main sets of proposed sustainable development goals (SDGs), shown in Table 2.1, reveals that transportation is not likely to be highlighted at the "goal" level. However, each of the three sets of SDGs includes transportation in a "target" under one or more of its goals (Table 4.2).

Both the Sustainable Development Solutions Network (SDSN 2014) and UN High-level Panel (UN 2013) focus on providing "universal access" to transportation, with the SDSN's targets making a distinction between urban versus rural access.[15] Universal access to transportation services is framed within the context of

[15] The urbanrural nexus (that can be characterized as consisting of flows of urbanrural people, goods, and services) was a key topic of conversation during the deliberations of the UN Open Working Group (OWG) on SDGs (IISD 2014). In particular, the OWG discussed the link that transportation provides between rural and urban areas and how the two are interdependent. The overriding concern was to ensure that any targets developed would not position urban and rural areas against one another, given their "common destiny" (IISD 2014, p. 7). It is noteworthy that the OWG choose not to differentiate between the urban and rural setting in their recommended targets (see Table 4.1).

enabling economic development and productivity. The High-level Panel also includes an explicit focus on doubling the rate of improvement in the energy efficiency of transportation.

The most extensive transportation target is provided by the UN Open Working Group (OWG), which highlights the need for "safe, affordable, accessible and sustainable transport systems for all" (UN 2014). Emphasis is given to expanding public transportation as a means of improving road safety and to meeting the needs of vulnerable groups. Like the High-level Panel, the OWG also calls for doubling the global rate of improvements in energy efficiency but does not specifically target or mention the transportation sector.

The draft SDGs and targets will surely evolve prior to the official release of the final set in 2015. Input from transportation groups, such as the Partnership on Sustainable, Low Carbon Transport (SLoCaT) and the World Resource Institute's (WRI's) EMBARQ initiative for sustainable transport and urban development, provides a valuable source of information on how sustainable transportation could be framed and measured. For example, SLoCaT's (2014b) *Results Framework on Sustainable Transport* presents six transportation-related targets[16] and supporting indicators that have been used as a foundation for developing recommendations on how the OWG's SDG targets could be refined (SLoCaT 2014a). SLoCaT (2014a) supports the OWG's decision to not create a dedicated transportation SDG (given the cross-sectoral nature of transportation) and focuses its attention on how transportation could be mainstreamed into the targets supporting eight of the OWG's 17 proposed SDGs. SLoCaT also presents the following argument for retaining Goal 11 of the OWG's proposed SDGs (see Table 4.2) given its direct link to transportation: "If it were decided to mainstream the cities and human settlements SDG into other SDGs like energy, water, health or education, it would become very difficult, if not impossible, for the transport sector to take meaningful guidance from the SDG framework in the development of the transport sector" (SLoCaT 2014b, p. 1).

The work of SLoCaT and other international organizations provides some insight into the complex process of creating the SDGs and their supporting targets. The need for a manageable set of targets for the SDGs means there is little space for extended lists of objectives with associated indicators. Thus, not all of the factors that have been associated with sustainable transportation can be incorporated in the SDG framework. An important question is what factors are currently not covered by the proposed targets?

Table 4.3 combines the core elements of the CST and European Council's definitions with several other notable definitions to create an overview of factors that could be included in a comprehensive definition of sustainable transportation. The table helps identify the factors that fall outside the proposed SDG

[16] The six targets focus on improving rural access, improving urban access, improving national access and regional connectivity, improving road safety and security, reducing air pollution, and reducing emissions (SLoCat 2014a).

**Table 4.3** Components of a comprehensive definition of sustainable transportation

| Dimension | Topic | A sustainable transportation system ... |
|---|---|---|
| Environment | Health and Environmental Damage | Minimizes activities that cause serious public health concerns and damage to the environment[a, b, d] |
| | Noise | Minimizes the production of noise[b–e] |
| | Land Use | Minimizes the use of land[c, e] |
| | Emissions and Waste | Limits emissions and waste to levels within the planet's ability to absorb them and does not aggravate adverse global phenomena including climate change, stratospheric ozone depletion, and the spread of persistent organic pollutants[b–e] |
| | Renewable Resources | Ensures that renewable resources are managed and used in ways that do not diminish the capacity of ecological systems to continue providing these resources[a–e] |
| | Nonrenewable Resources | Ensures that nonrenewable resources are used at or below the rate of development of renewable substitutes[a–e] |
| | Energy | Is powered by renewable energy sources |
| | Recycling | Reuses and recycles its components[c] |
| Society | Access | Provides access to goods, resources, and services while reducing the need to travel[a, c, e] |
| | Safety | Operates safely[a, c, e] and ensures the secure movement of people and goods |
| | Intragenerational Equity | Promotes equity between societies and groups within the current generation[c, e], specifically in relation to concerns for environmental justice |
| | Intergenerational Equity | Promotes equity between generations[c, e] |
| Economy | Affordability | Is affordable[a, c, e] |
| | Efficiency | Operates efficiently to support a competitive economy[a, c, e] |
| | Social Cost | Ensures that users pay the full social and environmental costs for their transportation decisions[a] |
| | Employment | Provides meaningful and well-paid employment opportunities |

[a]U.K. Round Table on Sustainable Development. (1996). *Defining a sustainable transport sector.* London
[b]OECD. (1997). *Towards sustainable transportation* (the Vancouver Conference), Paris
[c]The Centre for Sustainable Transportation. (1997). *Definition and vision of sustainable transportation.* Ontario
[d]OECD. (2000). *Environmentally sustainable transport (EST).* Paris
[e]European Council. (2001). Council resolution on *Integrating environment and sustainable development into transport policy.* Luxembourg. See also Barrella et al. (2010) and Amekudzi et al. (2011)

transportation-related targets (Table 4.2), which emphasize accessibility, energy efficiency, safety, affordability, public transportation, and the needs of vulnerable groups. While the framing of safety and affordability is comparable between the SDG targets and Table 4.3, there are several nuanced differences between the other overlapping factors that are worth highlighting.

Ensuring access to transportation services is included in both the SDG targets and Table 4.3, but there is an important difference between how the term is used. The SDG targets speak of access to transportation primarily in the context of developing countries where many people have little ability to travel or send their products to markets. In contrast, the concept of access in Table 4.3 emerged from developed countries where the discourse tends to focus on the different functions and activities people should have access to and not necessarily on providing more transportation. In this context, accessibility could be improved through integrated land use planning (proximity access) or enhanced telecommunication services (virtual accessibility), which could reduce the need to travel.

With regard to energy, the SDG targets advocate energy efficiency improvements, whereas Table 4.3 advocates a transition to renewable energy resources while ensuring that the capacity of ecological systems to continue providing these resources is not diminished. One of the proposed SDG targets relating to transportation highlights the need to provide special attention to vulnerable groups, whereas Table 4.3 emphasizes the need to promote intragenerational equity and environmental justice. These three examples indicate that while there is overlap between the general transportation-related factors included in the SDG targets and Table 4.3, the different ways in which the factors are framed mean that the actions they influence could vary considerably. This observation highlights how critical frameworks are to defining a problem and creating a system to measure and manage progress in addressing this problem (see Chap. 7).

The factors that the SDG targets do not explicitly or implicitly cover include transportation-related environmental and public health concerns, noise, land use, and recycling of waste.

### 4.2.6 The Dimensions of Sustainability Applied to Transportation

As should be evident from the above discussion, the core principles of sustainable development—i.e., meeting human needs and improving quality of life (QoL); living within the earth's ecological carrying capacity; living off ecological interest rather than consuming natural capital; and protecting future generations (Beatley 1995; WCED 1987; Rees 1995; Daly 1991a; Costanza and Daly 1992; Holdren et al. 1995)—have been incorporated to varying degrees in conceptualizations of sustainable transportation. There is an international consensus that the concept of sustainable transportation can be defined under the three dimensions of environment, equity/society, and economy. These dimensions are often referred to as the Three E's (Hall 2006), the three pillars, or the triple bottom line (TBL). The latter framing (TBL) originated in the business world, although it is now frequently used by transportation agencies and practitioners. While the labeling of each aspect can vary, the underlying objective is to ensure that environmental and social considerations are balanced or considered alongside economic outcomes.

The term the "Three E's" was first used in the mid-1970s in discussions on the topics of the *Economy*, the *Environment*, and *Energy*. During the 1990s, Energy became an intrinsic part of the Environment and was replaced by *Ethics* (or Equity) as society gradually became aware that a movement toward a sustainable future could not occur without a transformation of individual priorities and values (Kidder 1990). The notion was that the environment and the economy are shaped by our ethics—our sense of right and wrong—and that incorporating ethics into decisions might begin to alter the past objectives of growth, accumulation, and excess toward new objectives of sustainability, sharing, and restraint. The emphasis given to energy efficiency improvements in the proposed sets of SDGs signals that energy considerations will remain central to efforts to promote sustainable development. Thus, it may be useful to once again separate energy from the environment, in the same way that separating employment from the economy enables the development of policies that can co-optimize multiple objectives at once (Ashford and Hall 2011). The latter case raises the question of whether solutions designed to promote green growth—defined as economic growth that has been *decoupled* from growth in environmental pollutants—also creates well-paid and meaningful jobs. A failure to explicitly consider employment could lead to the displacement of jobs via technological upgrading enabled by policies to green industrial performance (Ashford et al. 2012). While somewhat unwieldy, the "Five E's" of Environment, Energy, Economy, Employment, and Equity may enable the creation of "multi-purpose" policies/solutions that have less unintended consequences. It could be argued that the single-purpose design of policies—such as creating a policy to "only" improve environmental quality—is one reason why progress toward sustainable development has been slow and difficult to achieve (Ashford and Hall 2011).

In response to the proliferation of factors included in the concept of sustainable transportation (see Table 4.3), Holden et al. (2013) argue for a return to Brundtland's formulation of sustainable development to reduce the risk of the concept becoming too diluted and, hence, ineffective. Black (1996, p. 151) previously provided a Brundtland-inspired definition of sustainable transportation as satisfying "current transport and mobility needs without compromising the ability of future generations to meet these needs" (Black 1996, p. 151).[17] The problem with

[17] The Bogota Declaration that was adopted at the Regional Sustainable Transport Forum in Bogota, held on June 23–24, 2011, provides a more nuanced version of this definition. It defines sustainable transportation as "the provision of services and infrastructure for the mobility of people and goods needed for economic and social development and improved quality of life and competitiveness. These services and transport infrastructure provide secure, reliable, economical, efficient, equitable and affordable access to all, while mitigating the negative impacts on health and the environment locally and globally, in the short, medium and long term without compromising the development of future generations" (FTS 2011, p. 1).

this type of definition is that it does not provide transportation professionals with any clear guidance on how to design transportation policies and programs to promote sustainable development (Black 1996, 2010).

Holden et al. (2013) propose the adoption of four main dimensions of sustainable development that can be derived from *Our Common Future* and applied to transportation—(1) safeguarding long-term ecological sustainability, (2) satisfying basic human needs, (3) promoting intragenerational equity, and (4) promoting intergenerational equity. By focusing on passenger transportation, Holden et al. (2013) developed an indicator with an associated target for each dimension. These indicators and targets are as follows: the daily per capita energy consumption for passenger transportation should not exceed 5.6 kW h (dimension 1); the minimum daily per capita travel distance by motorized transportation should be 9.2 km (dimension 2); the minimum value for the Public Transport Accessibility Level (PTAL) indicator should be three, which relates to moderate accessibility (dimension 3); and minimum renewable energy share for transportation should be 15 % (dimension 4). Regardless of whether one agrees that these indicators effectively measure each dimension, they do provide a starting point for developing transportation policies that can be linked with the Brundtland formulation of sustainable development. This back-to-basics approach also positions the transportation sector alongside other sectors in the pursuit of sustainable development.

If the transportation sector is considered in the broader context of sustainable development, one might question whether the sectoral-focus implied by the term sustainable transportation is too narrow and constraining. Indeed, if taken at face value, it implies that the transportation system can be made sustainable in its own right, possibly without the need to consider other sectors—such as energy, manufacturing, and housing/land use. This argument touches upon the ongoing debate about transportation and sustainability—i.e., "whether it is about sustainable transportation, transportation sustainability, or transportation in support of a sustainable society" (McVoy et al. 2010, p. 3). When reviewing the foundational texts on sustainable development, the transportation system is positioned within the broader context of sustainable development. This holistic perspective puts a spotlight on the contribution of transportation and other sectors to unsustainable development, which has important implications in terms of framing the problem and potential solutions.

The previous sections introduce important transportation-relevant text within the Stockholm and Rio Declarations, Agenda 21, the Rio+20 conference report, and the post-2015 agenda, which have shaped the definitions and principles of sustainable transportation. Much of the work focusing on the transportation sector at the international level—specifically work driven by the UN—continues to call for, and build on, the objectives set out in the two UN declarations and Agenda 21 (ECOSOC 2001a, b). An interesting characteristic of this work is that the topic of transportation is treated as a subset of other topics such as human settlements or energy (WEHAB Working Group 2002; UNCHS 2001). The explicit recognition of "sustainable transport" in the Rio+20 conference report and the post-2015 development agenda, however, signals that transportation is likely to have greater prominence in the future international sustainable development agenda.

**Discussion Topics**

- Is it possible to make significant progress toward sustainable development without a national or international framework in place to promote this objective? In the absence of such a policy framework, what options exist for transportation agencies to promote sustainable development?
- Is it viable for the transportation sector to take the lead in promoting sustainable development, or would this be equivalent to the tail (i.e., the transportation sector) trying to wag the dog (i.e., development)?
- While the definitions and principles of sustainable transportation are valuable, how should we deal with competing objectives? For example, achieving a safe or accessible transportation system may conflict with the objective of providing affordable services. Is there a hierarchy to the principles/factors? How would such a hierarchy be decided upon?
- Can you identify examples from national, regional, or local policy where a transportation "need" has been defined? How important is this need to social progress within a sustainable transportation paradigm?

## 4.3 Holistic and Sector-Specific Definitions of Sustainable Transportation

An important question raised previously is whether it is beneficial to develop transportation policies and programs from a sustainable development (i.e., holistic) rather than a sustainable transportation (i.e., transportation-centered) perspective. We believe both perspectives are important.

The holistic view is important since it defines the boundaries (the macro-ecological limits) within which all sectors must collectively operate. Three approaches that lend themselves to the holistic perspective are the capital model of sustainable development, ecological economics, and the notion of macroecology. The holistic view also invites a broader consideration of QoL and expands the analysis of equity to include distributional impacts of transportation between regions and nations over time. In contrast, the transportation-centered view is important since it provides *sector-specific* objectives that guide the development of transportation policies and programs. It should be acknowledged, however, that some decisions are more relevant and appropriate within a transportation-centered view (e.g., route alignment once a decision to build infrastructure has been taken) while others command a more integrated cross-sectoral approach (e.g., investments in welfare support to improve the QoL for low income families, where transportation has an important role but is one of many factors). While some decisions would fare better under some framings than others, a fair assumption is that the transportation-centered view is linked with institutional mandates that narrowly focus agencies on their own core remits. Thus, the existence of a national sustainable development policy/strategy that permits/encourages cross-sectoral partnerships may be an essential step in realizing a holistic approach to development. These issues are explored further in Chap. 5.

The holistic view of development highlights the importance of knowing whether existing sectors (such as energy, transportation, agriculture, etc.) are operating within sustainable parameters. An interesting framework for considering this holistic view is the capital model of sustainable development. This model works by identifying the current stocks of natural, manufactured, human, financial, and social capital and determines how these should be maintained or invested in for future generations.[18] While these stocks of capital cover the physical, social, and virtual domains—the core building blocks of development—this discussion is primarily interested in the physical domain that is captured by natural, manufactured, and human capital.

Natural capital can only be protected and enhanced if there is a mechanism to monitor and set limits to resource usage and pollution levels. These limits can be defined in terms of maximum sustainable yield, carrying capacity, critical levels, quality standards, vulnerability, resilience, fragility, etc. (Munn 1989; Nijkamp 1994). The limits become the operating parameters within which all sectors must collectively function. A sector's actions can only be described as "sustainable" (with regard to natural capital) if the sector is operating within its allotment of pollution rights and resource usage rates. The allocation of such would either need to be set by government or determined using market mechanisms,[19] which is where ecological economics can be applied.

Daly's (1991b) notion of a steady-state economy provides a useful conceptual framework for considering how production and consumption cycles exist within ecosystem limits (see Sect. 2.3) (Fig. 4.1). In Fig. 4.1, the transportation system is considered as one of many physical systems—such as energy, communications, industrial/manufacturing, etc.—that collectively support production and consumption cycles. For substantive progress to be made toward sustainable development, it is the cumulative impact from *all* the systems/sectors that matters, not whether one particular system is sustainable in its own right, ignoring the issue of whether this outcome is even possible. Consider the current debate on decarbonizing economies in order to dramatically reduce greenhouse gas emissions. The decision about whether and how fast to electrify the passenger car fleet depends critically on the costs of such technology and the extent to which the energy sector is itself being

[18] In practice, however, only the first four of these types of capital are considered in any detail due to the difficulty in measuring social capital.

[19] The elegance of market mechanisms is that a government would not be required to determine how the burden of staying within ecological and resource-use limits should be divided between sectors. Instead, the trading mechanisms would (theoretically) allocate these burdens in the most economically efficient manner.

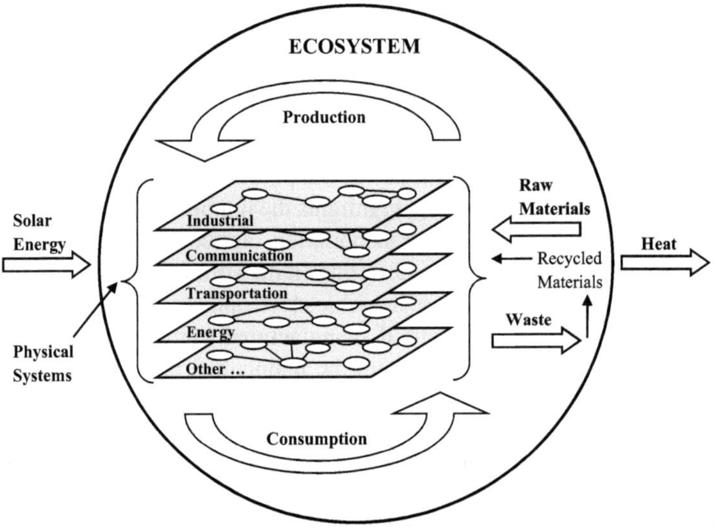


**Fig. 4.1** The transportation system as one of many systems supporting production and consumption cycles within ecosystem limits. *Source*: Adapted from Daly (1991b, p. 181)

decarbonized. The opportunities for earlier deeper cuts in other sectors may be more cost effective. The decisions are integrated and not independent.

The idea of maintaining/improving natural capital is conceptually straightforward; however, setting limits to pollution levels and resource usage is likely to be difficult from both a scientific and political perspective. If the same principle of maintaining/improving capital is applied to manufactured and human capital the analysis becomes far more complicated. Identifying acceptable and objective ways to "value" the preservation or development of these forms of capital is extremely difficult—perhaps impossible.[20] A rational conclusion, therefore, is to focus on maintaining natural capital and to reorient inputs to development on renewable resource streams. Such an approach is adopted by the Natural Capital Project, that focuses its analysis on "quantifying the values of natural capital in clear, credible, and practical ways," to enable decision-makers "to quantify natural capital in biophysical, socio-economic and other dimensions, to visualize the benefits delivered today and in the future, to assess the tradeoffs associated with alternative choices, and to integrate conservation and human development aims."[21]

[20] The tight interconnection between natural and manufactured capital means that if the use of natural capital is constrained, so too is the development of manufactured capital. However, to what extent the development of manufactured capital is constrained will depend upon whether a weak or strong form of sustainable development is applied (see Sect. 2.3).

[21] Source: The Natural Capital Project, http://www.naturalcapitalproject.org/ (accessed on April 19, 2015).

While the definitions of sustainable transportation recognize the need to limit pollution and the use of resources to sustainable levels, they often do not explicitly recognize the role of other sectors in achieving this objective. One way to adjust the existing definitions of sustainable transportation is to recognize the need for the transportation sector to coordinate (or at best integrate) its decision-making processes with those of other sectors. From a macroecological perspective, this means recognizing that each sector is embedded within a larger system and actions taken in one subsystem may have an impact on the larger system.

Given the above context, a close read of the European Council's definition of sustainable transportation reveals its transportation-centered focus. For example, it states that a sustainable transportation "limits emissions and waste within the planet's ability to absorb them, uses renewable resources at or below their rates of generation, and, uses non-renewable resources at or below the rates of development of renewable substitutes ..." (European Council 2001, pp. 15–16), without reference to the cross-sectoral collaboration needed to realize this objective. Preceding this statement by the term "in coordination with other sectors," would position the transportation sector alongside other sectors in the macroecosystem. Adjusting the definition in this way presents an explicit requirement for the transportation sector to work with other sectors to solve problems associated with the environment. It makes inter-sector cooperation a primary agenda item for transportation agencies, providing legitimacy to action taken in this area. Of course, until institutional mandates are updated to enable such cooperation, agencies in all sectors of the economy will need to work within their existing institutional missions in innovative and creative ways. There is also the question of whether agencies would be willing to work with new agencies, especially if this action required a cultural shift in the organization. Overcoming the inertia of existing planning and decision-making processes may be a difficult challenge requiring dedicated time and attention.

While the above focus is on the environmental aspect of the definition, it would be equally important for sectors to work together when addressing the social and economic dimensions as well. For example, following the recession that started in 2008/2009 there has been significant stimulus money spent on infrastructure to generate employment, although this is only one of many ways in which the labor market could have been stimulated. Further, strategies targeted at only one sector may inadvertently create problems in other sectors given the complex interrelationships that exist between transportation, energy, agriculture, etc. (see Sect. 2.5). In the social dimension, actions taken to promote development within a sector must not lead to growing inequality in other sectors.

While challenges clearly exist to advancing a more integrated and cross-sectional approach, if making progress toward sustainable development is the primary objective, such action will be essential. In Chap. 5, we provide a foundation for thinking about the governance of transportation and how this presents challenges and opportunities for linking/integrating policy within the transportation sector and across sectors. The US FHWA's *Transportation Planning for Sustainability Guidebook* provides the following comment regarding the potential value of adopting a multi-sector perspective on sustainable development.

**Table 4.4** Advantages and disadvantages of adopting a transportation-centered or holistic view of sustainable development

| Sustainable transportation perspective (*the transportation-centered view; Three E's*) | Sustainable development perspective (*the holistic view; global perspective*) |
|---|---|
| *Advantage* – Provides sector-specific objectives and principles that guide the development of transportation policies and programs | *Advantage* – Highlights the need to establish a national framework/policy to address sustainable development that can encourage sectors to coordinate/integrate their activities |
| *Disadvantage* – Does not explicitly connect impacts from the transportation sector with those from other sectors. Thus, transportation tends to be considered in a vacuum | *Disadvantage* – Does not provide detailed sector-specific objectives and principles to guide the development of transportation policies and programs |
| *System Perspective* – Single system | *System Perspective* – Multiple interconnected systems |

> There are several examples of international efforts to address sustainability, and a great deal of them have a broader scope because they are legally authorized and have allocated funding to address sustainability for entire nations and even regions. New Zealand (NZ) and the United Kingdom (UK) have national strategies for sustainable transportation. In the case of the UK, this strategy is part of a broader national sustainable development strategy involving a number of sectors and institutions, e.g., energy and the environment. The European Union (EU) has also developed a sustainable development strategy having a transportation component. Nations and regions that invest in the development of broader sustainable development visions, goals and objectives are likely to develop more comprehensive solutions involving multiple sectors and several institutions with related functions. They are also more likely to identify confounding effects of policies that may be good for one sector, but not particularly effective for another, thus motivating agencies to work together to achieve systemic and enduring solutions (Amekudzi et al. 2011).

Table 4.4 summarizes the advantages and disadvantages of adopting a transportation-centered versus holistic approach to sustainable development. We believe that the advantages of both approaches are needed. The transportation-centered approach provides focused objectives and principles that guide the development of transportation policies and programs. The holistic perspective then links these initiatives to the global ecosystem and asks whether the transportation sector's contribution to sustainable development is sufficient to realize substantive change. It also opens a dialogue for a discussion about the most cost effective way to realize a desired objective—i.e., which sector or sectors are best positioned to lead change in a specific area. Thus, the underlying question we raise in this book is whether current best practices are sufficient to make substantive progress on addressing the major environmental challenges we face.

Connecting the perspectives in this way is not a simple task and would require leadership, resources, and an environment where new approaches can be tried and tested, and failure is an acceptable part of the process. It would also present society with difficult decisions or trade-offs that lie at the heart of sustainable development.

For example, if a transportation service is to be constrained to keep emissions within a predefined limit, society would be confronted with a decision as to whether to curtail its mobility and economic activity or search for ways to maintain the level of service by seeking emission reductions in another sector, which in turn may impact other non-transportation services. The process of initiating such a conversation that is supported by objective analyses could spur innovation by revealing future opportunities and markets for innovative solutions. The act of constraining a problem has the effect of opening up the "design space" of the planner, analyst, designer, engineer, decision-maker, etc. who can develop targeted solutions. Thus, the macroecological constraints that the holistic perspective provides may be an essential factor in the development of innovative (perhaps, disruptive) solutions that may or may not be led by the transportation sector.

**Discussion Topics**

- How important is it to have a definition that explicitly requires sectors to work together? Is this a necessary condition for progress to be made toward sustainable development?
- What types of decisions require a more comprehensive cross-sectoral approach and which require a more transportation-centric approach? How does this work in practice?
- What examples can you find where a definition or policy explicitly calls for sectors to work together to address a complex societal problem? How did the various government agencies and stakeholders respond to this call?

## 4.4 Conclusions

This chapter has explored how the concept of sustainable transportation evolved from the foundational principles of sustainable development. These general principles are global in scope and are not constrained by sectoral boundaries. Thus, when translating these principles to the transportation sector, it is important to place them within a broader sectoral and macroecological context. The underlying theme of this chapter is the importance of integrated, multi-sectoral solutions to the sustainability challenge ahead (Dernbach 2003; Ashford and Hall 2011). As David Banister, a prominent transportation academic, has commented, "Much of the decision-making process is carried out at all levels of government within a sectoral framework. Sustainable development is all embracing and requires new thinking so that cross-sectoral decisions can be made" (Banister 2005, p. 3). Hence, by taking a critical look at existing principles/definitions of sustainable transportation, this chapter highlights an approach by which the transportation-centered versus holistic views of sustainable development can be considered to take advantage of the strengths of both approaches. This combination means that system-level sustainable transportation indicators need to be expressed within the context of regional/national/global indicators—that is, they indicate the transportation sector's contribution to the environmental, social, and economic problems faced. These indicators need to go beyond a traditional "silo" approach to transportation performance measurement.

Encouraging transportation professionals to consider both the transportation-centered and holistic approach to sustainability establishes a framework where critical questions can be asked such as whether current sustainable transportation policies and programs are sufficient to make substantive progress in critical areas of concern. Without the holistic perspective, it is possible to frame almost any progress, no matter how small, as advancing sustainability objectives. The challenge we pose to transportation agencies and professionals is to go one step further and ask whether these changes are sufficient to make a substantive difference at the regional or global (macroecological) scale. Implicit in this approach is the need to adopt a mutli-sector/agency view, which will require the creation of an environment where new institutional and organizational relationships can be tried and tested. Research on the question of how transportation agencies might respond to future demands for sustainable transportation services has already begun (Booz Allen Hamilton 2014), but much more work is needed to find the strategies that work in all settings. Performance measurement frameworks will be central to understanding, guiding, and communicating the contribution the transportation sector makes toward sustainable development.

The following chapter (Chap. 5) takes an important look at the *governance* of the transportation system. In particular, it discusses the key elements in transportation decision-making. Chapter 6 then describes the process of creating useful indicators and how indicators can be used for different applications in the planning, delivery, and operation of transportation services. Chapter 7 builds on Chaps. 5 and 6 and describes a structured process (or framework) that will enable transportation organizations to incorporate sustainable development principles into their practices.

## References


Amekudzi, A. A., Meyer, M. D., Ross, C. L., & Barrella, E. (2011). *Transportation planning and sustainability guidebook*. Washington, DC: U.S. Department of Transportation. Available at http://www.fhwa.dot.gov/hep/climate/sustainability/sustain00.cfm

Ashford, N. A., & Hall, R. P. (2011). *Technology, globalization, and sustainable development: Transforming the industrial state*. New Haven, CT: Yale University Press.

Ashford, N. A., Hall, R. P., & Ashford, R. (2012). Addressing the crisis in employment and consumer demand: reconciliation with environmental and financial sustainability. *The European Financial Review,* October-November, 63–68.

Banister, D. (2005). *Unsustainable transport: City transport in the new century*. London: Routledge.

Barrella, E., Amekudzi, A. A., Meyer, M. D., Ross, C. L., & Turchetta, D. (2010). Best practices and common approaches for considering sustainability at U.S. state transportation agencies. *Transportation Research Record, 2174*, 10–18.

Beatley, T. (1995). The many meanings of sustainability: Introduction to a special issue of JPL. *Journal of Planning Literature, 9*, 339–342.

Benfield, F. K., & Replogle, M. (2002). Transportation. In J. C. Dernbach (Ed.), *Stumbling towards sustainability* (pp. 647–665). Washington, DC: Environmental Law Institute.

Black, W. R. (1996). Sustainable transportation: A U.S. perspective. *Journal of Transport Geography, 4*, 151–159.

Black, W. R. (2010). *Sustainable transportation: Problems and solutions*. New York: Guilford Press.

Booz Allen Hamilton. (2014). *National Cooperative Highway Research Program Report 750: Strategic issues facing transportation, Volume 4: Sustainability as an organizing principle for transportation agencies*. McLean, VA: Booz Allen Hamilton.

Brunner, R. D., & Klein, R. (1999). Harvesting experience: A reappraisal of the U.S. Climate Change Action Plan. *Policy Sciences, 32*(2), 133–161.

Caid, N., Crist, P., Gilbert, R., & Wiederkehr, P. (2002). Environmentally sustainable transport: Concept, goal and strategy—The OECD's EST Project. *Proceedings of the Institution of Civil Engineers, Transport, 153*, 219–226.

Centre for Sustainable Transportation. (1997). *Definition and vision of sustainable transportation*. Ontario: The Centre for Sustainable Transportation.

Clinton, W. J., & Gore, A. (1993). *The climate change action plan*. Washington, DC: The White House. http://gcrio.gcrio.org/USCCAP/toc.html

Costanza, R., & Daly, H. E. (1992). Natural capital and sustainable development. *Conservation Biology, 6*, 37–45.

Daly, H. E. (1991a). Elements of environmental macroeconomics. In R. Costanza & L. Wainger (Eds.), *Ecological economics: The science and management of sustainability* (pp. 32–46). New York: Columbia University Press.

Daly, H. E. (1991b). *Steady-state economics*. Washington, DC: Island Press.

Dernbach, J. C. (2002). National governance. In J. C. Dernbach (Ed.), *Stumbling toward sustainability* (pp. 723–743). Washington, DC: Environmental Law Institute.

Dernbach, J. C. (2003). Achieving sustainable development: The centrality and multiple facets of integrated decisionmaking. *Indiana Journal of Global Legal Studies, 10*, 247–285.

Dernbach, J. C., & Bernstein, S. (2003). Pursuing sustainable communities: Looking back, looking forward. *Urban Lawyer, 35*, 495–529.

European Commission (EC). (1992). *Green paper on the impact of transport on the environment: A community strategy for sustainable mobility*. Brussels: European Commission.

European Council. (2001, April 4–5). *2340th Council meeting—Transport/telecommunications—Luxembourg*. Belgium: Council of the European Union.

Foro de Transporte Sostenible (FTS). (2011). *Bogotá declaration*. Bogotá: Foro de Transporte Sostenible (FTS). http://www.uncrdlac.org/fts/BogotaDeclaration.pdf

Gudmundsson, H., & Hojer, M. (1996). Sustainable development principles and their implications for transport. *Ecological Economics, 19*, 269–282.

Hahn, R. W., Olmstead, S. M., & Stavins, R. N. (2003). Environmental regulation in the 1990s: A retrospective analysis. *Harvard Environmental Law Review, 27*, 377–415.

Hall, R. P. (2006). *Understanding and applying the concept of sustainable development to transportation planning and decision-making in the U.S.* Engineering Systems Division, Massachusetts Institute of Technology, Cambridge. http://esd.mit.edu/students/esdphd/dissertations/hall_ralph.pdf

Holden, E., Linnerud, K., & Banister, D. (2013). Sustainable passenger transport: Back to Brundtland. *Transportation Research Part A, 54*, 67–77.

Holdren, J. P., Daily, C., & Ehrlich, P. R. (1995). The meaning of sustainability: Biogeophysical aspects. In M. Munasinghe & W. Shearer (Eds.), *Defining and measuring sustainability: The biogeophysical foundations*. Washington, DC: World Bank. Distributed for the United Nations University.

International Institute for Sustainable Development (IISD). (2014). Summary of the seventh session of the UN General Assembly Open Working Group on Sustainable Development Goals: 6-10. *Earth Negotiations Bulletin, 32*(7), 1–18.

Kidder, R. M. (1990). Perspectives: The three E's of the 1990s. *The Christian Science Monitor.*

Lyons, W. M. (2000, December 13–14). The U.S. Transportation Equity Act for the 21st century and the Clean Air Act Amendments—An innovative framework for transportation & environmental policy. In *ECMT-OECD Workshop on overcoming institutional barriers to implementing sustainable urban travel policies*, Madrid. http://www.oecd.org/cem/UrbTrav/Workshops/InstBarriers/Lyons.pdf

McVoy, G. R., Nelson, D. A., Krekeler, P., Kolb, E., & Gritsavage, J. S. (2010, November 14–17). Moving towards sustainability: New York State Department of Transportation's GreenLITES Story. In *Green streets and highways 2010: An interactive conference on the state of the art and how to achieve sustainable outcomes*, Denver, CO.

Munn, R. E. (1989). Towards sustainable development: An environmental perspective. In F. Archibugi & P. Nijkamp (Eds.), *Economy and ecology: Towards sustainable development* (pp. 42–72). Dordrecht: Kluwer.

Nijkamp, P. (1994). Roads toward environmentally sustainable transport. *Transportation Research, Part A, 28*, 261–271.

Organisation for Economic Co-operation and Development (OECD). (1996). Pollution prevention and control. Environmental criteria for environmentally sustainable transport. Report on phase 1 of the project on Environmentally Sustainable Transport (EST). Paris: Organisation for Economic Co-operation and Development (OECD).

Organisation for Economic Co-operation and Development (OECD). (1997, March 24–27). *OECD proceedings. Towards sustainable transportation. The Vancouver conference*. Paris: Organisation for Economic Co-operation and Development (OECD).

Organisation for Economic Co-operation and Development (OECD). (2000). Environmentally sustainable transport: Futures, strategies and best practice. Synthesis report of the OECD project on Environmentally Sustainable Transport EST. Paris: Organisation for Economic Co-operation and Development (OECD).

Pollard, T. (2009). Transportation: Challenges and choices. In J. Dernbach (Ed.), *Agenda for a sustainable America* (pp. 365–379). Washington, DC: Environmental Law Institute.

President's Council on Sustainable Development (PCSD). (1996a). *Energy and transportation: Task Force Report.* Washington, DC: President's Council on Sustainable Development, Energy and Transportation Task Force.

President's Council on Sustainable Development (PCSD). (1996b). *Sustainable America: A new consensus for prosperity, opportunity, and a healthy environment for the future.* Washington, DC: President's Council on Sustainable Development (PCSD).

President's Council on Sustainable Development (PCSD). (1999). *Towards a sustainable America: Advancing prosperity, opportunity, and a healthy environment for the 21st century.* Washington, DC: President's Council on Sustainable Development (PCSD).

Rees, W. E. (1995). Achieving sustainability: Reform or transformation? *Journal of Planning Literature, 9*, 343–361.

Short, J. (1995). Freight transport as an environmental problem: Direct policy measures for road freight can be effective in reducing environmental harm. *World Transport Policy and Practice, 1*(2), 7–10.

SLoCaT. (2014a). Results framework for sustainable transport. Partnership on Sustainable Low Carbon Transport (SLoCaT). http://www.slocat.net/

SLoCaT. (2014b). Integration of sustainable transport in OWG 12 document on SDGs and targets. Partnership on Sustainable Low Carbon Transport (SLoCaT). http://www.slocat.net/

Sperling, D., & Gordon, D. (2010). *Two billion cars: Driving toward sustainability.* Oxford: Oxford University Press.

Spyke, N. P. (2005). Heeding the call: Making a matter of Pennsylvania law. *Penn State Law Review, 109*, 729–782.

Sustainable Development Solutions Network (SDSN). (2014, April 11). *Proposed sustainable development goals (SDGs) and targets*. New York: Sustainable Development Solutions Network.

Transportation Research Board (TRB). (1997). *Toward a sustainable future: Addressing the long-term effects of motor vehicle transportation on climate and ecology*. Washington, DC: National Academy Press.

UK Round Table on Sustainable Development (UKRTSD). (1996). *Defining a sustainable transport sector*. London: UK Sustainable Development Commission.

United Nations (UN). (1993). *Earth summit: Agenda 21, the United Nations programme of action from Rio*. New York: United Nations.

United Nations (UN). (2012). *The future we want. A/CONF.216/L.1*. New York: United Nations.

United Nations (UN). (2013). *A new global partnership: Eradicate poverty and transform economies through sustainable development. The report of the high-level panel of eminent persons on the post-2015 development agenda*. New York: United Nations.

United Nations (UN). (2014). *Open Working Group proposal for sustainable development goals. Full report of the Open Working Group of the General Assembly on Sustainable Development Goals is issued as document A/68/970*. New York: United Nations.

United Nations Centre for Human Settlements (UNCHS). (2001). *The role of urban transport in sustainable human settlements. Background paper no. 7*. New York: United Nations, Commission on Sustainable Development.

United Nations Economic and Social Council (ECOSOC). (2001a). *Energy and transport. Report of the Secretary-General*. New York: United Nations.

United Nations Economic and Social Council (ECOSOC). (2001b). *Transport. Report of the Secretary-General*. New York: United Nations.

WEHAB Working Group. (2002). *A framework for action on energy*. New York: United Nations. http://www.johannesburgsummit.org/html/documents/wehab_papers.html

Weiner, E. (1997). *Urban transportation planning in the United States: An historical overview* (5th ed.). Washington, DC: Office of the Secretary of Transportation.

World Commission on Environment and Development (WCED). (1987). *Our common future*. Oxford: Oxford University Press.

Yevdokimov, Y. V. (2004). *Sustainable transport in Canada*. New Brunswick: University of New Brunswick. http://www.unb.ca/transpo/documents/SustainableTransportationinCanada...04.pdf

Zietsman, J., & Rilett, L. R. (2002). *Sustainable transportation: Conceptualization and performance measures* (p. 149). Austin, TX: University of Texas.

## Additional Readings

Banister, D. (2008). The sustainable mobility paradigm. *Transport Policy, 15*, 73–80.

Forman, R. T. T., Sperling, D., Bissonette, J. A., Clevenger, A. P., Cutshall, C. D., Dale, V. H., et al. (2003). *Road ecology: Science and solutions*. Washington, DC: Island Press.

Han, S., Kim, Y. H., & Lee, S. (2009). *Green growth in transportation: International views on the transition*. 2009 KOTI-EWI Special report. Gyeonggi-do: The Korea Transport Institute.

Holden, E., Linnerud, K., & Banister, D. (2013). Sustainable passenger transport: Back to Brundtland. Transportation Research Part A, 54, 67–77.

Ieda, H. (Ed.). (2010). *Sustainable urban transport in an Asian context*. New York: Springer.

Ugboaja, P. C., & Ukpere, W. I. (2011). Determining economic sustainability of Nigerian National Transport Policy. *African Journal of Business Management, 5*(8), 3139–3147.

# Governance and Decision-Making in Transportation

# 5

## 5.1 Purpose and Content

The transportation system is developed and managed by a wide range of governmental actors and agencies. As transportation crosses administrative boundaries, there is a need for coordination of policies and actions between the local, regional, national, and international level. In Chaps. 2–4, the importance of coordinating (or better still, integrating) policy areas within any given level of government to promote sustainable development is identified. Further complexity is added when one considers that the transportation system is owned and operated by a mixture of public and private operators acting within a framework set by different levels of government. The challenge of steering and coordinating this complex system is referred to as governance.

This chapter reviews the concept of governance and identifies the need to consider state and non-state actors as well as formal and informal practices in considering how to steer future policy development. It sets out the case for state intervention in transportation. An analysis of existing institutional arrangements and an in-depth case study are used to identify several characteristics of governance systems that are considered to promote planning for sustainable transportation.

Given that the rate of institutional change is typically slow, if the transportation sector is to play a role in contributing to a more sustainable future, we suggest that it is necessary to consider the principles of sustainable development in all levels of transportation decision-making and provide constructive ways of connecting these processes to those in other sectors. This requires reflection on the interplay between politics, institutions, and the use of information to inform and influence decisions. Planning for sustainable development is more than a technical endeavor.

The chapter concludes by identifying key common stages in decision-making processes which are adopted in the public and private realm of the transportation sector. Two broad domains of the transportation planning process—"planning" and "delivery" are defined. This distinction reflects both the different roles of the domains and the different tools which are applied in the two domains. Within



H. Gudmundsson et al., *Sustainable Transportation*, Springer Texts in Business and Economics, DOI 10.1007/978-3-662-46924-8_5

each domain, we also identify a series of functional areas which range from long-term strategic planning to maintenance. This structure provides an important first building block for a framework to integrate decisions about sustainable transportation.

## 5.2 Governance

Treib et al. (2007, p. 3) summarize governance as the "steering and co-ordination of interdependent (usually collective) actors based on institutionalized rule systems (Benz 2004: 25)." There are three important dimensions of government that emerge from this definition (Treib et al. 2007):

1. The rules, laws, and a series of formal or informal practices (or customs) through which systems are governed. Lynn et al. (2001, p. 7) construe governance broadly as "regimes of laws, rules, judicial decisions and administrative practices that constrain, prescribe and enable the provision of publicly supported goods and services." Rosenau (1992, p. 4) refers to governance as "systems of rules, goal oriented activities and purposive behavior."
2. The network of actors that influence decision-making. This moves beyond a notion that decision-making is purely a matter for the state and accepts that it is dependent on inter-organizational networks (Rhodes 1997). Stoker (1998, p. 17) points out the importance of both state and non-state actors, noting that "the boundaries between and within the public and private sectors have become blurred. ... The essence of governance is its focus on governance mechanisms that do not rest on recourse to the authority and sanctions of government the interactive relationship between and within governmental and non-governmental forces."
3. The development of policies to steer progress toward a set of policy objectives, where governance is seen as a "mode of political steering" (Héritier 2002, p. 185). To understand the steering of policy, it is necessary to develop a more nuanced understanding of what policy is. Hall (1993, p. 278) identified policy-making as a process typically involving three elements: overarching policy goals that guide policy, the instruments or techniques used in seeking to attain these goals, and the precise setting of the instruments. Hall (1993) argues that it is more common to see the adjustment of settings of instruments (e.g., the level or structure of an environmental tax) and the instruments used (e.g., the introduction of the EU emissions trading scheme) than it is to see a reevaluation of policy goals.

It is relatively straightforward to translate these generic governance principles to the transportation sector. The transportation system has many rules and conventions. These include matters such as which side of the road to drive on, the minimum safe standard of vehicle operation, the safe curvature of road alignments for different speeds, the tolerance for alcohol in the blood when driving, and the acceptability of low speeds in residential areas. The transportation system is also comprised of a range of state and non-state actors. While the state is normally the main funder of new infrastructure, there has been an increase in privately financed roads with the costs recouped through tolls levied on users. Different levels of state involvement also exist in the management and operation of public transportation services. In the UK, for example, outside of London, bus services are provided by a privatized and deregulated market. In contrast, in the USA, bus services are operated by metropolitan authorities. Thus, there is clearly a mix of public and private sector actors to engage with. Decision-making competencies also exist at different levels from the EU or US Federal level through to nation-state or state level (respectively) and levels below including regional and local. This complex multilevel environment makes a difference to transportation policy development and delivery—see Bache and Flinders (2004), Marsden and May (2006), MacKinnon et al. (2008), and Kern and Bulkeley (2009). Finally, the use of transportation policy as a steering mechanism has been a clear theme over time. Dudley and Richardson (2000), for example, look at the rise and fall of the roads building agenda in the UK. Preston (2003) looks at the role of privatization and deregulation of the bus industry and the myriad potential management models. Marsden et al. (2011) look at the extent to which climate change has changed transport policy.

Governance is about how the system is steered and by whom. If a more sustainable transportation system is to become a reality, then sustainable development principles need to be fully integrated in the governance process—across all actors, through legislation at different scales, and in the day-to-day decision-making practices across the system. While governance consists of far more than the actions of a single government, the government is a key actor. The next section reviews the case for state intervention in transportation.

## 5.3 The Role of Governance in Transportation

While the principal components of governance are evident in transportation, there remain questions as to why and to what extent the state should intervene in the transportation system. Knowles et al. (2008) identify the complexity of the mobility patterns (Sect. 3.2) and their fundamental importance to society as a major driver for action. Transportation is an important area of social policy. However, this is only one of many reasons for state intervention. Shaw et al. (2008) discuss two broad schools of thought on the overarching rationale for managing transportation. The first is the neo-liberal approach (Peck 2001) which suggests that transportation is treated as a market and that the system should evolve through market principles. The second is the "welfare model" which is typically more prevalent in continental Europe than in the UK or USA where the transportation system is owned or regulated by the state to promote a series of social policy goals (Ranci 2011;

Shaw et al. 2008). In reality, a mix of these two approaches is usually adopted. Shaw et al. (2008, p. 65) note that "the state is involved in the regulation and management of transport activities because the conditions rarely exist for an entirely—or even largely—free market in transport to function." The major reasons for intervention are reviewed below.

- Basic standards of operation are required in order for users to know what to expect on the network. The development of a Highway Code for example is synonymous with the need for pedestrians, cyclists, and drivers to know how to perform in the many different circumstances they will face on the highway network. Cowie (2010) defines such interventions as "qualitative regulation."
- The conditions for a perfectly competitive market do not exist. Most transport markets tend toward conditions which lead to small numbers of dominant operators (oligopolies or monopolies). These structures drive up prices and lead to undersupply of transportation, generating economic inefficiency, and can demand higher subsidies for operation (Cowie 2010, p. 175). Regulation is, therefore, required to establish a "second-best" approach to the operation of the transportation market which can take a variety of forms (e.g., regulation of privatized industries, franchising of services, and/or regulation of quality).
- There are a range of "externalities" (see Sect. 3.4) where the users of the transportation system do not experience the full costs of their decisions to travel and therefore "overconsume" travel relative to the societal preference. For example, drivers ignore the delays that they cause to other drivers and do not experience the emissions they produce from their own vehicles during the journey (if their travel is relatively uncongested). These external costs are quite significant and undermine economic, environmental, and social objectives (Mallard and Glaister 2008).
- There are problems of coordination, where it is not always in the commercial interests of transportation operators to integrate services, while for the user this is clearly advantageous (White 2008). This situation is seen in the UK where local and national government acts to provide an integrated bus-rail timetable and information service while each operator considers only its own routes.
- There is a conflict between solutions which are affordable and effective in the short run, but which will make it difficult to respond to longer-term challenges. Social trends such as aging and environmental trends such as climate change may require actions now which are not commercially viable but which will potentially reduce monetary costs and environmental and social impacts in the long run (Stern et al. 2006). In such cases, the state must intervene to enable or incentivize such activities.
- Major infrastructure schemes can be extremely expensive (e.g., estimates of the cost of the Boston "Big Dig" project were US$14 billion while the first 120 miles of a new proposed High Speed Rail line in England is forecast to cost around £16 billion—around US$24 billion). The state has access to capital funding at interest rates which are below the costs at which companies can borrow. While some private sector toll roads and networks have been constructed,

public transport schemes typically cover their operating costs but not construction, leaving a requirement for state intervention at some level (Abelson 2005).

- In the absence of state intervention, it is difficult for new technologies to flourish and establish themselves in the socio-technical regime (Geels 2002; Ieromonachou et al. 2004). The strength of the incumbent technology and the economies of scale of industry matter. For example, since the early unit costs of electric vehicles are extremely high and there is, as yet, no commercial incentive to invest in a network of charging points, the state will need to intervene if it wishes to accelerate the deployment of such technologies. The state may also intervene to fund basic research that holds potential to advance the next generation of transportation technologies/services.
- Social equity concerns in the politics of most countries are poorly represented in neoclassical economics and in commercial decision-making. It is politically and socially desirable to provide some form of service for rural communities or communities with mobility impairments (Farrington and Farrington 2005). The costs of a failure to provide good levels of accessibility to all in society can be quite significant (SEU 2003) and typically fall back on the state in the form of welfare support, health costs, and lower productivity.
- Finally, as set out in Chap. 3, the transportation system needs to be conceptualized as part of a much broader set of systems. Decisions made in other sectors can have significant influence on travel needs which cannot always be met by users. For example, specialization of health care in major centers may make access too difficult for some (Stead 2008). Transportation also relies on energy and is therefore intrinsically linked with the energy system. Decisions about the costs and benefits of actions related to energy futures for the transportation system cannot be made by transportation actors in isolation.

By simply taking account of the known limitations of free market principles for managing transportation, the arguments for state action in the transportation sector appear compelling. The role of government becomes more important when notions of sustainable development are considered. Sustainable development demands that the key externalities are tackled and that equity concerns are fully integrated into decision-making both for current generations and with a much longer-term view for future generations. As we argue in Chap. 4, transportation cannot be sustainable in and of itself, but instead must be part of a sustainable policy system that is integrated with other policy streams. That said, it also seems clear that the transportation sector will need to adopt considerable technological innovations and to operate in a more integrated manner that makes public transportation, walking, and cycling attractive for more journeys if resource consumption is to be effectively driven down. Both of these are unlikely to happen fast enough without state intervention.

A note of caution is also necessary. State intervention does not always lead to more efficient outcomes and identifying effective interventions is more difficult than identifying the need to intervene. Poorly targeted subsidy and regulatory capture by interest groups can serve to undermine policy goals and waste public resources.

Similarly, an entirely state-owned and operated transportation system can be subject to poor discipline on costs and supply choices (Cowie 2010). We do not seek to promote one extreme or the other but rather to underline the need to steer or govern the actions of the various state and non-state actors toward a common goal of sustainable development.

A topic related to the role of governance in transportation that we do not discuss in detail in this book is public–private partnerships (PPPs). PPPs are often viewed as a way to reduce the inefficiencies of state-owned transportation systems. While PPPs are a useful tool to deliver parts of the transportation system, they are put into place to meet specific needs that are identified based on existing goals and current operating environments. We therefore suggest that PPPs are not a solution or panacea for overcoming the challenges of funding complex, long-term cross-sectoral challenges such as sustainable development, but rather a mechanism that can work toward sustainability only in an organizational setting that is aligned toward sustainability.

## 5.4 Institutional Structures in Transportation

The literature on multilevel governance provides a useful framework through which to consider the different elements of institutional structures (Bache and Flinders 2004). Elected governmental bodies that are required to operate across a range of policy areas are referred to as Type I institutions. They have a clearly defined geographical area of remit (their electoral boundaries) and there is a nesting of levels from supranational (EU) or Federal (US) through nation-state (EU) or State (US) down to a local level. There may be many different configurations in between these levels which reflect particular countries' governance structures (MacKinnon et al. 2008). An example would be the extent to which regional governance structures are important and democratically embedded (Pangbourne 2010). Type II institutions are functionally specific but can operate across different spatial scales. Examples would include environmental lobby groups (e.g., The Royal Society for the Protection of Birds in the UK or the Sierra Club in the USA) or arm's-length agencies that are not directly democratically accountable (such as the UK Energy Savings Trust). In addition to these actors, there are a range of private sector operators and interests that will act at the different levels of government (e.g., lobbying politicians).

Transportation systems are organized in different ways in different countries and they work to different sets of norms and philosophies (Table 5.1 discusses institutional settings for selected European countries).

These structures also change over time within a country, and this has been particularly evident with the adoption of neo-liberal principles and the moves to increased marketization and private sector ownership described in Sect. 5.2 (Dudley and Richardson 2000). Formal institutional structures, particularly those with an electoral mandate, are slow to change (Low and Astle 2009). It must also be remembered that the institutional structures that do exist often need to meet the needs of a range of policy areas. It may not make sense to organize care for the

**Table 5.1** Classification of institutional settings of transport sectors

| | Finland | Germany | Greece | United Kingdom |
|---|---|---|---|---|
| Centralization | Moderate decentralization | Moderate decentralization | Moderate decentralization | Moderate decentralization |
| Participation and consultation | Moderate level | Moderate level | Moderate level | High level |
| Coordination across modes | Fragmented | Fragmented | Moderate coordination | Moderate coordination |
| Interest group influence | Moderate influence | Strong influence | Moderate influence | Moderate influence |
| Conflict resolution philosophy | – | Highly efficient conflict resolution procedures | Moderately efficient conflict resolution procedures | – |
| Regulatory intervention | High degree of state control | High degree of state control | High degree of state control | Partially liberalized with overall control by state |
| Knowledge management/ information availability | High level of efficiency | High level of efficiency | Moderate level of efficiency | High level of efficiency |
| Quantification of policy targets | Moderate level of quantification | Low level of quantification | Moderate level of quantification | High level of quantification |
| Feedback and evaluation | Moderate level of feedback or evaluation efficiency | Moderate level of feedback or evaluation efficiency | Moderate level of feedback or evaluation efficiency | High level of feedback and evaluation efficiency |

*Source*: Zografos et al. (2005)

elderly or refuse collection, for example, on the same scale as transportation (which might be organized to support/provide travel to work). While the formal institutional structures are only one component of the governance process, with rules and norms that influence behavior and power of actors within a network, there is evidence that they make a difference to the delivery of transportation policies (Legacy et al. 2012; Marsden and May 2006). The formal structures serve to outline the remit of organizations (e.g., to regulate the rail network infrastructure) or governmental departments (e.g., whether transportation is integrated with planning and the environment (Beecroft 2002) or whether a particular modal remit is stated). They have an influence on the nature of the networks of stakeholders that operate and the power structures that they operate within (Legacy et al. 2012; Lodge 2003). The case study below illustrates a number of tensions observable in transportation governance.

### 5.4.1 Case Study: The English Multimodal Studies

Most transportation powers reside with the governments of the countries comprising the UK (England, Scotland, Northern Ireland, and Wales) following devolution in 1999 (Shaw et al. 2006). In 1998, following the election of a new government which promised an integrated approach to transportation policy (DETR 1998), the then Department for Environment, Transport, and the Regions established a series of 22 "multimodal studies" for England at a cost of £32 million (US$47.7 million). Having inherited a roads program consisting of 147 schemes, the government approved around 50 schemes and referred the rest for decisions in the context of these new studies. The studies covered a range of urban and inter-urban contexts and took a fully multimodal view as to how to solve the worst congestion problems in the country. The studies were to be assessed against a range of policy objectives:

- Integration;
- Safety;
- Economy;
- Environmental impact; and
- Accessibility.

The structure of the studies and the subsequent delivery arrangements are set out in Fig. 5.1.

Figure 5.1 shows three important layers of governance within the system. First, the studies were commissioned by the national government even though some of the solutions proposed related to routes of only regional or local importance or local policies. The studies were managed by the national government's regional offices in close consultation with the unelected regional assemblies that were developing their own regional planning guidance and regional transportation strategies. The regional assemblies had the opportunity to comment on the final outcomes of the studies and to make additional recommendations, which occurred on two occasions (Marsden 2005). The local authorities had prepared Local Transport Plans for the period 2001–2006 and these fed into the regional planning processes and were part of the evidence base available to the multimodal studies teams. The teams themselves were consultants working to a steering group of local and regional stakeholders as well as the main national infrastructure planning agencies of the Highways Agency (responsible for major national roads) and the Strategic Rail Authority (SRA) (responsible for national rail enhancements). There appeared to be no one agency capable of running these studies.

The studies recommended a mixture of public transportation investments, road construction, behavioral adaptation, and strong demand management measures. Only when all of these elements were in place was there a real prospect for keeping congestion levels below levels seen in 2000 (Fig. 5.2). The importance of delivering an integrated package was underlined by many of the study teams (HoC 2003). In total, £28 billion (US$41.7 billion) of investment was recommended from 2001 to 2031 with more than two-thirds of this investment occurring by 2021. Sixty-two

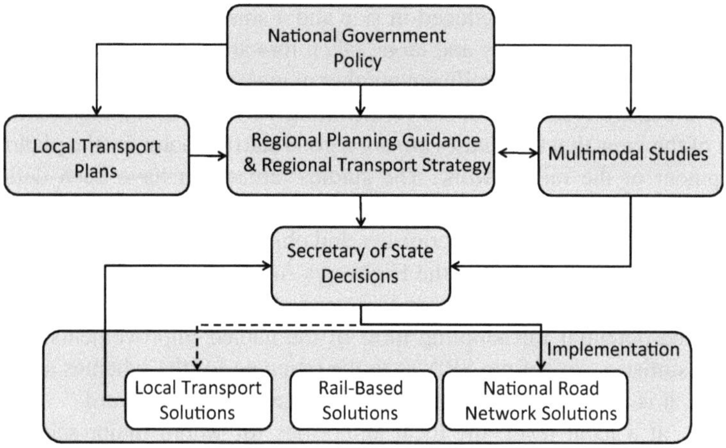


**Fig. 5.1** Overview of multimodal studies planning and implementation

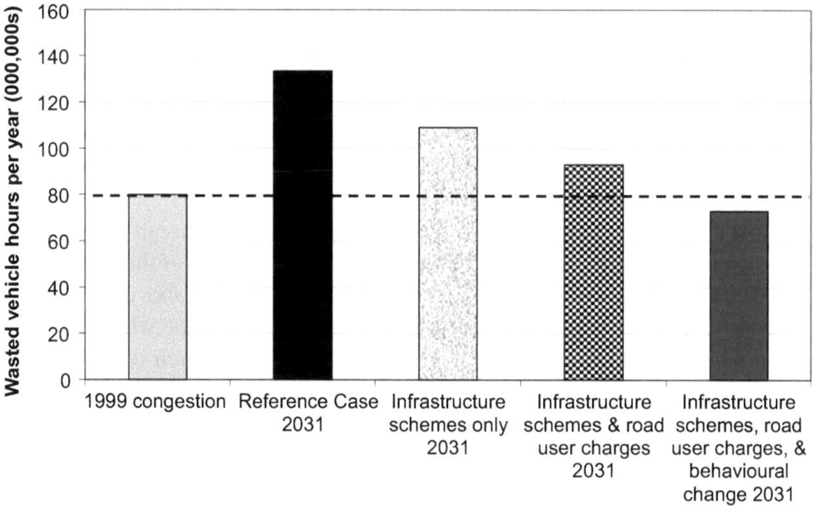


**Fig. 5.2** West midlands area urban multimodal study findings (based on HoC 2003)

percent of the investment was for public transport schemes with one-third (34 %) for roads-based schemes. Given that the studies were commissioned to replace a list of potential road schemes, it could be argued that they demonstrate that truly integrated planning is possible. The institutional framework for delivery, however, complicated the picture considerably.

The Highways Agency is an arm's-length delivery agency for the English government. It had a budget assigned for road construction and was essentially awaiting decisions from the multimodal studies to populate its forward program.

While some schemes were reduced in size and a small number were rejected, the main road proposals were, by and large, taken forward (Shaw et al. 2006).

The SRA already had a significant number of major schemes to be taken forward and had allocated its budget for the forthcoming 10 years. It commented that "The SRA is of the view that the studies are not a good starting point for the planning and development of the rail network. The studies' areas that have been defined are drawn largely to fit unresolved issues on the highway network" (HoC 2003, p. 53). Essentially the SRA had its own priorities and objectives which were not consistent with those of the studies. While the Highways Agency was instructed to accept or advance the design of most of the recommended schemes, the SRA was asked to consider the potential for adopting most of the named improvements. Given the rather pessimistic view of the SRA as to the rationale for the schemes and the lack of budget, it is unsurprising that few were ultimately taken forward.

Finally, at a local level, the local authorities for whom major schemes were recommended then had to develop a case which went back to the Secretary of State for further approval, alongside other local schemes which were not part of the multimodal studies process. This may not have been so problematic had the overall financial planning assumptions of the 22 studies somehow been capped to an affordable level. As things stood, however, each study recommended its preferred package in isolation. The Secretary of State conceded to Parliament that "Had I accepted everything in every single multi-modal study that came my way already I would probably have spent more than I would get for 20 years never mind ten years" (HoC 2003, para 141). There was clearly no realistic prospect of all of these local schemes being funded.

A final element of integration worth reflecting on is the nature of the assessment process. While the multimodal studies undertook the appraisal of the schemes as an integrated whole, each of the three delivery streams described above conducted the appraisal of their own specific schemes in isolation. Schemes which formed an integral part of a package may not necessarily look good on their own and so both the delivery assessment and financial processes were not joined up. Shaw et al. (2006, p. 576) concluded that "the MMS process has arguably been reduced to little more than a road builder's charter."

## Discussion Topics

- In England, Transportation has operated as a stand-alone government department and as a department integrated with energy and/or planning at various times over recent decades. What are the advantages and disadvantages of such integration?
- Pick a scale of governance from local to national. Draw a map of all of the stakeholders involved in transportation policy (both state and non-state). Identify who is involved in the planning, delivery, and funding of one of the following:
  - A major new road scheme;

  - A new urban commuter rail line; or
  - The purchase of fuel cell buses.
- Discuss the implications of the actor networks involved.

It is not the purpose of this book to review optimal institutional structures, if indeed such a concept exists. This book focuses on providing the necessary tools and understandings to allow decision makers to link transportation with other policy areas across different spatial scales. Whatever the structures in place, there is a need to find ways of making sustainable development matter at all stages in the decision-making process. However, it is possible to identify the following general principles that promote governance for sustainable transportation:

- There should be shared policy objectives, preferably developed across different levels and scales of governance;
- Within the different levels and scales of governance, there should be integration across sectors/government activity areas, particularly between land use planning, environmental protection, and transportation;
- Within transportation, multimodal planning and implementation is preferable to a separate modal structure;
- Transportation should be viewed as part of broader social and environmental policy such that equity concerns are fully integrated into the planning paradigm; in that regard, stakeholder coordination and public participation should reflect in all elements of the transportation decision-making process (discussed in the next section).
- Incentives for innovation in new technologies should work in ways which are mutually supportive across different sectors such as energy and transportation;
- The state should have policies and/or programs that can meaningfully influence the provision, quality, and integration of all modes of transportation; and
- Prices for externalities should be consistently applied to all forms of activity.

While we identify these principles, the slow pace of institutional change demands that we consider how best to integrate sustainability concerns into the decision-making structures that exist today. We begin this process by outlining some components of the decision-making process that are broadly common in different contexts around which we build our decision-support framework.

## 5.5 Key Elements in Transportation Decision-Making Processes

Decision-making in transportation can take many different forms and this is reflected to a degree in the literature that reports on it. Meyer and Miller (2001) identify five different approaches to decision-making:

- The rational actor model is where alternatives are selected to attain some set of predetermined goals and objectives in a utility-maximizing manner;
- The "satisficing" model is where the first solution to meet a core set of criteria is adopted;
- The incremental approach, which focuses on moving away from problems rather than on moving toward objectives;
- The organizational process model which suggests that decisions are reached based on the influence of different parts of the organization and typical practice; and
- The political bargaining model that involves multiple actors in reaching resolutions which, it is suggested, leads to difficult decisions being postponed.

Innes and Gruber (2005) identify four different decision-making forms which they suggest coexist: technical/bureaucratic, political influence, social movement, and collaborative. They observed that "Each style tended to be associated with different types of outcomes, though this was not explicit in discussion. The political planners divided resources among players, whereas the collaborative and the social movement planners were associated with strategies designed to benefit the region as a whole" (Innes and Gruber 2005, p. 177).

Emberger et al. (2008) identified three different approaches:

- Vision-led—which tends to be the result of a political leader or influential policy entrepreneur pushing through a vision for a city (e.g., Curitiba or Bogota);
- Plan-led—which tends to take the form of a more rational or systems-based analysis by technocrats where there is a process of setting objectives, assessing problems, developing potential interventions, evaluation of those interventions against the objectives, implementation, and then monitoring (see May 2003); and
- Consensus-led—where different stakeholders are engaged throughout the process to try and agree on the objectives, problems, potential solutions, and preferred priorities.

Banister (2002, p. 130) suggests that for transportation planning more rational or "systems analysis procedures have reigned supreme throughout the last 40 years with alternative approaches either being ignored or marginalized," although, examples of alternative approaches are also evident—e.g., see Banister et al. (2008) and Jones et al. (2009).

### 5.5.1 Information and Governance

The discussion above points to a multitude of potential approaches to decision-making. It is essential to observe that, even in the rational planning model, the decisions about what the goals are, what information will be used to ascribe progress toward those goals, and what solutions might therefore be included in the potential solution space need to be made by someone or by a group of people. Decision support is not distinct from the process of decision-making in some idealized scientific sense, but deeply enmeshed within it. As Gao et al. (2013, p. 232) note for example, "the choice and use of indicators is not only technical and science-led, but also a value-laden social process, and thus concerns public participation and political judgement." Understanding the importance of the interplay between politics and practice in decision-making is one of the rationales underpinning the desire to write this book.

In Chaps. 2 and 4, the definition of sustainable development and sustainable transportation was treated with some care. Clarity in the overall objectives is important for the transparency of decision-making processes and to the joining together of agendas. As Rein and Schön (1993, p. 161) note, when agreement on top level objectives is not clearly made "people may talk past one another, unaware of their actual disagreements. It is only in the everyday business of making and analyzing policy that the clash between frames becomes clearly evident."

While clear overarching definitions of sustainable transportation are important, they are not a guarantee of more sustainable decision-making processes. Gudmundsson and Sørensen (2013, p. 43) explored the deployment of indicators in sustainable transportation in Europe finding that, thus far, "'use' does not automatically mean 'influence' on policies or processes in more than a superficial manner." Tennøy (2010) posits explanations for the failure to make progress on sustainability issues even when there are shared objectives (and informational needs), identifying political considerations and different technical perspectives on the most effective courses of action as drivers of what gets implemented.

Political realities could paint a depressing picture for the role of sustainability decision-support tools. While identifying some very real challenges, the body of work to date also offers the prospect of improvement through better alignment of the decision-support tools in the more complex governance environment set out in Sect. 5.3. Rhodes (1996) recognizes this complexity through his influential work on the need to govern through networks, with very few agencies now solely in charge of direction and delivery. Hezri and Dovers (2006, p. 88) identify the implications of this for decision-making and information management noting that "in the interests of accountability and efficiency, the decision process broadens, away from simple coercive mechanisms, towards consensus building. … In governance, the utility of indicators as a policy tool whose traditional role was to fulfil the instrumental need of rationality must, in the new reality, enhance 'steering', 'mapping', and 'weaving' (see Parsons 2004)." So, decision-support tools might be the glue to bring together the many hands that are at play in implementation. Holden (2013), explicitly looking at a sample of (non-transport) sustainability decision-support tools, identifies the importance of the fit of the indicators to the processes they are being used to influence as an explanator of their relevance to the decision-making process. Indicators can be used as a positive part of the decision-making process to enhance transparency and accountability as with the congestion charge trial evaluation in Stockholm (Gudmundsson et al. 2009). Equally, their use can be driven by the expediency of what can be measured (Marsden and Snell 2009), by a desire to demonstrate progress toward a narrow target (Hood 2006), and by the type of questions which provide the backdrop for their application (Schön and Rein 1995). This will matter in different ways to different applications of indicators.

Understanding the fit of indicators to decision support requires a greater appreciation of the range of ways in which information can be used. Chapter 6, therefore, explores this idea, from "describing how things are" through "deciding on what to do" to "learning how to improve." This short foray into decision-making and information is to make readers aware of the need to combine the political with the scientific understanding of what indicators are used for and to recognize that these debates and struggle are part of learning about what sustainable development and sustainable transportation mean to different agencies. Indicators can sometimes be a revealing way of making real the tensions that exist between actors (Holden 2013). In Chap. 6, we define a range of common indicator application areas that are then applied in the case studies in the final part of the book.

## 5.5.2 Decision-Making Domains

The discussion in Sect. 5.5.1 suggests that all decisions are subject to debate and contestation. This is perhaps truer of some types of decisions than others, however, as the burden of reevaluating the basis for routine decisions would paralyze policy making and delivery. Technical standards and ways of working are commonplace in the field of transportation. Decisions which do not have significant uncertainty and risk attached to them are less likely to have the basis on which they have been taken reevaluated (Iseki et al. 2007). Once a decision to adopt a particular transportation solution has been made, then the questions revolve around how rather than whether it will be implemented (this notion is explored further in the High Speed Rail case study in Chap. 9).

Our broad characterization is to suggest that there are two sides to the transportation process, those of "planning" and "delivery," and we refer to these two as broad **domains** within our framework. We then break the domains down into a series of **functional areas** which we deem to reside within the domains as shown in Fig. 5.3.

We see this characterization as being as applicable to the private sector as to the public sector, albeit for different decisions and with a different set of goals. Brief descriptions of our interpretation of these functional areas are provided below. Different indicators (Chap. 6) and different decision-making frameworks (Chap. 7) are likely to be at play, at least in part, across these quite distinct domains and functions. Strategic planning might, for example, be interested in the long-term economic growth benefits of a regional transportation plan, while a decision on whether to renew a road surface every 10 or 11 years would necessarily focus on the costs and environmental impacts.

| | Domain | |
|---|---|---|
| | **"Planning"** | **"Delivery"** |
| **Functional Areas** | Strategic Planning | Construction |
| | Programming | Operations |
| | Project Development | Maintenance |

**Fig. 5.3** Domains and functional areas in transportation planning

### 5.5.3 Planning Domain

Planning is a future-oriented reflective activity which typically operates on a fixed periodic cycle. In the USA, for example, the state MPOs are required to prepare a long-range transportation plan every 5 years if they are in attainment with the National Ambient Air Quality Standards (NAAQS) or 3 years if not. The MPO must also create a financially constrained 3-year transportation improvement program to implement the long-range transportation plan.[1] In England, local governments produce a 15-year strategy with a shorter 3–5 year implementation plan.[2] The overall outcomes of the planning domain, as we define it, are a series of projects or policies ready for implementation. We break this down into three functional areas.

**1. Strategic (or Long-Range) Planning**

The systems planning function is the space in which top-level strategy is developed. It requires an understanding of the context of the organization and how it relates to other organizations both in terms of its operations, its geography, and its recent history of action. Two further stages then typically occur whereby vision statements and key goals are established and problems are identified. Vision statements should be developed by stakeholders who are to "own" the vision. The goals should also have strong stakeholder input although organizations are often subject to goals driven by other external agencies such as the Federal or national government (for example on environmental standards). Some example vision statements and goals are shown in Table 5.2.

**Discussion Topics**

- Using Table 5.2 or drawing from three different types of agencies from your own context, examine each agency's vision and goals and identify:

[1] See the Metroplan Orlando website for an example of the long-range transportation plan (LRTP) and the transportation improvement program (TIP), among other programmatic and planning documents, http://www.metroplanorlando.com/plans/ (accessed on April 20, 2015).

[2] See the West Yorkshire Local Transportation Plan Partnership for an example of a local transport plan (LTP), http://www.wymetro.com/wyltp/ (accessed on April 20, 2015).

**Table 5.2** Visions and goals

| Organization | Vision | Goals |
|---|---|---|
| Federal Aviation Administration www.faa.gov (Federal/Modal) | To provide the safest, most efficient aviation system in the world | • *Next level of Safety*. By achieving the lowest possible accident rate and always improving safety, all users of our aviation system can arrive safely at their destinations. We will advance aviation safety worldwide<br>• *Workplace of Choice*. We will create a workplace of choice marked by integrity, fairness, diversity, accountability, safety, and innovation. Our workforce will have the skills, abilities, and support systems required to achieve and sustain NextGen<br>• *Delivering Aviation Access through Innovation*. Enhance the flying experience of the traveling public and other users by improved access to and increased capacity of the nation's aviation system. Ensure airport and airspace capacity are more efficient, predictable, cost-effective, and matched to public needs<br>• *Sustaining our Future*. To develop and operate an aviation system that reduces aviation's environmental and energy impacts to a level that does not constrain growth and is a model for sustainability<br>• *Improved Global Performance through Collaboration*. Achieve enhanced safety, efficiency, and sustainability of aviation around the world. Provide leadership in collaborative standard setting and the creation of a seamless global aviation system |
| UK Highways Agency (national motorway and major trunk road operator) www.highways.gov.uk (National/Modal) | To be "The world's leading road operator" | • *We provide a service that our customers can trust*. Before setting out, road users will be able to make more informed journey planning decisions. . . . Trusted on-road communications will inform road users on the move, and our Traffic Officer Service will manage traffic more effectively<br>• *We set the standard for delivery*. This will mean we are known for first class service delivery, to time, and budget. We will be relied upon for efficient implementation, preserving value for money through effective procurement and contract management<br>• *We deliver sustainable solutions* . . . We will be . . . actively contributing to sound planning decisions, so that operation of our network supports housing growth and economic recovery. All the while, we will reduce our contribution to climate change by lowering our carbon emissions, and we will mitigate our impact on the environment and the communities adjacent to our network |

| | | |
|---|---|---|
| | | • *Our roads are the safest in the world.* We will continue to reduce deaths and serious injuries on our network. . . . We will challenge complacency about road safety. . . . In partnership with our stakeholders and suppliers we will design and deliver the safest roads in the world<br>• *Our network is a dynamic and resilient asset.* We will sustain the long term integrity and accessibility of our roads through effective management of a flexible, yet ageing asset. . . . We will reduce the whole life costs of improving and maintaining our network . . . we will enhance our growing reputation for keeping the network in operation during severe weather |
| West Yorkshire Metropolitan Integrated Transport Authority www.wyltp.com (Metropolitan/Integrated) | Working together to ensure that West Yorkshire's transport system connects people and places in ways that support the economy, the environment, and quality of life | • *Economy.* To improve connectivity to support economic activity and growth in West Yorkshire and the Leeds City Region<br>• *Low Carbon.* To make substantial progress towards a low carbon, sustainable transport system for West Yorkshire, while recognising transport's contribution to national carbon reduction plans<br>• *Quality of Life.* To enhance the quality of life of people living in, working in, and visiting West Yorkshire |
| California Department of Transportation (State Department of Transportation) California Transportation Plan (CTP) 2025 (*currently being updated to CTP 2040*) www.californiatransportationplan2040.org | California has a safe, sustainable, world-class transportation system that provides for the mobility and accessibility of people, goods, services, and information through an integrated, multimodal network that is developed through collaboration and achieves a Prosperous Economy, a Quality Environment, and Social Equity | • *Goal 1. Improve Mobility and Accessibility*: Expanding the system and enhancing modal choices and connectivity to meet the State's future transportation demands<br>• *Goal 2. Preserve the Transportation System*: Maintaining and rehabilitating California's extensive transportation system to preserve it for future generations<br>• *Goal 3. Support the Economy*: Ensuring the State's continued economic vitality by securing the resources needed to maintain, manage, and enhance the transportation system, while providing a well-organized and managed goods movement system<br>• *Goal 4. Enhance Public Safety and Security*: Ensuring the safety and security of people, goods, services, and information in all modes of transportation<br>• *Goal 5. Reflect Community Values*: Finding transportation solutions that balance and integrate community values with transportation safety and performance, and encourage public involvement in transportation decisions<br>• *Goal 6. Enhance the Environment*: Planning and providing transportation services while protecting our environment, wildlife, and historical and cultural assets |

  - How well the Agency's vision and goals map to sustainable development principles;
  - How might the objectives that do not easily map to sustainable development principles influence Agency behavior; and
  - How the nature of the Agency affects the type of objectives set.
- How necessary is it to align the goals of these different agencies? If it is necessary, how easy would it be to do so?

Problem analysis can be conducted through data-led exercises, using forecasts or by more user-led exercises. It seeks to understand the likely trends in goal attainment and the underlying drivers for attainment or non-attainment as a basis for identifying solutions. The process of goal setting and problem analysis is typically an iterative process over time, with key problems informing the top-line goals. If the problem analysis identifies a failure to meet goals either today or at some point within the planning period, then this acts as the stimulus for identifying interventions at the next stage in the process (Marsden et al. 2011; Rose 2005). Figure 5.4 illustrates the outcome of a problem analysis in West Yorkshire. The changes above the dotted line together lead to a series of impacts (shaded box) which will impact negatively on the key objectives (below the dotted line).

Recent research in city-to-city learning in Europe and North America suggests that this is far from a logical and well-ordered process (Marsden et al. 2012a, b). It typically relies on a mixture of past experience as well as selective learning from other innovations in trusted peer organizations (Marsden et al. 2011). A range of potential solutions are typically assessed against the goals either individually or in packages (May et al. 2006). Although the boundaries of the different functions are

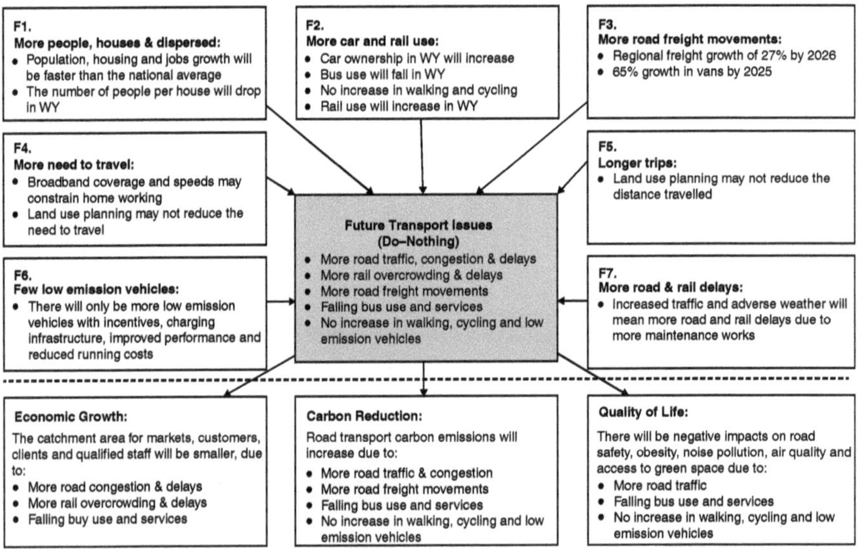


**Fig. 5.4** Problem analysis from West Yorkshire local transport plan 3 (LTP3)

artificial and somewhat fuzzy, it is at this point that we consider the programming function to take place.

**2. Programming**

The strategic planning process can help identify a range of critical needs and potential solutions for transportation. Some of these solutions are implementable in the short term (such as recalibrating traffic signals) while others may take a decade or more to realize (such as a new road or public transportation scheme). Similarly, some solutions may have an identifiable funding stream (such as a tolled bridge crossing where the toll pays for the construction), whereas others may not (such as new Light Rapid Transit system where the ticket revenue will cover operational but not construction costs). The programming function is where the realities of implementation meet the aspirations of strategy. The role of the organizations involved is to attempt to phase the introduction of projects such that they meet the overall goals of the strategy while acknowledging what can feasibly be introduced. Programming is an ongoing action which is revisited much more frequently than strategic planning. Budgets are often set for 1-year periods and there is a requirement to spend against the budget (or to risk being seen as not needing it). As such, where implementation problems emerge with one project or policy, the programming function helps to determine which are next in the list that can be brought forward.

To illustrate the distinction between strategic planning and programming, consider a metropolitan authority developing an integrated transportation strategy for the coming 15 years. The West Yorkshire Local Transport Plan identified 26 different proposals, each of which contained several initiatives (e.g., the capacity enhancements proposal included projects for the road and rail network and a new mass rapid transit scheme). The weak economic climate in the first implementation period 2011–2014 had two impacts on the schemes brought forward for the programming stage. First, the recession had reduced the demand for car travel and provided some additional time before capacity enhancements were required in some areas. Secondly, it limited the likely availability of capital funds to support major new transportation projects. The focus was therefore modified to be on better information, integrated ticketing, increased walking and cycling, and making better use of existing resources (through behavior change and better management).

In the programming stage, policies and projects have to be designed to a reasonable degree of sophistication to allow the estimation of costs and likely impacts. For example, the London congestion charging scheme had been analyzed with an anticipated set of prospective charge levels and a cordon location (ROCOL 1998). The final congestion charge, details of hours of operation, exemptions, and the way in which payment and enforcement would actually work were not clear and this becomes a matter of project development.

**3. Project Development**

At the programming stage, lots of different projects and policies exist which are prioritized for implementation. As described above, the level of detail of each proposal has to be such that a reasoned decision can be taken on bringing it forward with respect to likely benefits and impacts and potential cost. However, it is not feasible to design all of the projects in a program up to "shovel ready" specification if they are not in a position to be implemented. Such an approach would not allow innovations in delivery methods or the adoption of new solutions to current problems to be utilized. The project development function therefore takes the policies and programs outlined in the programming stage and prepares them to the point where they are ready for construction or another type of implementation. This stage involves the assessment and final decision on matters such as route alignments, vehicle design, operational capacity, fares and payment methods, and preferred technology. Decisions will also be made through the design process as to the balance between initial project and whole life-cycle costs. The approach to procurement and the extent to which the design is allowed to evolve as the project is delivered will be determined. Environmental Impact Assessments or Strategic Environmental Assessments will be required to understand what, if any, impacts of the policy or project may require mitigation and the outcomes will need to be included within the project delivery plan.

### 5.5.4 Delivery Domain

The delivery domain covers the implementation of projects and policies and their subsequent operation and maintenance (including decisions to decommission or stop projects and policies). While major new infrastructure costs can be quite significant, the maintenance requirements of the existing large-scale infrastructures are also significant. In England, the Highways Agency spends around one-third of its budget on maintenance and a further 15 % on operations (HA 2011). Similarly, in the USA, 15 % of total state highway disbursements went to maintenance costs, which was second only to capital outlays for construction and rehabilitation projects (FHWA 2011).

Since many of the actions taken in the delivery domain reflect the broader goals and outcomes of the planning domain, the extent to which those goals fully encompass sustainability can impact how sustainability is addressed. At the same time, there are actions within the functional areas listed here that can also have a potentially significant impact on sustainable development. The choice and management of the materials used in construction and maintenance projects has direct impacts on resource consumption. As noted in Zietsman et al. (2011), construction and staging footprint, erosion control practices, use of renewable and recyclable materials, minimizing environmental impacts, and social disruption can all impact sustainability. Operational decisions on when to switch street lighting on and off have impacts on road safety and on energy consumption. Decisions on how often to maintain infrastructure, and to what standard, have direct implications on the resources required, the energy used, and the costs.

It is also worth noting that it is not just infrastructure that requires implementation, operational management, and maintenance; policies do as well. For example, policies on road vehicle taxation in the UK have been overhauled in recent years to take account of the need to link the amount motorists pay to the environmental quality of the vehicle. Such changes require design, consultation, implementation, and reappraisal as circumstances change. Similarly, in the USA, the Corporate Average Fuel Economy (CAFE) Standards need to be continually monitored and adjusted as new technologies emerge.

The three functional areas that we distinguish within the delivery domain are:

**1. Construction**

This function describes the period in time where the project or policy moves from the end stages of project development to being operational. While many decisions regarding the design have already been taken, there is typically scope to procure materials and to manage the materials effectively to minimize resource consumption, energy use, emissions, and disruption caused during the construction of the project. This function concludes when the project opens.

**2. Operation**

This function describes the day-to-day running of the system. This includes a wide range of elements (see Fig. 3.3) which are put in place to keep the infrastructure functioning to some predefined set of operational goals. This can include decisions on the level of policing to be provided, how to ensure collective transport systems run relative to their schedules, or the target speed at which to try and operate traffic on a particular route. Decisions are increasingly taken in real time, taking account of the greater possibilities that information and communication technologies afford. Examples include variable speed limit control on motorways and ramp metering which both respond to high traffic levels and seek to smooth flows (Hegyi et al. 2005). High occupancy toll lanes can have variable tolls which are responsive to levels of delay on the surrounding road network (Dahlgren 2002).

**3. Maintenance**

This function describes the process of intervening to adjust a project or policy to allow it to continue functioning in line with the agreed design principles. Infrastructure systems wear out with use and require timely maintenance to avoid the need for costly replacement. The strategy for maintenance should be directly linked to the design and construction phase so that the whole life of the asset is considered in decision-making. However, the budgets for construction and maintenance are typically held separately and this can lead to different decision-making processes from those initially planned.

On the policy front, policies similarly require maintenance to reflect changing circumstances and to ensure that changes to other policies do not render them inconsistent or irrelevant (e.g., the increase in the London congestion charge from £5 to £8 to £10 over an 8 year period to keep congestion levels down). Similarly, in the USA, there is discussion regarding more sustainable alternatives to the current fuel tax policy such as the use of mileage-based user fees in light of federal gas tax revenues being unable to keep pace with inflation and improved vehicle fuel economy (Burris et al. 2013).

## 5.6 Conclusions

There is a strong case for state intervention in transportation to correct for imperfections in market structures, to address externalities that are not currently paid for by users, and to properly integrate both the long-term agenda demanded by sustainable development and also the need to incorporate social equity as a central theme in decision-making.

State intervention is, however, only one part of the picture. The "state" actually comprises many different levels that are not necessarily aligned and coordinated in their objectives and remits. Within any particular level, there are potential conflicts and synergies between different policy-making areas such as transportation and land use and also within transportation (such as the competition between different modes for available funding). The state is also only one of many actors with an important stake in the funding and delivery of transportation services. This chapter has considered the notion of governance and how the state works within a network of actors to try and steer policy direction. The chapter also acknowledges that a range of formal institutional structures exist as well as informal decision-making practices in different contexts. It is also necessary to be aware that information as part of decision-making processes is not "value neutral," but can be an important part of the exertion of power between organization and over processes. If transportation is to play a full role in contributing to a more sustainable future, we suggest that it is necessary to integrate the principles of sustainable development in the decision-making processes at all scales and to provide constructive ways of joining up those processes.

In order to begin considering how these decision-making processes might be joined, this chapter has identified some key common stages in decision-making processes which are adopted within transportation both in the public and private sector. Two broad domains of the transportation planning process—"planning" and "delivery" are defined. This distinction reflects both the different roles of the domains and the different tools which are applied in the two domains. Within each of these domains, we also identify a series of functional areas which are the operational approaches adopted ranging from long-term strategic planning to maintenance. This provides an important first building block for a framework to integrate decisions about sustainable transportation. Information is critically important to each of these functions, and Chap. 6 begins the process of exploring how information can be used in decision-making.

## References


Abelson, P. (2005). Financing transport infrastructure: Public finance issues. In K. J. Button & D. A. Hensher (Eds.), *Handbook of transport strategy, policy and institutions* (pp. 359–373). Amsterdam: Elsevier.

Bache, I., & Flinders, M. V. (2004). Multi-level governance: Conclusions and implications. In M. V. Flinders & I. Bache (Eds.), *Multi-level governance* (pp. 195–206). Oxford: Oxford University Press.

Banister, D. (2002). *Transport planning* (2nd ed.). London: E&FN Spon. ISBN 0-415-26172-4.

Banister, D., Hickman, R., & Stead, D. (2008). Looking over the horizon: Visioning and backcasting. In A. Perrels, V. Himanen, & M. Lee-Gosselin (Eds.), *Building blocks for sustainable transport: Obstacles, trends, solutions*. Bingley, UK: Emerald.

Beecroft, M. (2002). From DETR to DfT via DTLR, what are the potential implications for transport planning of these changes in departmental organisation? Accessed June 25, 2014, from www.tps.org.uk/files/Main/Library/2002/0102beecroft.pdf.

Benz, A. (2004). Einleitung: Governance—Modebegriff oder nützliches sozialwissenschaftliches Konzept? In A. Benz (Ed.), *Governance—Regieren in komplexen Regelsystemen: Eine Einführung* (pp. 11–28). Wiesbaden: VS Verlag für Sozialwissenschaften.

Burris, M., Lee, S., Geiselbrecht, T., & Baker, R. (2013, October). Equity evaluation of sustainable mileage based user fees. Report to the Southwest Region University Transportation Center (Report 600451-00007010).

Cowie, J. (2010). *The economics of transport: A theoretical and applied perspective*. Oxon: Routledge.

Dahlgren, J. (2002). High-occupancy/toll lanes: Where should they be implemented? *Transportation Research Part A: Policy and Practice, 36*(3), 239–255.

DETR. (1998). *A new deal for transport: Better for everyone*. London: Department of Environment, Transport and the Regions.

Dudley, G., & Richardson, J. (2000). *Why does policy change? Lessons from British Transport Policy 1945-1999*. London: Routledge.

Emberger, G., Pfaffenbichler, P., Jaensiriak, S., & Timms, P. (2008). "Ideal" decision-making processes for transport planning: A comparison between Europe and South East Asia. *Transport Policy, 15*(6), 341–349.

Farrington, J., & Farrington, C. (2005). Rural accessibility, social inclusion and social justice: Towards conceptualization. *Journal of Transport Geography, 13*, 1–12.

FHWA. (2011). US Federal Highway Administration, Our Nations Highways 2011. Accessed April 20, 2015, from http://www.fhwa.dot.gov/policyinformation/pubs/hf/pl11028/onh2011.pdf.

Gao, J., Kørnøv, L., & Christensen, P. (2013). The politics of strategic environmental assessment indicators: Weak recognition found in Chinese guidelines. *Impact Assessment and Project Appraisal, 31*(3), 232–237.

Geels, F. W. (2002). Technological transitions as evolutionary reconfiguration processes: A multi-level perspective and a case-study. *Research Policy, 31*, 1257–1274.

Gudmundsson, H., Ericsson, E., Hugosson, M. B., & Rosqvist, L. S. (2009). Framing the role of decision support in the case of Stockholm Congestion Charging Trial. *Transportation Research Part A, 43*, 258–268.

Gudmundsson, H., & Sørensen, C. H. (2013). Some use—Little influence? On the roles of indicators in European sustainable transport policy. *Ecological Indicators, 35*, 43–51.

HA. (2011). Highways Agency Business Plan 2011–2012. Accessed June 10, 2014, from www.highways.gov.uk.

Hall, P. (1993). Policy paradigms, social learning, and the state the case of economic policymaking in Britain. *Comparative Politics, 25*, 275–296.

Hegyi, A., De Schutter, B., & Hellendoorn, H. (2005). Model predictive control for optimal coordination of ramp metering and variable speed limits. *Transportation Research Part C: Emerging Technologies, 13*(3), 185–209.

Héritier, A. (2002). New modes of governance in Europe: Policy-making without legislating? In A. Héritier (Ed.), *Common goods. Reinventing European and international governance.* Lanham, MD: Rowman & Littlefield.

Hezri, A. A., & Dovers, S. R. (2006). Sustainability indicators, policy and governance: Issues for ecological economics. *Ecological Economics, 60*, 86–99.

HoC. (2003). Jam tomorrow?: The multi-modal study investment plans. Evidence to the Third report of the House of Commons Transport Committee (HC 38-II). TSO, London.

Holden, M. (2013). Sustainability indicator systems within urban governance: Usability analysis of sustainability indicator systems as boundary objects. *Ecological Economics, 32*, 89–96.

Hood, C. (2006). Gaming in targetworld: The targets approach to managing British Public Services. *Public Administration Review, 66*(4), 515–521.

Ieromonachou, P., Potter, S., & Enoch, M. (2004). Adapting strategic niche management for evaluating radical transport policies—The case of the Durham Road Access Charging Scheme. *International Journal of Transport Management, 2*(2), 75–87.

Innes, J. E., & Gruber, J. (2005). Planning styles in conflict: The Metropolitan Transportation Commission. *Journal of the American Planning Association, 71*(2), 177–188.

Iseki, H., Yoh, A., & Taylor, B. (2007). Are smart cards the smart way to go? Examining their adoption by US transit agencies. *Transportation Research Record, 1992*, 45–53.

Jones, P. J., Kelly, C. E., May, A. D., & Cinderby, S. (2009). Innovative approaches to option generation. *European Journal of Transport and Infrastructure Research, 9*(3), 237–258.

Kern, K., & Bulkeley, H. (2009). Cities, Europeanization and multi-level governance: Governing climate change through transnational municipal networks. *Journal of Common Market Studies, 47*, 309–332.

Knowles, R. D., Shaw, J., & Docherty, I. (Eds.). (2008). *Transport geographies: Mobilities, flows and spaces.* Malden, MA: Blackwell.

Legacy, C., Curtis, C., & Sturup, S. (2012). Is there a good governance model for the delivery of contemporary transport policy and practice? An examination of Melbourne and Perth. *Transport Policy, 19*(1), 1–18.

Lodge, M. (2003). Institutional choice and policy transfer: Reforming British and German railway regulation. *Governance, 16*(2), 159–178.

Low, N., & Astle, R. (2009). Path dependence in urban transport: An institutional analysis of urban passenger transport in Melbourne, Australia, 1956-2006. *Transport Policy, 16*(2), 47–58.

Lynn, L. E., Jr., Heinrich, C. J., & Hill, C. J. (2001). *Improving governance: A new logic for empirical research.* Washington, DC: Georgetown University Press.

MacKinnon, D., Shaw, J., & Docherty, I. (2008). *Diverging mobilities? Devolution, transport and policy innovation. Current research in urban and regional studies.* Oxford: Elsevier. Series Ed. Tiedsell.

Mallard, G., & Glaister, S. (2008). *Transport economics: Theory, application and policy.* Basingstoke: Macmillan.

Marsden, G. (2005). The multi modal study transport investment plans. Proceedings of Institution of Civil Engineers. *Transport, 158*(2), 75–87.

Marsden, G., & May, A. D. (2006). Do institutional arrangements make a difference to transport policy and implementation? Lessons for Great Britain. *Environment and Planning C: Government and Policy, 24*(5), 771–790.

Marsden, G., & Snell, C. (2009). The role of indicators, targets and monitoring in decision-support for transport. *European Journal of Transport Infrastructure Research, 9*(3), 219–236.

Marsden, G., Frick, K. T., May, A. D., & Deakin, E. (2011). How do cities approach policy innovation and policy learning? A study of 30 policies in Northern Europe and North America. *Transport Policy, 18*, 501–512.

Marsden, G., Bache, I., & Kelly, C. (2012a). A policy perspective. In T. Ryley & L. Chapman (Eds.), *Transport and climate change* (Transport & sustainability book series). Bingley, UK: Emerald.

Marsden, G., Frick, K., May, A. D., & Deakin, E. (2012b). Bounded rationality in policy learning amongst cities: Lessons from the transport sector. *Environment and Planning: Part A, 44*(4), 905–920.

May, A. D. (2003). Transport policy. In C. A. O'Flaherty (Ed.), *Transport planning and traffic engineering* (pp. 42–76). London: Elsevier Butterworth Heinemann. Chapter 3.

May, A. D., Kelly, C. E., & Shepherd, S. P. (2006). The principles of integration in urban transport strategies. *Transport Policy, 13*(4), 319–327.

Meyer, M. D., & Miller, E. J. (2001). *Urban transportation planning* (2nd ed.). Singapore: McGraw-Hill. ISBN 0-07-120000-2.

Pangbourne, K. (2010). *The changing geography of Scottish transport governance*. PhD thesis, University of Aberdeen.

Parsons, W. (2004). Not just steering but weaving: Relevant knowledge and the craft of building policy capacity and coherence. *Australian Journal of Public Administration, 63*(1), 43–57.

Peck, J. (2001). Neoliberalizing states: Thin policies hard outcomes. *Progress in Human Geography, 25*(3), 445–455.

Preston, J. (2003). A 'thoroughbred' in the making? The bus industry under labour. In I. Docherty & J. Shaw (Eds.), *A new deal for transport? The UK's struggle with the sustainable transport agenda*. Oxford: Blackwell.

Ranci, C. (2011). Competitiveness and social cohesion in Western European cities. *Urban Studies, 48*(13), 2789–2804.

Rein, M., & Schön, D. (1993). Reframing policy discourse. In F. Fischer & J. Forester (Eds.), *The argumentative turn in policy analysis and planning*. London: UCL Press.

Rhodes, R. A. W. (1996). The new governance: Governing without government. *Political Studies, 44*, 652–667.

Rhodes, R. (1997). *Understanding governance: Policy networks, reflexivity and accountability*. Buckingham: Open University Press.

Road Charging Options for London. (1998). Road charging options for London: A technical report. Government Office for the South.

Rose, R. (2005). *Learning from comparative public policy: A practical guide*. Oxon: Routledge.

Rosenau, J. (1992). Governance, order and change in world politics. In J. Rosenau & E.-O. Czempiel (Eds.), *Governance without government: Order and change in world politics* (pp. 1–30). Cambridge: Cambridge University Press.

Schön, D., & Rein, M. (1995). *Frame reflection: Toward the resolution of intractable policy controversies: Toward the resolution of intractable policy controversies*. New York: Basic Books.

Shaw, J., Hunter, C., & Gray, D. (2006). Disintegrated transport policy: The multimodal studies process in England. *Environment and Planning C: Government and Policy, 24*, 575–596.

Shaw, J., Knowles, R., & Docherty, I. (2008). Transport governance and ownership. In R. Knowles, J. Shaw, & I. Docherty (Eds.), *Transport geographies: Mobilities, flows and spaces* (pp. 62–80). Oxford: Blackwell.

Social Exclusion Unit. (2003). *Making the connections: Final report on transport and social exclusion*. London: SEU.

Stead, D. (2008). Institutional aspects of integrating transport, environment and health policies. *Transport Policy, 15*(3), 139–148.

Stern, N., Peters, S., Bakhshi, V., Bowen, A., Cameron, C., Catovsky, S., et al. (2006). *Stern review: The economics of climate change*. London: HM Treasury.

Stoker, G. (1998). Governance as theory: Five propositions. *International Journal of Social Science, 50*(155), 17–28.

Tennøy, A. (2010). Why we fail to reduce urban road traffic volumes: Does it matter how planners frame the problem? *Transport Policy, 17*, 216–223.

Treib, O., Bähr, H., & Falkner, G. (2007). Modes of governance: Towards a conceptual clarification. *Journal of European Public Policy, 14*(1), 1–20.
White, P. (2008). *Public transport: Its planning, management and operation*. Oxon: Routledge.
Zietsman, J., Ramani, T., Potter, J., DeFlorio, J., & Reeder, V. (2011). A guidebook for sustainability performance measurement for transportation agencies, NCHRP Report 708. Transportation Research Board of the National Academies, Washington, DC.
Zografos, K. G., May, A. D., Marsden, G., Kallioinen, J., Tegner, H. (2004) *Surveys of transport institutional issues in Europe*. Transport Institutions in the Policy Process, Deliverable 3, European Commission DG TREN.

## Additional Readings

Akram, W., Hine, J., & Berry, J. (2013, August 31–September 1). Transport governance, structures and policy implementation: A methodological framework. In *Proceedings of the ITRN2011*, University College, Cork.
Finger, M., & Holvad, T. (Eds.). (2013). *Regulating transport in Europe*. Cheltenham: Edward Elgar.
Marshall, T. (2013). Planning at the national level in Europe in relation to major infrastructure. *European Planning Studies, 19*(5), 887–905.
Meyer, M. D. (Ed.). (2009). *Transportation planning handbook* (3rd ed.). Washington, DC: Institute of Transportation Engineers.
Stopher, P., & Stanley, J. (Eds.). (2014). *Introduction to transport policy: A public policy view*. Cheltenham: Edward Elgar.
Weiner, E. (2012). *Urban transportation planning in the United States: History, policy, and practice* (4th ed.). New York: Springer.

# Indicators

# 6

## 6.1 Purpose and Content

The ideas and concepts of sustainability need to be given operational forms if they are to influence and count in the governance of transportation systems. Such a prerogative has been acknowledged by many policy bodies and scholars over the last two decades, and the term *indicators* is often evoked as an important element in this respect (Jeon and Amekudzi 2005; Joumard and Gudmundsson 2010). The purpose of this chapter is to introduce the concept of an indicator as a key informational device for both planning and delivery (see Chap. 5) with regard to sustainability in transportation, and to present various ways to develop and use indicators. However, a key problem for making sustainability count in transportation with the use of indicators is the diversity of values, scientific disciplines, and planning tasks that are involved. This creates a number of challenges for developing indicators that will be effective and resonate with a broad array of stakeholders and contexts. It also undermines the idea that one "general set of sustainable transportation indicators" can be found, which is why we have not pursued this task in this book.

This chapter will first provide a definition of an indicator that is subsequently applied throughout the remainder of the book. The chapter will also discuss other related concepts such as performance measures, indices, and benchmarks. The chapter will then introduce basic types of indicators and what kinds of informational support they can provide, and will also draw attention to some of the limitations indicators may suffer from. An important question is how to distinguish a good indicator from a poor one. The chapter will present a set of *criteria* to identify, assess, and select suitable indicators from a measurement as well as management point of view. These criteria are exemplified for the hypothetical situation of an agency selecting indicators for a low-carbon transportation plan.

Another important aspect is to distinguish different functions and roles of indicators in the governance of transportation. One would not necessarily use the same indicator to rationally analyze a problem, to support political negotiations, to



H. Gudmundsson et al., *Sustainable Transportation*, Springer Texts in Business and Economics, DOI 10.1007/978-3-662-46924-8_6

manage an organizational process, or to deliver a project, even if all examples made reference to sustainability. The chapter introduces the concept of *indicator applications* to capture this phenomenon. The chapter will identify different indicator application types, such as description, diagnosis, forecasting, decision support, and accountability, and introduce what each of them may require for the adequate design and selection of indicators.

This chapter provides an important progression from the previous chapters that have explored concepts, systems, and impacts of transportation and sustainability, essentially addressing *what* needs to be measured to make sustainability count. This chapter is about *how* and *why* to measure using indicators. The chapter forms a stepping stone from the previous chapters to Chap. 7, which moves beyond individual indicators to *frameworks* that connect indicators together in comprehensive systems for planning and decision support. A framework should connect substance, procedure, and intention of measurement, or integrate the "what," "how," and "why" of indicating sustainability in transportation. Chapters 6 and 7 will together provide essential building blocks not only for measuring sustainability in transportation, but also for understanding what is done in actual planning and delivery processes with the compromises and experimentation these involve. This will be demonstrated in the four case studies that follow after Chap. 7, where the provided definitions and typologies will be used to describe and critically examine real-world applications within different fields of transportation governance. The case study chapters will not only look into what is being measured with operational indicators in each case (and what is not), but also how the indicator application reflects constraints embedded in the framework and governance context. To build the foundations for such analysis the basic features of the indicator tool need to be elaborated first, which is the topic for the following sections.

## 6.2 What Is an Indicator?

### 6.2.1 Terminology and Definitions

An *indicator* (from the Greek word "indicare" meaning to point out, to announce, to give notice of) is used to measure or evaluate a particular characteristic of interest. For example, with an instrument or gauge, like a speedometer (Fig. 6.1), the needle on the scale *indicates* a vehicle's speed, with some degree of accuracy. Specifically, the indicator is to be understood as the *variable* that is measured and displayed by the needle, in this case, the speed of the vehicle, measured in kilometers (or miles) per hour (km/h or mph). The speed is shown as a position of the needle on the display, but it could also be illustrated in other ways, for example, with digits, colors, images, or even sounds.

In this book, we will use the following definition of an indicator (see also Fig. 6.2 and Table 6.1 for key related terminology):

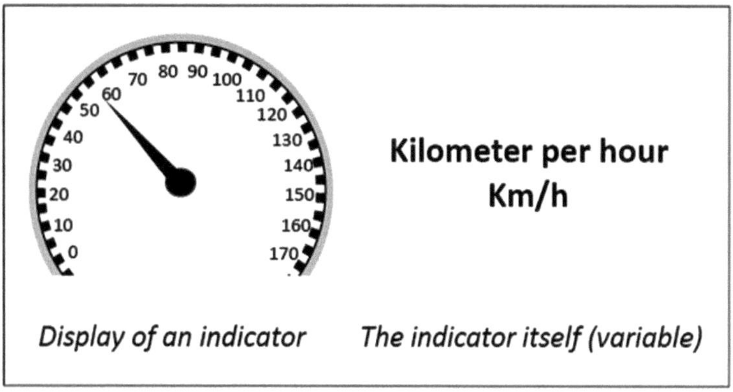


**Fig. 6.1** A speedometer reading as indicator of speed

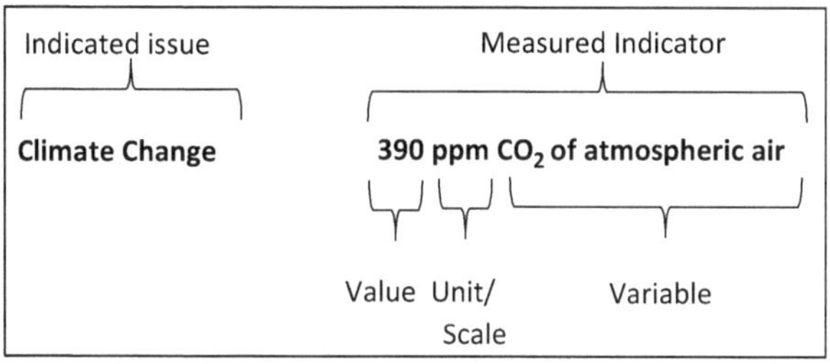


**Fig. 6.2** Indicator terminology illustrated

*An indicator is a variable, or a combination of variables, selected to represent a certain wider issue or characteristic of interest.*

This definition underscores the importance of indicators to reflect a specific issue or characteristic (such as sustainability) or problem (such as traffic safety). The term "wider" is used since it is often not possible to directly measure an issue or problem. Here wider means "beyond what can be fully captured in a single measure." The word "selected" is included since indicators are never pure, or value-neutral representations; they are selected for various reasons, and therefore inevitably have subjective aspects to them, hidden or not.

The core element of an indicator is the unit by which the indicator is measured. The variable must have a clear conceptual link to the phenomenon (such as "speed" being a key dimension of traveling somewhere in space and time) to be its indicator. A variable is again defined as an operational representation (say, km/h) of an attribute (say, speed), which can assume different values (say, from 0 to 200 km/h). An indicator is often a quantitative variable such as speed, concentration, or cost, but it need not be. An example of a qualitative indicator could be: "Level of perception of on-street livability" among citizens (May et al. 2010, p. 61), where the

**Table 6.1** Key terminology for indicator components

| Term | Definition | Examples |
|---|---|---|
| Indicator | An indicator is a variable, or a combination of variables, selected to represent a certain wider issue or characteristic of interest | Concentration of carbon dioxide in the atmosphere, as an indicator of climate change<br>Number of fatal crashes as an indicator of road safety |
| Variable | A variable is a measurable value or attribute that may vary over time or space. Indicators are, by definition, variables. However, not all variables are indicators, unless they are selected to represent a certain wider issue or characteristic of interest | In the example above, the concentration of carbon dioxide in the atmosphere is both an indicator and a variable, whereas the concentration of oxygen in the atmosphere is a variable, but not an indicator of climate change |
| Unit/ Measurement scale | A unit or measurement scale is the way in which an indicator or variable is measured or categorized<br>Measurement scales are usually classified as nominal, ordinal, interval, and ratio, and can range from scientific measurement units to customized scales developed for a specific context | In the above example, the unit of measurement is parts per million (ppm) of carbon dioxide<br>In the context of roadway operations, "level of service" is a measurement scale devised to assign roads with a letter grade ranging from A to F based on its operational characteristics |
| Value | A value is the magnitude associated with a variable or an indicator, represented by a number, figure, symbol, etc. A value is meaningless if not connected to a variable or an indicator. A value, when associated with an indicator in a specific context, produces a full indicator | "400" is the value of the indicator, if the concentration of carbon dioxide in the atmosphere is quantified at 400 using the unit "parts per million" (ppm) |

units would be, for example, "high" and "low," rather than numbers. Indicators are sometimes derived from a scientific theory about the observed system (for example, "Parts Per Million (ppm) Carbon Dioxide in the atmosphere" for global warming), but it may also be fed directly from popular or political concerns such as "accident black spots" on the road network or "gender balance" among the senior leadership of a transportation agency; the latter variable is used in the country of Sweden. Note that each of these variables can assume different values, either continuous (e.g., between 0 and 600 ppm $CO_2$) or discrete ("male" or "female"). In all these cases, when indicators are used to measure or describe a specific context, they have a value associated with them. In this book, we say the indicator in such a case has been measured—regardless of whether the associated variable is quantitative or qualitative.

Table 6.1 provides definitions and examples of the key indicator terminology that is used in this book. There are several other terms outside of those in Table 6.1 that are used in the context of indicators and measurement. These include parameters, metrics, indices, performance measures, etc. Performance measures and broader concepts of performance measurement and management are addressed in Chap. 7, in relation to their use in specific frameworks. The terms "indicator" and "performance measure" are sometimes used interchangeably, and in this book we distinguish between the two terms on the basis of use and intent. Broadly speaking, performance measures can be viewed as indicators that are used in an organizational, goal-oriented setting.

A *parameter* in the context of this book can be considered the same as a *variable*. A *metric* is a broad term that is often used interchangeably with *variables, parameters, or indicators*; however, it is also used to signify specific aspects relating to *units or measurement scale*. The term *index* is generally used to describe a composite indicator (i.e., one developed to combine multiple indicators/parameters) that is developed for a context-specific application. A grade point average, used to measure students' performance taking into account scores on various subjects is an example of an index. A transportation example could be a "roadway safety index" developed by a local transportation agency to take into account roadway geometry, crash history, and other factors to evaluate priorities for safety improvements.

### 6.2.2 Why Are Indicators Needed?

In general, indicators are selected to provide a clear measure of something considered important. For example, if traffic safety is to be measured, a typical indicator for this issue could be "the number of people killed or seriously injured in traffic." Indicators such as this are used in a broad range of situations, from describing the current condition of a transport system, to predicting its future outcomes, to monitoring the results of projects, programs, or policies over time. Indicators are used by experts building scientific models as well as by planners, decision makers, and the general public. However, it is not always the same indicator that is preferred by different groups or in different situations.

Indicators can be used to illustrate many different aspects of a system or a problem, and they may cater to a broad variety of planning and policy making situations from day-to-day action to broad political debates. Indicators are important, as they provide focused information on the key issues involved. Moving forward from planning to decision-making, to implementation, to monitoring and evaluation of actions is almost impossible without indicators. This type of use of indicators in a planning and organizational context, termed as performance measurement and management, is discussed further in Chap. 7.

### 6.2.3 What to Indicate?

An indicator is used to *represent* something important (in our terminology this is the "issue"), like a key feature of a system (e.g., its capacity) or a critical concern associated with the system like its safety or environmental impact. The type of indicators to use, how many of them to employ, and the level of detail of the indicators are very important, and are often dependent on the *context* of their use.

Different decision-making processes rely on different quantities and types of indicators. The needs of the end user and the overall context of the indicators' use are therefore important in selecting what to indicate. If we consider the example of a speedometer gauge, for instance, it is used to provide a more accurate measurement of a vehicle's speed than the driver's own perception. In the same vein, if we contrast the informational needs of an individual car driver with that of the pilot of a large airplane or the captain of a battleship in action, we can expect that the car driver has, or should have, fewer significant concerns than the others; hence, the car driver has fewer variables (i.e., indicators) to consider and fewer needles to look at.

From a *planner's* or a *manager's* point of view, however, information of the current activities of an individual vehicle or vessel may not be of immediate value. He or she needs indicators of the condition of the larger system being planned for, which could include issues such as congestion on the road network, the need for maintenance of rail tracks, the demand for transit services in a city, or the emissions of carbon dioxide from aircraft operating over an entire continent. Information on the individual speeds of all of the vehicles in the network at any one moment would be too much information. Several variables can instead be defined to measure the concerns at a planning/managerial level, and several types of communication instruments (for example, a report, a computer screen, or even a needle) can be used to display the current or future status of those indicators.

### 6.2.4 Disciplinary Approaches to Indicators

While we provided a working definition of an indicator in the context of this book, it is also important to understand the approaches to the use of indicators in the different fields of science and management. An "indicator" tends to mean something slightly different in each scientific discipline or domain. For example, in biology, an indicator has been defined as "an organism that can be used to determine the concentration of a chemical in the environment" (Parker 2003, p. 1005). Here the presence of the indicator organism reveals a significant property of the studied environment. The capacity of indicators to "show what is hidden from sight" (in this case, the chemical) is an important and common one.

In the field of ecology, a related definition of an indicator refers to an organism or ecological community so strictly associated with particular environmental conditions that its presence is indicative of the existence of these conditions (Merriam-Webster 2013). Again the indicator is an observable entity (an organism, such as lichens growing on trees) that demonstrates the presence of a condition of interest (e.g., low acid content in the air and precipitation that are necessary conditions for lichens to exist), which may not be easily observable.

In the social sciences, an indicator has been defined as a "variable that is directly associated with a latent variable, such that differences in the values of the latent variable mirror differences in the values of the indicator" (Bollen 2004, p. 7283). The term "latent" here also refers to a phenomenon that cannot easily be directly observed, as is often the case in social sciences (think of concepts like "freedom" or "happiness"). One needs to look for measurable indicators that somehow reflect such conditions and increase or decrease in the same way as the desired condition (i.e., "mirror it") when the situation changes.

In economics, indicators have been defined as any of a group of statistical values (such as level of employment) that taken together give an indication of the health of the economy (Merriam-Webster 2013). Here the "healthy economy" is the latent, or hidden, concept that the indicators must seek to illustrate, using approximations such as Gross Domestic Product (GDP), the employment rate, the debt-to-savings ratio, etc. In this example, it is noted that several indicators, rather than just one, will often be needed to "capture" the larger, hidden concern.

Moving from the sciences to environmental planning and management, we find more specific indicator concepts. An example is the following definition that has been used by the US Environmental Protection Agency, according to which an indicator is:

> a numerical value derived from actual measurements of a pressure, ambient condition, exposure, or human health or ecological condition in a specified geographic domain, whose trends over time represent or draw attention to underlying trends in the condition of the environment (USEPA 2006, p. 2).

This definition reflects that environmental problems are complex, but have a certain structure. Different indicators may be needed to represent different aspects of a problem such as its cause (the "pressure") or effects ("human health"). Also, the "geographic domain" will suggest different indicators (e.g., whether the region is mountainous, coastal, or urban). The above definition maintains the general notion of indicators pointing to underlying "hidden" trends.

Across all these examples is the idea of using indicators to represent something that is not directly observable. In some definitions, like the classical natural science definitions, the indicators' representation is assumed to be "strong" (i.e., it is used to *determine* something that exists), while in other cases (e.g., social science), the representation may be weaker (the indicator *suggests* something that may be latent or hidden[1]); this is generally because many societal phenomena are not as predictable or well understood as certain natural phenomena, and the indicators are therefore less clear cut.

This also reinforces the issue that an indicator is never able to provide a complete description of a system or a planning concern. The temperature, for example, can be measured very accurately with a thermometer (in the "natural science" domain).

[1] The term "proxy" is often used to denote that indicators only provide an approximate description of an issue. It is not an exact term, but it highlights the important aspect that the indicator always needs to be interpreted by someone to form a conclusion about the problem.

However, it only gives some indication of the concern, whether it is the health of a person (when used in the medical context) or the potential need for a local government agency to call in its snow clearing personnel (in the context of public works management). This is where the use of multiple indicators, often in the form of a systematic framework, comes into play. Chapter 7 deals further with the concept of frameworks.

**Discussion Topics**

- An indicator measures *something* of interest using a selected variable. The variable needs to represent the "something" in a reasonably accurate way. Typically, one can define more than one indicator for a particular issue. Not all indicators are equally good measures of every aspect of the problem. Try to think of several possible indicators of the issues listed below. Consider if the indicator should describe the severity of the problem itself, or if the indicator should measure the factors that may cause it to occur. Do not worry too much about the scientific or other evidence that may be required at this point.
  - Traffic safety in general
  - Traffic safety near a school
  - Congestion on a road network in a city
  - Global warming
  - Benefits of transport infrastructure investments
- How do the "boundary" assumptions that you have used affect the indicator set you selected? Look at one of the issues and see what the implications are on the number of indicators required to drawing a narrower and broader boundary around the problem.

## 6.3 Indicators for Sustainability and Sustainable Transportation

Transportation and sustainability are areas that can benefit from the use of indicators. *Transportation* systems are large, complex, and highly dynamic entities, whose attributes may vary significantly over time, and change even by the second (see Chap. 3). For a transportation manager, it is crucial to find appropriate indicators to monitor system conditions to avoid effects like accidents, congestion, and delays. In the long run, transportation systems have significant impacts on environmental, social, and economic conditions, for example, through their influence on toxic air pollution, road accidents, local environmental disturbance, land-take, and congestion. It can be highly worthwhile to try to predict and monitor such trends using appropriate indicators, in order to prepare for the future or to reach for specified goals.

*Sustainability* is a notion that particularly calls for the use of indicators, since it is a concept that is hard to observe or measure directly and may need to be reflected by several indirect measures. The overarching and complex nature of sustainability makes it difficult to keep every aspect of sustainability in focus all of the time. An important role of indicator selection therefore is to seek out a few clear and representative variables that really matter in a specific context. Another important application for sustainability indicators is to identify "*un-sustainability*." For example, it is of critical importance to be able to trace irreversible ecosystem impacts for timely intervention.

As discussed in Chap. 2 the notion of *sustainable development* promotes a holistic perspective grounded in a set of principles. Those principles are not limited to the environmental aspects of human activity. Ever since the Earth Summit in Rio de Janeiro in 1992, attempts have been made to identify indicators to help gauge progress toward sustainability more generally. These tend to follow two broad approaches. One emphasizes the need for a multitude of different indicators for environmental, social, economic, and institutional dimensions of development and covers everything from causes to effects (Moldan and Billharz 1997). The other is concerned with more direct and complete indicators of sustainability that seek to measure in one number the "reproducibility of the way a given society" behaves (Opschoor and Reijnders 1991). In other words, these indicators provide an answer (e.g., "yes" or "no") as to whether development is sustainable or not. A broad definition of sustainability indicators that seeks to bridge both approaches describes them as:

> ... quantitative measures of human wellbeing, economic activity, and natural processes and conditions; they are needed to sense the degree to which human activity may continue or expand in the future (Lee 2001, p. 7045).

According to MacLaren (1996), sustainability indicators further need to support an integrated, forward looking assessment that can also capture distributional effects. The sustainability scientist Gilberto Gallopin broadly describes indicators as variables that summarize or otherwise simplify relevant information, make visible or perceptible phenomena of interest, and quantify, measure, and communicate knowledge (Gallopin 1996).

In other words, indicators are essential tools to make the notion of sustainability measurable. The demand for sustainability indicators (for transportation and in general) is high, and the use of these indicators in various planning and policy contexts has become increasingly common. The remainder of this chapter covers topics related to the types and applications of indicators in general.

## 6.4 Types of Indicators

Indicators can take many forms and can produce different kinds of information about the same issue, entity, problem, or process. The *types* of indicators distinguish between different *ways* of conveying information about a topic but do not specify a particular topic per se. For example, an "environmental" indicator is not necessarily a different *type* of indicator than an "economic" one. In contrast, a qualitative indicator is a different type of indicator than a quantitative one, because it communicates differently. Beyond this basic distinction, four indicator typologies will be introduced, according to three different ways of conveying information, namely:

- The different *dimensions* in which the indicator moves (time, space);
- The different *complexity* of the messages conveyed by the indicator;
- The different *positions* of the indicator before or after the events it indicates; and
- The different *stages* in a process that the indicator can support.

The *indicator typologies* discussed in this section should not be confused with *indicator applications* (i.e., the purpose for which the indicators are used). This concept of indicator applications is introduced in a separate section of this chapter. However, it should be noted that the types of indicators, to a certain extent, have a relationship to the indicator application (i.e., certain types of indicators lend themselves better to certain uses). Therefore, the discussion of indicator typologies in this section occasionally references the end use/application of indicators.

### 6.4.1 Indicator Typology: Based on Dimension

The first typology involves the dimensions of time and space, as illustrated in Fig. 6.3. The most simple indicator in this typology is an observation of one variable (e.g., outdoor air temperature), at one time (e.g., now), and one place (e.g., your current location).

One may call this a "single-point" indicator. The measurement of temperature is an example of this type of indicator. However, indicators used in planning or management would more often employ time series (comparing a situation over time) and/or cross-sectional indicators (comparing a situation across entities such as cities, or groups, or traffic routes). These types of indicators are obviously more dynamic and interesting than the single point. The cross-sectional indicator is a comparison that is synchronic in time, across entities. This is opposed to diachronic comparison (meaning changing over time) for one entity, which is the time-series indicator.

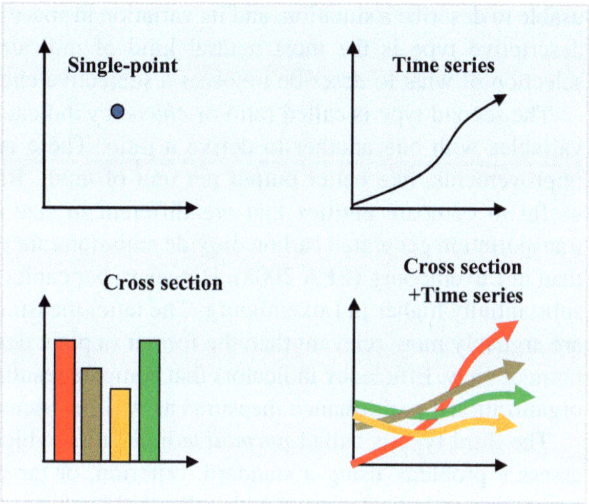


**Fig. 6.3** Indicator types by time/space dimension (*arrows* show change over time. *Colors* represent different entities)

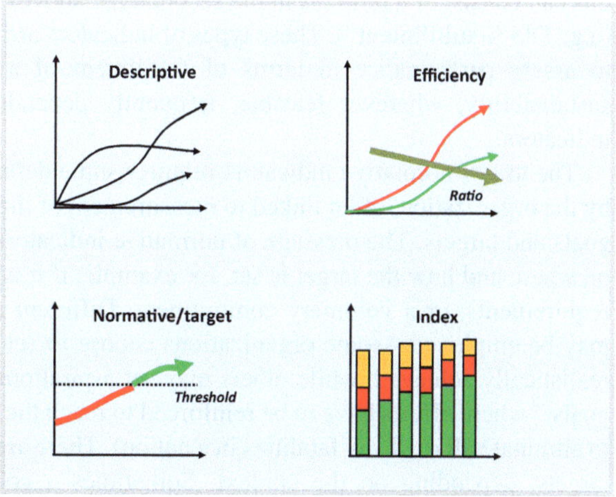


**Fig. 6.4** Four indicator types (inspired by EEA 1999) (*arrows* show change over time. *Colors* show different items measured)

### 6.4.2 Indicator Typology: Based on Message and Purpose

While the indicators shown in Fig. 6.3 can be used for almost any kind of information, the second typology that is adapted from the European Environment Agency (EEA 1999) is more specific. This typology distinguishes between four different ways to convey messages by building different components into the indicators. They are illustrated in diagrammatic form in Fig. 6.4.

The first type consists of *descriptive* indicators, used to give a clear illustration of a condition using a particular variable. The number of cars using a freeway or the tons of emissions of air pollutants from their tailpipes are examples of this type of indicator. All of the four examples in the previous typology (Fig. 6.3) are directly usable to describe a situation, and its variation in space and/or time. One can say the descriptive type is the most neutral kind of indicator, although, of course, the selection of what to describe involves a subjective choice.

The second type is called ratio or *efficiency* indicators. They divide at least two variables with one another to derive a ratio. These are helpful to assess relative improvements, like better output per unit of input. Ratio type indicators are also useful to compare entities that are different in size or otherwise. For example, transportation generated carbon dioxide emissions are two times higher in Denmark than in Luxembourg (EEA 2008). However, per capita or per vehicle emissions are substantially higher in Luxembourg. The latter measures (the efficiency indicators) are arguably more relevant than the former (a plain descriptive one), from a policy point of view. Efficiency indicators that compare results with costs are often used as organizational performance measures as will be discussed in Chap. 7.

The third type is called *normative* indicators, which are indicators that help to assess a problem, using a standard, criterion, or target as a reference point, like comparing an indicator trend with a threshold value or target line. An example is an indicator of the gap between a traffic safety goal and the actual number of injuries; this can consist of a ratio (as in the second type above) with a target reference point (e.g., "88 % fulfillment"). These types of indicators are therefore particularly useful to assess performance in terms of fulfillment of a goal or target. Measuring sustainability, wherever feasible, frequently depends on the use of normative indicators.

The use of normative indicators requires some definition of targets/benchmarks by the organization, often linked to measurement of their performance compared to goals and targets. The message of normative indicators therefore depends strongly on where and how the target is set, for example, if it is based in hard science, legal requirements, or a voluntary commitment. Different approaches to target setting may be employed—some organizations choose to set goals or targets that can be realistically achieved, while others may set aspirational targets (so-called "stretch goals," where efforts have to be reinforced to reach the target—in the case of a goal to eliminate all roadway fatalities in a nation). There are many approaches to setting targets, depending on the context. Sometimes a collaborative process may be required (as in the case of Barbour et al. 2011), while in other cases, scientifically determined criteria may drive the target setting (for example, an indicator of air pollution levels based on scientific evidence of when detrimental health impacts occur). Concepts related to target setting and their influence on indicator selection and use are discussed further in Sect. 6.5, dealing with the "target relevance" criterion for indicator selection.

The fourth type is referred to as an index, or "total welfare," aggregate, or composite indicator. The term index was previously defined as a composite indicator, combining multiple indicators/parameters developed for a context-specific application. These indicators are composed of several qualitatively different sub-indicators, which are usually measured in qualitatively different units (such as tons, area, and money). This means they cannot just be added together but have to be aggregated using an appropriate type of metric. Usually, the aggregation process involves some form of *scaling/normalization* of the indicators (to express them on a common basis) and *weighting/prioritization* to reflect the relative importance of the components.

The most frequently used method to construct a composite indicator is to assign weights to each sub-indicator and then add the weighted values, but many more sophisticated methods exist (Nardo et al. 2005). A well-known example is the "Human Development Index" (HDI) created by the United Nations (UNDP 2014). This index is composed of three units measuring income, health status, and literacy of individuals. These measures are first normalized on an index scale of 0–100, and then combined into one composite allowing country ranking on a common HDI scale. Another approach is to normalize each variable using a normative target value, and then average the variables. This produces a unit-less measure of the "distance to target" for each indicator. An average distance for all variables can be calculated, which is effectively an aggregated expression of progress overall, a "normative index" combining this and the former type of indicator. An example of this method has been used to compare the overall environmental performance for all the countries in Europe (EEA 2005).

The aggregation of indicators to produce an index or composite indicator is often linked to the application of multi-criteria decision-making (MCDM) methods. MCDM is an entire field of study in itself that goes beyond the scope of this book. However, there are several examples and discussions of the aggregation of indicators in the transportation and sustainability area that touch upon the use of MCDM methods—for example, Jeon et al. (2013) present aggregated sustainability indices along key sustainability dimensions, Ramani et al. (2010) discuss approaches to the normalization/scaling of indicators in the development of an aggregate sustainability index, and Castillo and Pitfield (2010) present a methodology for deriving priority weights for application to a suite of sustainability criteria.

However, indicator aggregation is a process that should be undertaken with care, and with an understanding of the pros and cons of employing this approach. Joumard and Gudmundsson (2010) provide a comprehensive discussion of aggregating indicators and related issues. Broadly speaking, an aggregate indicator can help provide a convenient single snapshot of a complex issue and allow for ease of communication with decision makers and stakeholders. At the same time, aggregation of indicators runs the risk of masking certain elements that may have appeared as a concern if the indicators/components were examined individually. Moreover, composite indicators can lack transparency if the methodologies used to develop them are not clearly outlined. Another issue with aggregation is the phenomenon called aggregation bias resulting from spatial and temporal aggregation. Zietsman and Rilett (2001) showed that vehicle emissions estimates can vary by as much as 20 %, depending on how the vehicle miles of travel and speed data are aggregated or depending on the type of data collection methodology used.

### 6.4.3 Indicator Typology: Based on Timeframe and Position

The third kind of typology concerns the temporal position of the indicator in relation to what it indicates. The key distinction here is between *leading* and *lagging* indicators. Leading indicators predict future changes in the phenomenon or entity of interest. Market analysts are particularly on the lookout for leading indicators. The purpose of a leading indicator is "to predict the peaks and troughs of the swings . . . sufficiently far in advance that it is possible to react to the extreme events they represent" (Seip and McKnown 2007, p. 277). A leading indicator with regard to sustainability could be the concentration of carbon dioxide in the atmosphere, which is used to predict potential climate change impacts in the future. Lagging indicators occur after the fact. They can help confirm that a certain pattern is occurring or if the desired outcome was produced as a result of a certain effort and not just coincidental (referred to as the "attribution problem"). For example, did the information campaign to increase cycling in a city lead to more people taking up cycling on a permanent basis? This objective could only be assessed with a time lag indicator. Analysis of a series of lagging indicators may help identify what caused a certain outcome if this is not obvious. For example, they can help identify if increasing fuel consumption is the result of changes in the number of vehicles, the annual distance traveled per vehicle, the average technical fuel efficiency of the cars, or any combination of these or other factors. A particular indicator may not be uniquely leading or lagging. For example, the price difference between a pure fossil gasoline and a fossil/bioenergy blend fuel may guide the consumer in making his or her choice (leading); the same difference of price may also serve ex post to help explain the observed pattern (lagging). On the other hand, it may be found that indicators believed to lead future changes did not in fact influence events (Mearns 2009) and were not leading after all. Figure 6.5 attempts to illustrate the difference between leading and lagging indicators.

To think in terms of leading and lagging indicators can be enlightening, but caution must be exercised, since real cause and effect chains are complex and may not be fully understood. A "leading" indicator may become misleading if one relies blindly on it and other events interfere with the actual trend. A simple example of

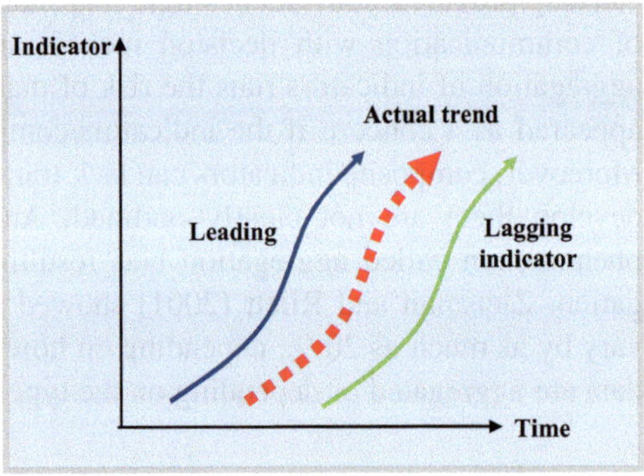


**Fig. 6.5** Leading and lagging indicators. *Note*: The actual trend is still unknown when the leading indicator is considered. In contrast, the trend has occurred but is hidden or poorly understood when the lagging indicator is considered

this would be an assumption that driving is safe if the speed is below a certain limit, without considering factors such as freezing road surfaces or drivers' alertness levels.

### 6.4.4 Indicator Typology: Based on an Organizational Production Process

This final commonly used indicator typology classifies indicators based on a combination of previously discussed factors, including the overall level, scale, and timeframe of their applicability in the process of producing or maintaining a certain service or product. This typology generally classifies indicators as either "input," "output," or "outcome" indicators depending on which stage of the process is being monitored (Kusek and Rist 2004). A process (or input) indicator directly relates to an activity conducted. Output and outcome indicators relate more to the impacts of a particular process or input, with output indicators being shorter-term and more tangible in terms of measurement when compared to outcome indicators. For example, if we are to consider investment in improving bicycle facilities in a city, an example of a related process/input indicator is the amount (in monetary units) spent on the program each year. An example of an output indicator could be the increase in kilometers of bicycle lanes constructed per year in the region, while an outcome indicator could be one related to public health improvements or automobile travel reductions in the region attributed to the investment. As with other indicator classifications, these distinctions are often quite fluid, though they provide a useful systematic basis to conceptualize and select indicators for an organization with specified services and goals. Combining these indicators allows the measurement of efficiency (input/output) and effectiveness (targets/outcomes). The importance of clear target setting for the strength of this approach is discussed in Sect. 6.5.3.

**Discussion Topics**

- What could be leading indicators for people choosing an environmentally friendly mode of travel to work such as light rail versus the car? Try to think of this in terms of an everyday choice situation versus a long-term planning situation. What does the traveler need to know to make an informed choice? What does the planner need to know?
- What lagging indicators could be used to assess the introduction of a light rail line extension? Which outcomes would be of interest and which factors could explain them?
- Note that in the planning situation the planner may need to consider if he or she is only concerned about the intended effects (e.g., did patronage increase, or did the investment pay off?) or if wider system effects are also dealt with (e.g., what happened to cycling and is the light rail fare, for example, an important leading indicator for potential cyclists?)

## 6.5 Developing and Selecting Indicators

The previous sections of this chapter dealt with the basics of what indicators are, why they are important in the context of sustainability and transportation, and the different ways of classifying the types of indicators. This section introduces some of the key issues involved in developing and selecting appropriate indicators considering both the measurement and governance aspects of indicators. The section will also introduce criteria to help distinguish more suitable from less suitable indicators.

### 6.5.1 Key Issues in Developing and Selecting Indicators

In several of the hypothetical examples discussed previously, the use of indicators is more complicated than the speedometer case that was described in the beginning of the chapter. There are three basic issues involved, which can be classified as *representational* issues (i.e., selecting indicators that truly represent the issue under consideration), *practical* issues (i.e., can the indicators actually be measured and tracked given the data and resources at hand), and *contextual* issues (i.e., whether the indicators selected are appropriate for the specific application and governance context).

In terms of *representational* issues, the ultimate concerns for decision makers often involve broad and complex areas such as economic development, social inclusion, or sustainability. There is no single variable that can measure all aspects of such broad topics, but neither can one simply collect and consider all possible variables related to a problem. Congleton and Sweetser (1992, p. 16) argue that the informational burden on decision-making from too much information "tends to reduce the efficacy of political institutions" and leads to stakeholder conflict and delay. In general, it is necessary to find the few indicators that capture the essence of the phenomenon for decision makers and discard the irrelevant ones.

To start selecting indicators, it is important to conceptualize the problem by focusing on a key concern and how can it be represented by a certain measure. It is necessary to have a clear problem definition as well as some knowledge of causes and effects of a problem to be able to identify really telling, representative indicators. Sources to help build a good conceptual foundation for an indicator include scientific theories about the problem, empirical research results, statistical analysis, expert judgment, and logic.

The second major issue deals with *practicality* of indicators. For example, the issue of *data quality and availability* is another major challenge that faces practitioners, planners, and analysts looking for indicators. Even if a suitable indicator set is identified to represent the problem at hand, each indicator may have its own data challenges associated with it. This could include logistical constraints such as availability of measurement equipment, data storage capacity, quality control, manpower, or forecasting models. Data quality and comparability over space or time are also considerations. A major obstacle can also be costs associated with data collection. Sometimes one can also start developing indicators from available data and see which problem or concept they can be used to represent. For example, data on gasoline sales can be used as a proxy for pollution from traffic or the number of complaints over traffic noise can be used as an indicator of noise nuisance. Another typical data problem can be secrecy or other kinds of limited access to information, even if the information does exist. Qualitative indicators require similar attention to the collection of reliable data/information and the construction of clear scales/measures, but may have lower data collection demands than quantitative indicators.

**Discussion Topics**

- Every indicator may involve one or more challenges in connection with the measuring, gathering, storing, or communication of data. Try to think of possible obstacles for collecting and verifying data for the following types of indicators, assuming that any conceptual issues with the indicator have been resolved:
  - Delay or congestion on a road network;
  - Climate change effects of transportation;
  - The economic benefits to companies generated by new infrastructure; and
  - The difference in access to a transportation system experienced by men and women, respectively.

The final problem relates to issues of *context*. Generally speaking, the role of indicators is not fulfilled just by representing a problem in a conceptually unambiguous way and using available data of good quality. The indicator should provide relevant, appropriate, resonant, even compelling information to be worthwhile reporting. The compatibility with a particular purpose or the specific *application* of an indicator is crucial for its usefulness. The relevancy of indicators in a particular context, to a particular decision maker, or to a certain group of stakeholders, is of course a matter of perspective. Indicators are also not completely neutral—there is always a subjective component to them. For example, in the reporting of roadway fatalities as a general indicator of safety, the specifics of the indicator selected can influence the perception of the results. In this case, the choice between reporting fatalities as a whole, fatalities normalized by vehicle mile of travel, or fatalities normalized on a per-capita basis may each provide a different picture of the indicator and overall performance.

**Discussion Topics**

- A variable needs to represent a phenomenon in a *relevant* way. Not all indicators are equally relevant for everyone, every situation, or every application. Try to think of possible indicators of relevance for the area of electromobility. Look at this issue and define the most interesting indicators from the following different perspectives:
  - You are a car dealer who is interested in maximizing your profit, and you are considering the electric vehicle market;
  - You are a green activist who wants to promote the most sustainable solutions for your city and the planet;
  - You are a politician who has promised to create 100,000 new jobs in your country within the next 5 years; and
  - You are an energy planner who needs to consider the development of energy infrastructure in your region for the next two decades.

### 6.5.2 Criteria for Developing and Selecting Appropriate Indicators

This section looks at what it takes to create a "good" indicator taking into account representational, practical, and contextual factors. A "good" indicator here means one that is appropriate and realistic for measuring and reporting a phenomenon of interest in a given context. A key tool in this effort is the use of *criteria* to assess the capacity of individual variables as "candidates" to serve as indicators.

In the following text, a limited set of ten criteria proposed for use in sustainable transportation planning by Joumard and Gudmundsson (2010) is considered. The criteria are divided into three categories that detail the *representational*, *practical*, and *contextual* challenges that were introduced previously. For each criterion, a definition is given along with some information on how to apply the criterion in the assessment of potential indicators. Detailed methodologies for validation of specific indicators are not discussed here but can be found in the technical literature (see Additional Readings).

To support the discussion in this section, a hypothetical situation is considered whereby a transportation agency would like to develop a long-term transportation plan for a city wishing to become "carbon neutral" by 2050. Within this context, there is an interest in developing an indicator of the impact of global warming. A range of variables that could be candidate indicators of global warming impact have been gleaned from various sources such as scientific literature about global warming, existing datasets, and concerns raised by stakeholders. The candidate variables assumed in this hypothetical case are limited but quite diverse, including ambient temperature, concentration of $CO_2$ in the atmosphere, emissions of greenhouse gases from transportation, and average fuel efficiency of vehicles in the vehicle fleet.

**Representational Criteria**

The three criteria in this group—validity, reliability, and sensitivity—all refer to how sound or accurate a representation of the problem a given variable provides. The main focus is on getting the variable right.

The *validity* criterion is most simply described as whether the variable actually measures the issue of interest (i.e., if there is a close causal correspondence between the indicator and the issue of interest). If we look at the candidate variables introduced above, the ambient temperature in a city is certainly a valid indicator of the local climate, but not of global warming impacts due to emissions from traffic. Science would show that the causal link between local traffic and local climate is negligible, even if global warming does involve increasing average temperature at the global scale. The emission of greenhouse gases from transportation is a more valid candidate to measure the issue of interest since there are well-established ways to calculate the contribution to global warming from such emissions. Fuel efficiency is a little less valid, since the global warming impact also depends on the particular fuel used (its carbon content) not just the efficiency of the car.

A valid indicator must be based on a conceptual model that justifies how the indicator and the issue are causally connected (e.g., through laws of nature or causal mechanisms confirmed in practice or identified in science). The model should be well accepted by the research community involved in the particular field (i.e., conceptual validity).Validity of indicators can sometimes be consolidated by statistical tests of the agreement between a prediction obtained from the indicator and other, more direct or "objective" measurements of the same phenomenon (i.e., predictive validity). Predictive validity without conceptual validity can however be misleading and should not be considered a substitute for it.

*Reliability* means that the indicator behaves in a controlled and predictable way when measured, that is, it produces the same value over repeated measurements. Reliable indicators allow different people to obtain the same results when operating the indicator. Reliability is, therefore, often more difficult to obtain for qualitative indicators that involve interpretation as part of the measurement process.

In our example on local climate change, we can say that variables like temperature and the concentration of $CO_2$ in the atmosphere are superior since the measurement methods for those variables are technically well established and reproducible. In contrast, emissions and fuel efficiency at the vehicle fleet level are calculated based on samples of vehicles and various assumptions in test driving and computer programs. Their reliability is likely to be somewhat lower than the former ones, but this would not overrule the stark difference in validity. A reliable indicator is *never* useful if it is not valid at the same time, while a valid indicator can be the best choice even if it is not very reliable. The reliability of ambient temperature measurements does not push this variable to the forefront considering its lack of validity in the present context.

Figure 6.6 provides a visual representation of indicators that are: (1) valid, but not reliable, (2) reliable, but not valid, and (3) both reliable and valid.

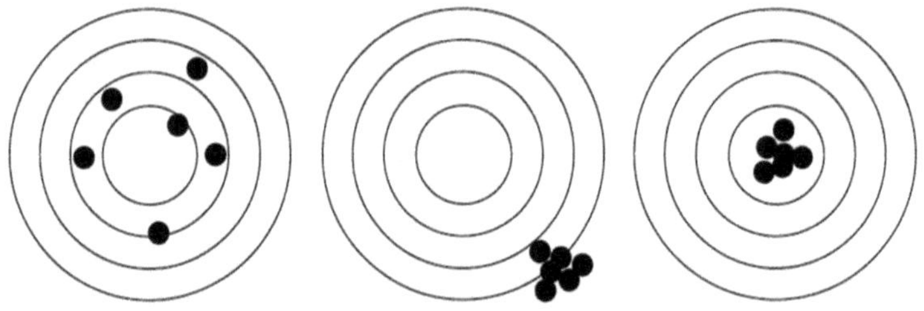


**Fig. 6.6** Representation of valid and reliable indicators

*Sensitivity* is really a variation of the validity and reliability criteria. It considers whether an indicator is able to discriminate any significant changes in the outcome of interest. Consider an example of tracking the impacts of local transportation policies on climate change. A temperature indicator is ruled out as an extremely insensitive indicator to changes in local transportation. Changes in $CO_2$ concentrations locally could be a result of changes in transportation emissions, but could just as well reflect a number of other changes in the local atmosphere (for example, as a result of point sources). A vehicle fuel efficiency variable has limited sensitivity since it would not capture effects associated with, for example, more travel or longer trips, changes in modal split toward lower or higher emission modes, etc. Only a transportation emissions indicator would have a high sensitivity, since it is almost a direct representation of the topic of interest.

What should be especially clear from these examples is that the definition of the problem or topic of interest is critical to selecting representative indicators. The problem/topic definition should be considered carefully, since the validly and sensitivity of particular indicators depend on it.

**Practical Criteria**

The three practical criteria—measurability, data availability, and ethical concerns—all refer to the operation of an indicator system and the associated monitoring. The main focus is on indicators in their capacity as *data values*. What does it require to collect, report, and compare the data to feed into the candidate variables? Is it doable, affordable, and does it raise any problems of an ethical nature to actually collect and publish the indicator?

A *measurable indicator* should be straightforward and relatively inexpensive to measure. In the example case, most of the variables are *measureable*. The measurement of ambient temperature is, of course, the most simple and inexpensive, compared to the relatively extensive systems required to calculate emissions and to monitor fuel efficiency. The cost and effort of measurement greatly depends on the level of ambition with regard to representation, as discussed above. Take fuel efficiency where an average value often can be obtained from national statistics or surveys. However, this is not so useful if there is great variation in the vehicle fleet across cities or (worse) if the concern of interest is not simply the aggregate emissions but to measure the effectiveness of a local program to promote fuel efficient driving patterns.

The boundary between "measurability" and "data availability" is fluid. Often something *could* in principle be measured, but not without excessive costs to establish a measurement program to make the data available. Indicators that are based on data that are readily available or can be made available at reasonable cost and time meet the *data availability* criterion. Again, the performance of various indicators according to these criteria will strongly depend on the specification of the problem to be indicated as well as the goals of the program which the measurement aims to support.

Finally, where applicable, an indicator must not violate any fundamental human rights and only use data that are consistent with established principles and ethics. This *ethical criterion* has been introduced in the context of human health assessment to ensure that health data collection does not violate privacy or other ethical concerns of people. Similar concerns might be appropriate with regard to other aspects of human and social activity (e.g., travel behavior, criminal records, etc.). An indicator should not be based on data that are offensive for people to report or could be used against them. In travel surveys, information is collected about travel activities including "private" information about people's choice of destinations, travel purposes, timing of trips, etc., on a certain day. The use of the data is often restricted by privacy safeguards, and users have to sign confidentiality agreements.

**Contextual Criteria**

As discussed previously, this issue deals with the suitability of the indicators in terms of *context of application*. This category has four criteria—transparency, interpretability, target relevance, and actionability. They are all applicable in a situation where the variables are selected, the data are collected, and the indicators are being *applied*. While these criteria can and should be addressed even in the indicator development process, the consequences of a poor choice only materialize forcefully at the application stage.

*Transparency* means that it is possible to understand and trust the way in which the indicators have been produced. This is sometimes an issue for decision makers especially if there is political disagreement over the course of action. It is easier to debate an opponent's position if it is strongly dependent on data for which he or she has no clue about the underlying metrics, or cannot refer to an accepted source for its verification. Transparency is sometimes an issue with more complex and "subjectivity-infused" variables such as monetized values of environmental damage or loss of human life. Transparency is associated with but not identical to simplicity. A simple indicator may be more attractive because it is easier to show how it is produced. However, complex indicators may also be transparent if the methodology is well justified, well defined, and well explained.

*Interpretability* refers to drawing clear conclusions on the basis of an indicator—i.e., whether it is clearly understood as showing an improvement or the opposite. Interpretability also depends on how the indicator is influenced by uncertainties. In the unlikely case the "ambient temperature" locally was used as an indicator for transportation impacts, it would be erroneous to interpret any movement in the indicator as relevant to the problem; it cannot be soundly interpreted. In the case of emissions, it is fairly obvious that a decrease is positive. The fuel efficiency indicator could in theory lead to misinterpretation, since an increase is desirable, while some may read increasing values on a graph as bad, especially in an environmental context where most reported information is about problems for which a decline is desired. There are real cases where policy makers have actually read the indicators in the wrong way—see, for example, Rousseaux (1994).

*Target relevance* means the variable measures something directly related to articulated goals, objectives, targets, or thresholds. If an impact is quantifiable, an indicator should make possible a comparison with any relevant threshold or reference value (standard, political target, etc.). If there are no quantified targets or thresholds, the indicator should be considered in terms of its relevance for non-quantified policy objectives or goals. Targets establish the direction where decision makers wish to go, but may also denote an absolute value to be achieved. Setting a target can be risky for elected officials who may see their support wane if the promised results do not materialize. In the hypothetical example, the city has a target to become "carbon neutral" in 2050. None of the variables directly measure this, but greenhouse gas emissions probably come the closest. As noted previously, it is easier to use indicators if clear and unambiguous aims and targets are formulated (see Sect. 6.5.3 on "SMART" criteria below), and this is not immediately the case for a term like "carbon neutral."

Finally, *actionability* refers to the level of control over the item that is measured with an indicator. An actionable indicator measures factors that can be changed or influenced directly by management or policy action. The indicator can be directly actionable by measuring a parameter that is also a policy variable (e.g., number of police controls to check vehicle emission control equipment), or indirectly by measuring something that can be influenced by policy (e.g., population exposure to air pollution above limit values). An indicator directly measuring the parameters of decisions (e.g., funding dispersed) is more actionable than indicators measuring the general environmental conditions (e.g., temperature rise of the atmosphere). Actionability is often constrained by the division of powers and factors like multilevel jurisdictions with distinct responsibilities.

## Discussion Topics

- You have been asked to develop an indicator to measure the congestion of roads in a metropolitan area. To ensure that you carefully think through the possible indicator options, develop a matrix that contains the possible indicators on one axis and the ten criteria shown below on the other axis. For each indicator, try to complete the cells in the matrix. After completing this task, could any of the indicators you selected be described as "good"? How is "good" defined in your case?
  **Representational criteria:** validity, reliability, and sensitivity.

**Practical criteria:** measurability, data availability, and ethical concerns.
**Contextual criteria:** transparency, interpretability, target relevance, and actionability.

This section has demonstrated the importance of taking into account the needs and perspectives of decision makers and other users of indicators when they are designed and selected. If the indicator is not transparent, target relevant, and actionable, it may end up being ignored in the planning or decision-making process, and will likely be discontinued.

While the proposed criteria can be helpful to assess individual indicators, it is important to recall that indicators are often not identified and applied one by one, but rather as an element in sets of indicators framed by an overall purpose and need for the particular indicator set. This aspect will be considered in Chap. 7, with its focus on broader frameworks and processes to build comprehensive indicator systems and sets.

### 6.5.3 "SMART" Criteria for Goals

In this section, we introduce a simplified approach that can be used to support the selection of context-appropriate indicators, known as the "SMART" criteria. SMART is an acronym for *Specific, Measurable, Attainable, Relevant, and Timely.* The relevance criteria, contextual criteria, and practical criteria for good indicators discussed in the previous section are consistent with the criteria presented here. The SMART criteria provide less in terms of scientific support but are easier to apply. They have been widely cited in management literature, even if the exact definition is disputed (Rubin 2002). The criteria originally refer to the process of setting goals that are realistic and useful for an organization, where indicators can be thought of as a tool to help in this process. However, similar considerations can apply to indicators themselves, as proposed by Broughton and Hampshire (1997) who apply them to indicators used in project assessment (see Table 6.2).

In practice, even the use of SMART measures may not be feasible in every situation, and compromises may have to be made in the choice of indicators used. Performance measurement involves organizational issues in addition to the ones concerning indicators themselves, including those such as who defines the desired performance levels, how can performance results be attributed to the actions of the organization, and how will various groups or individuals react to having their performance measured and reported in a certain way? The issues about attribution are especially crucial to the actionability and general usefulness of an indicator. Many organizations struggle with indicators over which they have very little control, such as emissions of greenhouse gases.

**Table 6.2** "SMART" indicators

| S<br>Specific | M<br>Measurable | A<br>Attainable | R<br>Relevant | T<br>Timely |
|---|---|---|---|---|
| Key indicators need to be specific and should relate to the conditions the project seeks to change | Each indicator should be measurable and hence requires a precise definition | The indicator must be attainable at reasonable cost using an appropriate collection method | Indicators should be relevant to the management information needs of the people who will use the data | An indicator needs to be collected and reported at the right time to influence many management decisions |

*Source*: Broughton and Hampshire (1997)

## 6.6 Application of Indicators

This chapter has so far discussed indicators, their importance, key definitions related to indicators, and disciplinary approaches to the use of indicators. We also presented indicator typologies and discussed indicator aggregation, target setting, criteria for good indicators, and other key issues relating to indicator selection and development. All of these issues have some relation to the *indicator application*, i.e., the ultimate use of an indicator or a set of indicators for a specific purpose. Eight key indicator applications are defined and discussed in this section. These applications are then applied in the case studies in Part II of the book.

Indicators can be applied for several different purposes and organizational functions, from long-range comprehensive planning to day-to-day decision-making. Chapter 5 introduced key objectives, procedures, and tasks typically involved in transportation planning and decision-making. It was noted that there are several levels of activity, encompassing (but not limited to) the planning and delivery domains. The work tasks and their associated information needs will not be of the same scale or dimension, even if sustainability serves as a common concern. In this section, we offer a typology of indicator applications, which define the use of indicators in connection with a wide variety of planning, decision support, and operational tasks. Table 6.3 provides an overview of the selected indicator applications.

Before discussing each of these applications, a note of caution is offered. Planning and decision-making for sustainability in transportation cannot be understood only from a functionalist or rationalist perspective. These activities take place in a social and political context and are often charged with different, perhaps conflicting, ideas and interests of decision makers, decision takers, stakeholders, and professionals as described in Chap. 5. Thus, indicators are not just tools to perform technical tasks but can also be "weapons" in political argumentation and struggle. As pointed out by Rydin (2002, p. 90) "indicators function inside the governance process, they are not exogenous factors parachuted in, which can act like a magic bullet causing decision-making to become instantly objective and

**Table 6.3** Overview of selected indicator applications

| Application | Leading question |
|---|---|
| Describe | What is going on? |
| Forecast | Where are we heading? |
| Review | How are we doing? |
| Diagnose | How did we get here? |
| Decide | What should we do? |
| Account | Who is responsible? |
| Learn | How do we do better? |
| Communicate | How do we tell others? |

scientific." Looking at indicators from a political science perspective suggests at least three different ways or models for interpreting their use in planning and decision-making: a "rational-positivist" model, emphasizing instrumental policy applications of indicators such as aiming to quantify objectives and measure results; a "discursive-constructivist" model, where indicators serve as mechanisms to frame problems, identify shared values, and build a common discourse rather than perform as precise technical tools; and a third "strategic" model, where the competition and bargaining for political power leaves little room for objective knowledge as offered by indicators, other than as ammunition in fighting over power or "turf" (Boulanger 2007).

While taking note of possible limitations or even distortions in the use of indicators implied by these different models, the focus will now turn to define a set of applications as they will typically occur on the working agenda of professionals involved in various levels and stages of planning, management, or operations with regard to transportation and sustainability. These applications can not only serve to guide an indicator selection process but can also be used to analyze and interpret practical cases as will be shown in Chaps. 8–11.

**Describe—What is going on?** This is the most basic application of indicators and is usually applied when a new, perhaps much debated issue emerges on the agenda. The main purpose of using indicators here is to help establish some idea about the magnitude and evolution of the problem being considered. Suitable indicators will provide overviews of key historical trends, or evidence of how a situation (e.g., accessibility for physically disabled citizens) varies across a geographical space. A search for leading indicators to help refine knowledge from descriptions toward possible actions may be one obvious extension of this application.

**Forecast—Where are we going?** An important aspect of sustainability and transportation planning is a focus on the future. Apart from preparing for future problems of congestion and pollution, the attachment of indicators to forecasts may also be useful in the evaluation of alternative policy scenarios. De Ceuster et al. (2006) illustrate the effectiveness of this approach for the Mid-Term Review of the European Union's Transport Policy in 2005. While it was not feasible to evaluate the current policy accomplishments due to data limitations, the use of models allowed the development of forecasts, which demonstrated a need to revise current policy priorities. A limitation is that it is usually only possible to create a limited set of forecast indicators, such as transport volumes and greenhouse gas emissions.

**Review—How are we doing?** Moving deeper into the core functions of indicators, it will often be desirable or even necessary to perform an assessment of the present or predicted future situation in order to prepare a course of action. Such an assessment is normative with regard to objectives, standards, benchmarks, or simply determining which direction of change is desirable; hence, some form of normative indicators is needed. Examples of questions addressed by the review application include: Will objectives be met? Is progress occurring? Does the program work? Or, are we becoming sustainable, in one sense or another? The Texas Department of Transportation for example uses a measure called "average pavement condition score" as one component in the review of a goal to preserve the value of its transportation assets (Ramani et al. 2011).

**Diagnose—How did we get here?** Policy making often proceeds in an incremental fashion, and is governed by the "art of the possible" or follows the intuition of policy entrepreneurs. Sometimes such policies are successful, but they may also utterly fail. Very often a need arises for evidence of "what works" or why something did not work. Analyses tracing correspondence between leading and lagging indicators, or between causes and effects, can be helpful. Such diagnostic efforts can help to unpack the factors behind a (possibly undesirable) present situation and also create a more solid basis for interventions to avoid it. An example is a decomposition analysis of $CO_2$ emissions from passenger cars in Greece and Denmark, which found that changes in vehicle ownership, fuel mix, engine capacity, and annual mileage all contributed to the increase in emissions in the two countries (Papagiannaki and Diakoulaki 2009). If the diagnosis can be linked to actionable policy variables—such as fuel type—it may be particularly helpful.

**Decide—What should we do?** Decision-making is formally the responsibility of elected officials or executives who act on their best judgment of a situation. However, it has been a common trend in transportation policies in most developed countries to apply technical decision support (DS) tools and procedures to parts of the process (Hayashi and Morisugi 2000). Cost–benefit analyses of infrastructure projects rely on a limited set of normative socioeconomic indicators, such as Net Present Value (NPV, see Chap. 7), where the final decision might be based on the project with the highest NPV. Indicators also play important roles in other decision support applications such as multi-criteria analysis, environmental impact assessment (EIA), and performance budgeting. The methodological challenges involved in selecting indicator variables and aggregating them to the appropriate degree differ greatly among various settings. For example, a cost–benefit analysis provides a more unambiguous basis for a decision about which construction project to choose if the compared projects share many similar features (Quinet 2011). In any project, small or large, the scrutiny of the indicators and their credibility tends to increase significantly when actual decisions draw near.

**Account—Who is responsible?** Over the last decade or so, performance management regimes have been adopted by or imposed on many national and local transportation agencies around the world. These have placed indicators in a central position as tools to operationalize strategic goals, monitor performance, and report results. An example is the Swedish Road Administration that was tasked with implementing cost-effective safety measures on the road network that would reduce the number of road fatalities by at least 20 % compared to 2006. The results of the program were later to be used in political negotiations or decisions on the allocation of future resources (Küchen and Nordman 2008). A key purpose of performance management with indicators is to allow "principals" to hold "agents" accountable for results; this applies to taxpayers with regard to elected officials as well as agency executives with regard to staff. The choice of indicators in a performance management regime can be perceived to have distorting effects on results if, for example, the performance indicators are not "SMART" (see Sect. 6.5.3) or if the set of available indicators is unbalanced.

**Learn—How can we do better?** Arguably a key aim of any performance management effort or indicator application is to build capacity to learn and improve. Improvements occur when results and experience are used to implement changes in the practices, procedures, or structures of the organization in a way that exploits experiences and enhances the capacity to perform in accordance with principles, goals, and capacities in the future—i.e., it is a learning process. Another way to foster learning was experienced in the region of Gothenburg Sweden, where a network of civil servants was able to promote multi-sector and multilevel collaboration and enhance the capacity for sustainable urban development (Polk 2010). An element in the strategy was a consensus formed around a specific definition of sustainability with associated indicators. However, a complex multi-actor network can also pose a challenge for consensus building and shared learning. The use of indicators to support continuous improvement is a long-term endeavor that is in no way guaranteed to succeed. In some studies, this "learning" approach is contrasted with the "accountability" approach as an alternative way to seek improvement in performance.

**Communicate—How do we tell others?** Communication is an essential and cross-cutting component in working with indicators, as it applies to and reinforces any (and all) of the other applications. The selection of variables to measure can communicate a strong focus on certain areas or a dedication to certain issues. The reporting indicator values is therefore essential for descriptions, assessment, diagnostics, accountability, and learning. The audience for the communication of indicators can vary from technical experts, to decision makers, to the general public and the press. The forms and media used to report the indicators should vary accordingly. There are a multitude of potential ways to communicate indicators, and only a small set of options have been exemplified in this chapter.

## 6.7 Conclusions

Indicators are variables that should be selected and designed for their ability to represent important characteristics of phenomena of interest. This chapter has used several examples of indicators and shown various ways to construct them, ranging from simple variables used to highlight a problem, to more complex metrics incorporating scientific models, aggregation functions, or spatial and temporal variations.

The chapter defined four indicator typologies by looking at the *dimensions* of indicators (e.g., time and space), the *complexity* of the messages conveyed (e.g., neutral, normative, composite), their *positions* (i.e., whether they lead or lag a phenomenon of interest), and the *stages* in a process that is measured (e.g., using input, output, or outcome indicators). Additional issues in organizational performance measurement such as "SMART" target setting were also covered. This whole array of indicator types will be useful or even indispensable to comprehensively support planning, management, and decision-making, as well as the broader governance processes affecting sustainability in transportation. The previous chapters jointly mapped out a wide range of sustainability and transportation topics to potentially measure, from climate change and nonrenewable resource use to mobility barriers and transportation facility costs and beyond. This chapter has, however, not converged around a particular indicator set for these topics, let alone one preferred indicator (or index) to measure the sustainability of transportation systems. Instead the focus has been to define typologies, criteria, and applications for indicators. There are several reasons why this approach has been taken.

As demonstrated in Chaps. 2–4, the sustainability of the transportation sector depends upon the conditions and interactions of a multitude of natural, social, and economic systems operating at different scales. These interactions are not so invariant (or well understood) as to allow a few universal control variables to be identified. The theoretical analysis suggests a comprehensive, inclusive approach to the measurement of sustainability based on general principles rather than pointing to a list of specific indicators. What needs to be measured and communicated in practice will differ with regard to time and space scales, boundaries of transportation system impacts, and different philosophical or political paradigms of sustainability.

The limited scope for a universal set of indicators for sustainable transportation becomes even clearer when the diversity of governance perspectives and the different domains and functional areas in transport decision-making are introduced as they were in Chap. 5. Even if policy makers, scientists, or other stakeholders could agree on the need to monitor factors like the energy consumption or costs of maintaining an expanding road network, the specific indicators to apply will strongly depend on the context of their application.

This is important in two ways. First, the consideration of context will help indicator developers determine the specific information needs and most appropriate indicators to represent an issue of interest. For example, different indicators would be needed to support project planning, monitor construction or maintenance activities, or for ex post evaluation of network performance. The indicator design and selection process can find support in the criteria introduced in Sect. 6.5, supporting the aspects of representation, practicality, and context relevance.

Secondly, the wider governance context may help understand why certain indicators and not others have been used in a particular situation. The role of knowledge in different governance models introduced in Chap. 5 is helpful to consider here. For example, indicators used to compare efficiency of road underpasses may be accepted more easily by a community of expert peers than indicators to determine the success or failure of a controversial policy measure such as free public transportation for all, or the abandoning of speed limits. Indicators are always embedded in a context of subjective choice and thus prone to be influenced by political, social, and psychological factors, which may overshadow technical measurement functions or at least provide additional clues to understand their real use and value. Political or philosophical disagreements can prevent any consensus emerging on how to measure and interpret the results of a policy and how then to proceed with further action. In this case, the need to indicate becomes a stumbling block rather than a stepping stone, due to possibly undisclosed controversies involved. As stated by Turnhout et al. (2007), indicators are objects that cross the boundary between science and policy making. They may help to bridge the gap between scientific and political worldviews, especially if criteria for representation, practice, and context are applied in the design process. However, there is no guarantee that this will always work. According to Turnhout et al. (2007) and Runhaar and Driessen (2007), issues that are "unstructured," in the sense of lacking agreement over policy ends and means are particularly challenging to define effective indicators and assessment procedures for.

For a systematic approach to what the context can mean for indicator development and use, this chapter has introduced the important notion of indicator *applications*. An application embodies the purpose to which the indicators are to be put, which is essential to observe if indicators are actually to be used and have influence on analysis, communication, or decision support. Section 6.6 introduced the following generic applications that can be used to characterize indicator use, as will be demonstrated in the case studies in Chaps. 8–11. Basic features of each application are summarized here:

- The *Describe* application is the most basic and is particularly relevant for introducing emerging or new issues with a low level of structure to a wide audience. With this application, it can be controversial to introduce indicators that incorporate strong normative elements or advanced scientific assumptions.

- The *Forecast* application is widely used in early planning or decision analysis. Forecasting assumes that the indicators will also be relevant in the future; they will often be connected to forecasting or simulation models that may narrow the scope of effects that can be forecasted. Uncertainty in the forecast may overshadow any problematic issues with the indicator itself.
- The *Review* application inevitably involves normative judgment. Indicators embedded in a normative interpretation framework or directly incorporating norms, thresholds, standards, or the like can provide strong support for action, but may also be rejected by those not sharing the implied norm. The normative elements used for review should not be "hidden" inside an indicator but made explicit.
- The *Diagnose* application aims to bring structure to the issue that is measured in terms of identifying cause and effect links and hence suitable versus less suitable interventions. Diagnosis is especially important for the development of representative indicators, but can also have a valuable communicative function to help decision makers or stakeholders understand the problem and agree to disregard possibly ineffective policy measures. This implies a rational use of information, which cannot necessarily be assumed.
- The *Decide* application involves the potentially most consequential use of indicators, if they are used to determine or inform decisions to construct, allow, prohibit, or fund transportation projects or activities. Indicators used for decision support should jointly reflect all relevant aims and concerns of those the decision applies to. Since indicators always come with some limitations or blind spots, in practice they should be supplemented with other types of information and explicit judgment.
- The *Account* application refers to real-time or ex post review of actions, which is particularly important in an organizational context. Simple accounting is to check if planned or prescribed actions have actually been conducted. If this fosters a strict control regime or results in mere "bean counting," it can have a negative or even distortive effect on the performance of those measured. It is more interesting, but also more challenging, to measure if actions have contributed toward achieving a desired outcome or goal. Thus, the application of accounting may be accompanied by diagnosis.
- The *Learn* application is often used as a rationale to develop a range of indicators to understand a specific situation or issue of interest. Learning can embody wider "enlightenment" aspects of indicators beyond the purely rational production of evidence; although such effects can be difficult to discern. Learning is often contrasted with the accounting approach, as a more productive way to inspire improvement than mere control. While learning can occur in any of the application types, it should be treated as a stand-alone application if the primary intent is to develop an understanding of how to manage a particular problem.
- The *Communication* application is attached to all other applications and plays a role in how successful they are. Indicators that are not shaped and conveyed in a way that is acknowledged by the intended recipients may fail. Enhanced communication can be a primary reason to use indicators as a tool in the first place. A

well-established indicator in a particular context (such as GDP or inflation rate for economic investment decisions) can potentially foster instant recognition and lead to immediate action among large groups of stakeholders.

These application areas are generic and comprehensive with regard to the way that indicators are used. They can also all play a role in making sustainability count within transportation. It is, however, very important to be aware that indicators may not always be applied as planned. There may be unintended and even negative effects from their use. Cousins (2004), Halachmi (2002), and Marsden et al. (2005) all point to possible misapplications of indicators if they are merely used to justify already adopted positions, or if they install "tunnel vision," where indicators draw the attention away from important issues that have not been quantified to the same degree. These types of pitfalls will be discussed and exemplified in the case study chapters. The generic applications should not, therefore, be considered exhaustive in terms of the specific uses to which indicators can be put in practice.

While this chapter has focused primarily on *individual* indicators, adequately addressing sustainability in transportation needs to take into account the interconnectedness of the transportation systems and transportation planning levels, and the full scope of sustainable development. It also needs to combine the contextual aspects (the "why" to measure as described by the indicator applications), with the "how" of the indicator criteria, and the "what" of the general and holistical concepts of sustainable development and transportation. This inevitably means that *systems* of indicators need to be devised. The next chapter embeds indicators in the wider notion of *frameworks* that serve as the final stepping stone toward the effective use of indicators for the measurement of transportation sustainability.

## References


Barbour, E., Newmark, G., & Deakin, E. (2011). Determining fair share regional targets for reduction of greenhouse gas emissions from transportation and land use: California's experience under Senate Bill 375. *Transportation Research Record: Journal of the Transportation Research Board, 2244*, 9–17.

Bollen, K. A. (2004). Indicators: Methodology. In N. J. Smelser & P. B. Baltes (Eds.), *International encyclopedia of the social & behavioral sciences* (pp. 7282–7287). Oxford: Elsevier.

Boulanger, P. M. (2007). Political uses of social indicators: Overview and application to sustainable development indicators. *International Journal of Sustainable Development, 10*(1–2), 14–32.

Broughton, B., & Hampshire, J. (1997). *Bridging the gap: A guide to monitoring and evaluating development project*. Canberra: Australian Council for Overseas Aid.

Castillo, H., & Pitfield, D. (2010). ELASTIC—A methodological framework for identifying and selecting sustainable transport indicators. *Transportation Research Part D, 15*(1), 79–188.

Congleton, R. D., & Sweetser, W. (1992). Political deadlocks and distributional information: The value of the veil. *Public Choice, 73*, 1–19.

Cousins, J. B. (2004). Commentary: Minimizing evaluation misuse as principled practice. *American Journal of Evaluation, 25*, 391–397.

De Ceuster, G., Logghe, S., & Van Herbruggen, B. (2006, September 18–20). *Indicator assessment of the objectives of the white paper on transport*. Presented at European Transport Conference, Strasbourg.

EEA. (1999). *Environmental indicators: Typology and overview*. Copenhagen: European Environment Agency.

EEA. (2005). The European environment—State and outlook 2005. Copenhagen: European Environment Agency.

EEA. (2008). Climate for a transport change TERM 2007: Indicators tracking transport and environment in the European Union. EEA Report No 1/2008. Copenhagen: European Environment Agency.

Gallopin, G. C. (1996). Environmental and sustainability indicators and the concept of situational indicators. A systems approach. *Environmental Modeling and Assessment, 1*, 101–117.

Halachmi, A. (2002). Performance measurement: A look at some possible dysfunctions. *Work Study, 51*(5), 230–239.

Hayashi, Y., & Morisugi, H. (2000). International comparison of background concept and methodology of transportation project appraisal. *Transport Policy, 7*, 73–88.

Jeon, C. M., & Amekudzi, A. M. (2005). Addressing sustainability in transportation systems: Definitions, indicators, and metrics. *Journal of Infrastructure Systems, 11*(1), 31–50.

Jeon, C. M., Amekudzi, A. A., & Guenseler, R. (2013). Sustainability assessment at the transportation planning level: Performance measures and indexes. *Transport Policy, 25*, 10–21.

Joumard, R., & Gudmundsson, H. (Eds.) (2010). Indicators of environmental sustainability in transport: An interdisciplinary approach to methods. *INRETS report*, Recherches R282, Bron, France, 422p. Accessed April 25, 2015, from http://cost356.inrets.fr

Küchen, T., & Nordman, K. (2008). Performance budgeting in Sweden. *OECD Journal on Budgeting, 8*(1), 1–11.

Kusek, J. Z., & Rist, R. C. (2004). *Ten steps to a results-based monitoring and evaluation system. A handbook for development practitioners*. Washington, DC: World Bank.

Lee, K. N. (2001). Human–environment relationship: Indicators. In N. J. Smelser & B. Baltes (Eds.), *International encyclopedia of the social and behavioural sciences* (pp. 7045–7050). Amsterdam: Elsevier.

Maclaren, V. W. (1996). Urban sustainability reporting. *Journal of the American Planning Association, 62*(2), 184–202.

Marsden, G., Kelly, C., Snell, C., & Forrester, J. (2005). Improved indicators for sustainable transport and planning. Sustainable transport indicators: Selection and use. DISTILLATE. Deliverable C1. Accessed April 25, 2015, from http://www.distillate.ac.uk/projects/project-c.php

May, A. D., Koh, A., Blackledge, D., & Fiorretto, M. (2010). Overcoming the barriers to implementing urban road user charging schemes. *European Journal of Transport and Travel Research, 2*(1), 53–68.

Mearns, K. (2009). From reactive to proactive—Can LPIs deliver? *Safety Science, 47*, 491–492.

Merriam-Webster. (2013). Merriam-Webster online: Dictionary and thesaurus. *Indicator*. Accessed April 25, 2015, from http://www.merriam-webster.com/dictionary/indicator

Moldan, B., & Billharz, S. (Eds.). (1997). *Sustainability indicators. Report on the project on indicators of sustainable development*. Chichester: Wiley.

Nardo, M., Saisana, M., Saltelli, A., Tarantola, S., Hoffman, A., & Giovannini, E. (2005). Handbook on constructing composite indicators: Methodology and users guide, 2005, OECD-JRC joint publication, OECD Statistics Working Paper, STD/DOC(2005)3.

Opschoor, H., & Reijnders, L. (1991). Towards sustainable development indicators. In H. Kuik & O. Verbruggen (Eds.), *In search of indicators of sustainable development* (pp. 7–27). Dordrecht: Kluwer.

Papagiannaki, K., & Diakoulaki, D. (2009). Decomposition analysis of $CO_2$ emissions from passenger cars: The cases of Greece and Denmark. *Energy Policy, 37*(8), 3259–3267.

Parker, S. P. (Ed.). (2003). *McGraw-Hill dictionary of scientific & technical terms*. Columbus, OH: McGraw-Hill.

Polk, M. (2010). Sustainability in practice: The interpretation of sustainable development in a regional planning arena for dialogue and learning in Western Sweden. *Planning Theory & Practice, 11*(4), 473–489.

Quinet, E. (2011). Cost–benefit indicators and transport programming. *Fiscal Studies, 32*(1), 145–175.

Ramani, T., Quadrifoglio, L., & Zietsman, J. (2010). Accounting for nonlinearity in the MCDM approach for a transportation planning application. *IEEE Transactions on Engineering Management, 57*(4), 702–710.

Ramani, T. L., Zietsman, J., Knowles, W. E., & Quadrifoglio, L. (2011). Sustainability enhancement tool for state departments of transportation using performance measurement. *Journal of Transportation Engineering, 137*(6), 404–415.

Rousseaux, P. (1994). Proposition de descripteurs de suivi de l'état de l'environnement. Lyon (France) (Proposal of indicators to monitor the state of the environment in Lyon). Observatoire des changements écologiques du Grand Lyon, Lyon, France, 39p.

Rubin, R. S. (2002). Will the real SMART goals please stand up? *The Industrial-Organizational Psychologist, 39*(4), 26–27.

Runhaar, H., & Driessen, P. P. J. (2007). What makes strategic environmental assessment successful environmental assessment? The role of context in the contribution of SEA to decision making. *Impact Assessment and Project Appraisal, 25*(1), 2–14.

Rydin, Y. (2002). *Indicators into action: Local sustainability indicator sets in their context*. London: The Pastille Consortium, London School of Economics.

Seip, K. L., & McKnown, R. (2007). The timing and accuracy of leading and lagging business cycle indicators: A new approach. *International Journal of Forecasting, 23*, 277–287.

Turnhout, E., Hisschemoller, M., & Eijsackers, H. (2007). Ecological indicators: Between the two fires of science and policy. *Ecological Indicators, 7*, 215–228.

UNDP. (2014). *Human Development Report 2014. Sustaining Human Progress: Reducing Vulnerabilities and Building Resilience*. New York: United Nations Development Programme.

US EPA. (2006). Report of the 2005 peer review of proposed indicators for the U.S. Environmental Protection Agency's Report on the Environment 2007 (ROE07) technical document. Final Report January 4, 2006. Accessed January 25, 2009, from http://www.epa.gov/ncea/ROEIndicators/prcomments/pdfs/1.cover-and-introduction.pdf

Zietsman, J., & Rilett, L. R. (2001). Analysis of aggregation bias in vehicular emission estimation using TRANSIMS output. *Transportation Research Record: Journal of the Transportation Research Board, 1750*, 56–63.

## Additional Readings

Franceschini, F., Galetto, M., & Maisano, D. (2007). *Management by measurement. Designing key indicators and performance measurement systems*. Berlin: Springer. 242p.

Geurs, K. T., & van Wee, B. (2004). Accessibility evaluation of land-use and transport strategies: Review and research directions. *Journal of Transport Geography, 12*(2), 127–140.

Hakkert, A. S., Gitelman, V., & Vis, M. A. (Eds.) (2007). Road safety performance indicators: Theory. Deliverable D3.6 of the EU FP6 project SafetyNet.

Meadows, D. (1998). *Indicators and information systems for sustainable development*. A Report to the Balaton Group. The Sustainability Institute, Hartland, VT.

# 7 Frameworks

## 7.1 Purpose and Content

The purpose of this chapter is to introduce the essential role of frameworks in measuring and managing sustainability in transportation. In Chap. 6 indicators were defined as key informational devices for planning and delivery with a number of distinct applications. Frameworks organize how combinations of indicators are selected and used. The chapter starts by defining and describing frameworks and why they are important. It then outlines a wide range of methods to "frame" indicators, from broad ideas and paradigms, to more specific systems adopted by organizations. The active process of "framing," i.e., setting up a framework to develop and apply indicators and performance measures will also be discussed. The chapter will illustrate a number of generic framework types as well as practical frameworks that are used in areas such as transportation appraisal, environmental planning, sustainability assessment, and performance measurement. These examples are each discussed in terms of how well they support thinking and acting upon sustainability in transportation, and how well they support the different indicator applications that were introduced in Chap. 6. The chapter will summarize the key features a framework should generally possess in order to provide optimal support to sustainability, without proposing one master framework to fit all situations. In the case studies discussed in Chaps. 8–11, detailed examples of the application of frameworks will be analyzed. The analysis will draw from the concepts and typologies presented in this chapter, and will serve to show how different contexts in terms of spatial scale, institutional setting, policy goals, or task portfolio can affect how sustainability is framed and measured.



H. Gudmundsson et al., *Sustainable Transportation*, Springer Texts in Business and Economics, DOI 10.1007/978-3-662-46924-8_7

## 7.2 Introduction to Frameworks

All indicators exist within a framework. Indicators are usually bundled together and linked to other information and management tools to serve an overall assessment purpose. For example, when indicators are used in an organizational, goal-oriented setting, they serve as performance measures in the review or accounting application. The number of crashes may be an indicator of road safety in general, whereas the number of annual crashes per million vehicle miles traveled can be used to measure the performance of an agency in reducing this number. The selection of indicators will be driven by the goals of the agency. This process of "framing" is therefore critical for how sustainability is made to "count." As Joanna Becker notes, "frameworks are useful tools to clarify the concepts of sustainable development and which indicators to use" (Becker 2007, p. 142).

But what is a "framework" and how can it provide a useful context for making a set of indicators count? At the most general level a "framework" is understood as *a way to organize information* according to an overall purpose or practice. A framework generally provides an outer boundary as well as an internal structure for deriving and using information. There are two ways to understand or visualize a "frame": as a "picture frame" that focuses attention on what is happening inside the frame rather than on the walls outside; or as a "building frame," like a set of steel bars, locking various elements of a house into their proper position (Polletta and Ho 2006). Each of these notions suggests a different perspective on the importance of frames, namely the *focus* (external frame) and *organization* (internal frame) of information, respectively. Frameworks usually do both to some degree—i.e., provide essential information and a context for planning and decision-making. Sustainability depends on how various elements of a system affect each other to create a total outcome (Jeon et al. 2013). Frameworks are therefore key in organizing and conceptualizing information and actions to inform the development of sustainable outcomes.

Frameworks come in different shapes and forms that serve various needs. Assmuth and Hilden (2008, p. 73) offer a useful distinction between "loosely structured or generic frameworks" versus "clearly structured or contextual frameworks with specific procedures." The former may be used in situations where problems are not well defined. Current examples of this could be concerns about the resilience of urban infrastructure with regard to flooding, or risks associated with road users' increased use of electronic devices during driving. In these cases, a loosely structured framework can help define basic concepts and map out existing information. In a clearly structured framework, knowledge has matured and sufficient data are systematically produced to manage a particular issue. Examples of structured frameworks include information systems for road pavement management or air traffic control. Here each indicator usually has a strict definition, clear function, and a specified action protocol.

In the philosophy of science, scholars have proposed a distinction between frameworks, theories, and models each having a gradually stronger ability to organize information, but also with a gradually narrower area of application

(Sabatier 1997). A framework points to a set of important variables and how they are related, but does not explain how they interact in detail (like a theory does) or estimate results (like a model does). Frameworks may link to, build on, incorporate, and even sometimes look like theories or models, but it is better to define them as something broader that directs how the information provided by the different types of knowledge should be produced and used.

**Discussion Topics**

- Think of an issue like "air pollution in the city." Which indicators could be used to describe the causes or input to the problem and the outcomes or impacts of the problem?
- How did you select your indicators? To what extent was your indicator selection based on a framework, theory, or model?
- Is a framework necessary to measure a construct such as sustainable transportation? Why not directly apply a theory or model?

There exist a number of well-known generic frameworks such as the "Balanced Scorecard" (used to manage performance in organizations), "Pressure-State-Response" (PSR) indicator systems (used widely in environmental assessment), and "The Natural Step" approach (applied by many companies to frame their sustainability work). Figure 7.1 provides a simplified overview of these generic frameworks, indicating their scope and areas for measurement. More in-depth descriptions on how frameworks have been applied to sustainable transportation will follow later in this chapter.

| **Balanced scorecard** | **P-S-R** | **The Natural Step** |
|---|---|---|
| ***Scope:*** | ***Scope:*** | ***Scope:*** |
| **Company** | **Country / Sector** | **Society / Organization** |
| ***Key areas measured:*** | ***Key areas measured:*** | ***Key areas measured:*** |
| • Internal business processes | • Pressures on the environment | • Substances extracted from the Earth's crust |
| • Creation of value for customers | • State or quality of the enviromnent | • Substances produced by society |
| • Financial data and performance | • Responses to environmental problems | • Harvesting of natural resources |
| • Learning and growth | | • Fair and efficient use of resources to meet basic human needs |

**Fig. 7.1** Simplified overview of three generic frameworks

Generic frameworks provide the overall direction, but they do not set targets or solve concrete data problems. Frameworks must therefore be adapted and applied to a particular case to be effective. Becker (2007, p. 130) says "A framework only provides a basis for discussion. The roadmap must still be compiled by those who are to follow it." Such "road map" or "practice" frameworks organize specific indicators and include procedures to collect, analyze, combine, and report on them. Some practice frameworks in transportation have become known brands, such as the 'Transport and Environment Reporting Mechanism' (TERM) indicator reports published by the European Environmental Agency (EEA 2011), or the "Gray Notebook" published by Washington State DOT in the USA (WSDOT 2011), whereas others remain relatively unknown due to their small-scale application.

Most practice frameworks are shaped by contextual factors such as the kind of organization using it, current policy goals, the task at hand, the knowledge of the involved professionals, and the need to feed certain data into assessment tools (such as monetized values for prioritization of projects). Hence, "framing" can derive from a range of sources that do not necessarily stem from one particular way of thinking; "frames are therefore not internally homogenous" as noted by Boezeman et al. (2010, p. 1757). Moreover the boundary between the framework and its environment can be difficult to draw. Generic frameworks may be easier to identify and review than practice ones.

## 7.3 Drivers of Frameworks

Frameworks of all kinds are important because they explicitly or implicitly suggest certain ways to think, organize, measure, and eventually act. The main advantage of frameworks is that they can provide organized ways to deal with communication needs: What should be measured and how and to whom should the results be reported? This is also their limitation, since all stakeholders may not share the same assumptions as those embedded in a particular framework. For example, decision makers frequently question the use of an economic method to quantify and present social or environmental impacts of transportation (Sager and Ravlum 2005; Bakker et al. 2010). One may say that the information is "overframed" for some stakeholders. Other stakeholders may, in contrast, not be able to cope with a loose framework reporting scores of disconnected indicators (Niemeijer 2002, p. 99). Here, the information is "underframed." However, "no framing" is rarely a good option for working with indicators either. Traditions, values, concealed mind-sets, or untested ideas will result in boundaries, structures, or blind spots, forming "invisible frameworks" and poor decision-making. Cobb and Rixford (1998) emphasize that a framework derived from a powerful storyline aiming to raise awareness about a critical problem may not produce indicators that are useful at *diagnosing* problems correctly or identifying the most appropriate countermeasures against them. Conversely, indicators that accurately measure traffic flows using complicated metrics may not be able to stir sufficient attention to ensure that the resulting congestion problems are really addressed. Hence, a framework is more than a convenient structure; it can bring focus, purpose, direction, clarity, and attention, but also limitations to what indicators can say and do. One could reasonably expect that the more explicit and well thought out the framework is, and the better it is balanced between generic and contextual elements, the more likely it is to produce relevant and influential information in a particular situation (Fischer et al. 2010). This is why frameworks and how they are developed and applied are so important.

There are three basic questions involved in building an indicator framework:

- "*Why*" is the information needed?—referring to the intention and application;
- "*What*" information is needed?—referring to the specific issues or impacts measured; and
- "*How*" is the information to be delivered?—referring to the framework operation.

Each framework would provide a different set of answers to these questions. Some typical examples are shown in Table 7.1.

In the study of frameworks, it has been observed that some appear to be dominated by either "what," "why," or "how" questions, or conceptual aspects, intentions, or procedures, respectively (Rametsteiner et al. 2011). The reasons for this may sometimes be found by looking at who developed the framework. Conceptually dominated frameworks often stem from scientists or strategists using theories for how a system is supposed to work. For example, an environmental monitoring system may be based on a theory of ecosystem taxonomy classes derived from biology, even if such classes are not recognized by policy makers or the public (Niemeijer 2002). At the other end of the spectrum lie frameworks that are intentionally derived from political needs or legal mandates (as noted in Chap. 5).

**Table 7.1** Examples of framing questions and answers

| | Why to indicate? | What to indicate? | How to indicate? |
|---|---|---|---|
| Asset management framework | Describe the infrastructure system's condition | Age and state of repair of each infrastructure system | Build a database that allows asset data to be stored and recovered |
| Environmental policy program | Review progress toward adopted environmental goals | Emissions of air and other pollutants | Report progress for policy reviews |
| Environmental problem analysis | Diagnose the causes of increasing waste accumulation | Sources and volumes of relevant waste/emissions | Define methods to allow statistical analysis |
| Project appraisal system | Decide among alternative project options | Project impacts, for example, on accessibility for those without a car | Combine GIS data with qualitative surveys of a target group |
| Performance reporting | Account for the implementation of programs | Implementation actions and processes | Provide incentives for accurate reporting |

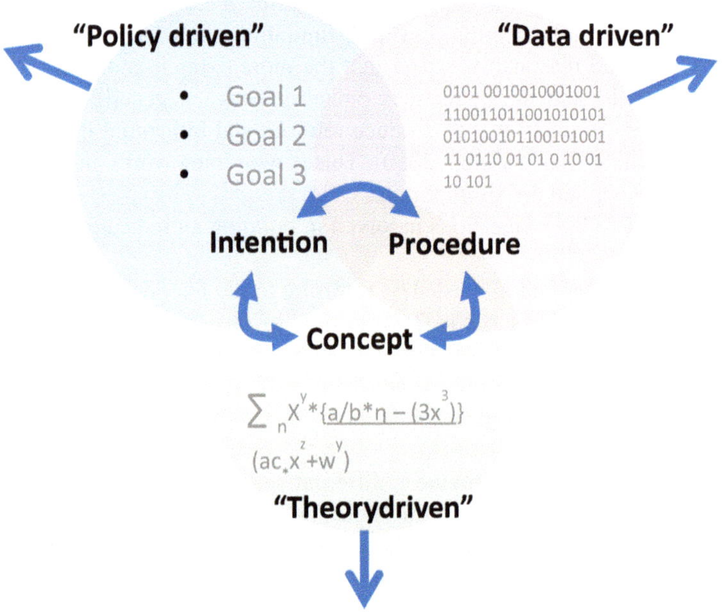


**Fig. 7.2** Key dimensions in building a framework

The structure of these frameworks is controlled by a set of policy goals that often lack a logical connection to each other and are not accompanied by scientific or operational definitions that facilitate their measurement. They may be built up incrementally over time with initial logic getting lost. A third situation where procedure dominates may arise if "bureaucrats" opt to report required information by exploiting what is already produced, regardless of the specific relevance. The result may be a "data-driven" framework (Niemeijer 2002). An example could be the use of an existing survey on transport modal split to report the results of a strategy to shift people from using cars to public transportation. These data may seem relevant but are not as useful if "new" public transit users are, for example, former cyclists or pedestrians or just more frequent users (Sørensen and Gudmundsson 2010).

Figure 7.2 illustrates key dimensions in building a framework. It shows the ideal situation of a mutual alignment between concepts (substance), procedures, and intentions but also highlights the risk of frameworks disintegrating due to excessive dominance of one particular dimension.

Looking at how these dimensions are aligned or not can help characterize a particular framework and why the information it contains may succeed or fail to make a difference for decision-making. The goal is to have a clear understanding of the three dimensions and ensure that an adequate balance between the dimensions is achieved when developing the framework.

## 7.4 Overarching Frameworks Relevant to Transportation and Sustainability

The following sections will review four categories of overarching frameworks that organize different kinds of existing planning and management practice. They include:

- *Transportation appraisal*, with a focus on "official" national frameworks for the assessment of transportation projects and plans;
- *Environmental policy review and reporting*, as conducted by bodies like the OECD and the European Environment Agency (EEA);
- *Sustainability assessment*, where a wide variety of methods and tools have been applied to evaluate economic, social, and environmental impacts; and
- *Performance management*, emphasizing the organizational level of decision-making, where performance indicators are used to report on results with regard to an agency's objectives and targets.

None of these overarching frameworks have the measurement of sustainable transportation as their exclusive purpose. Reviewing them, however, illustrates how institutionalized settings and conventional practice to some extent enable some but also constrain other aspects of measuring sustainability in transportation.

### 7.4.1 Transportation Appraisal[1] Frameworks

Transportation planning and management is characterized by the use of a range of appraisal and evaluation methodologies. Project appraisal, in particular, has a long history and some methodologies have an almost global reach although with significant variations in how they are applied (Mackie and Worsley 2013; Hayashi and Morisugi 2000). Giorgi and Tandon (2002) identified four dimensions along which many transportation assessment frameworks can be categorized, namely according to their:

- Analytic basis (paradigm, concept, theory, etc.);
- Level of decision-making (project, program, policy, etc.);
- Phase in the policy cycle (from agenda setting, to review of options, to evaluation); and
- Degree of technical formalization (use of mathematics versus more soft approaches).

[1] The concepts of Appraisal and Assessment are quite similar notions. Here, "Appraisal" is used for a more focused project review undertaken ex ante, while "Assessment" is used in a broader sense for different scales, scopes, and times.

It can be noted that "official" national transportation assessment practice in most countries is dominated by a tradition that combines these four dimensions in a particular manner, where the *analytic basis* is mostly welfare economic theory (i.e., with a view of optimizing resources to maximize social welfare), the *level of decisions* is mostly focused on infrastructure projects and programs, the *phases* of the policy cycle range from agenda setting to evaluation, and the methods used have a relatively *high degree of formalization* with extensive use of transport simulation and economic evaluation models (Bakker et al. 2010; Bristow and Nellthrop 2000).

**Cost–Benefit Analysis**

Cost–Benefit Analysis (CBA) of individual projects is the classic methodology required for project appraisal in a number of countries (Mackie and Worsley 2013; Browne and Ryan 2011; Bakker et al. 2010). Undertaking cost–benefit-based project appraisal requires substantial information about a project's expected impacts converted into monetary values if they are not already in that form. Some of the key steps where indicators are used or transformed include the following (DG Regio 2008; FHWA 2003; Quinet 2011):

- The identification of all relevant positive and negative effects of a given project;
- Monetization of significant external effects;
- Projection of future values calculated for all variables over the life span of the project;
- Discounting of future benefits compared with total costs, using a discount factor; and
- Deriving decision-making indicators such as a Benefit–Cost Ratio (CBR), Net Present Value (NPV), Net Benefits per year, and the Internal Rate of Return (IRR).

From a sustainability point of view, the standard economic approach encapsulated in benefit–cost frameworks represents a weak sustainability position (see Chap. 2), which assumes that all resources and impacts could in principle be compared and aggregated by their market or shadow value. This is a disputable assumption especially for large-scale projects that affect critical environmental resources in the future. National frameworks do try to address this problem in various ways. Some provide monetary values for environmental impacts, whereas others leave most environmental impacts out of the analysis and describe them qualitatively or as part of an Environmental Impact Statement (EIS), or post "flags" when there are particularly critical issues for policy makers to be aware of in the assessment of a project (Mackie and Worsley 2013). Some, instead, place restrictions on the degree of acceptable trade-offs among social, economic, and environmental impacts by, for example, setting a minimum score for each pillar (see Highways Agency 2013). Yet another approach is to apply Multi-Criteria Decision-Making (MCDM) by providing scaled scores and weights rather than an economic calculus, where projects are ranked instead of being assessed by an absolute number (Bickel et al. 2005). Mackie and Worsley (2013) demonstrate that countries differ as to which costs and impact indicators are included in economic appraisal frameworks.

Another important contested issue in economic project assessment is the use of a *discounting* factor for costs and benefits accrued in the future. Normal linear discounting imposes a much lower present valuation of impacts that may affect future generations (Joumard and Nicolas 2010). Some national appraisal frameworks have been adjusted in response to this issue by modifying the required discount rate (Arrow et al. 2012).

**Environmental Impact Assessment**

In addition to economic appraisal, most countries' transportation planning frameworks also require Environmental Impact Assessments (EIAs) of large-scale transportation projects, as well as Strategic Environmental Impact Assessments (SEAs) of higher level plans or programs (Fischer 2006). The EIA/SEA procedures seek to ensure that environmental effects (of potential significance for sustainability) have at least been described and considered before plans are adopted and economic decision support has led to final conclusions. Occasionally environmental assessments also help decide if the project or plan can proceed or if significant mitigating measures need to be included. Some countries have developed specific guidance for conducting EIA and SEA in the transportation sector (e.g., National Roads Authority of Ireland 2008). Assessment rules and guidance do not prescribe particular indicators to be used in EIA or SEA, but often highlight important aspects of environmental impacts that need to be measured such as permanent, cumulative, or irreversible effects.

A number of countries, states, and regions incorporate further elements to address sustainability in their transportation appraisal frameworks. Some US states for example apply detailed project rating tools to score and rank transportation projects according to how well they perform on a number of sustainability criteria. The rating can be used to prioritize among projects or enhance the design of the individual project (Samberg et al. 2011). This practice is exemplified in more detail in Chap. 10. At the other end of the scale some national frameworks include, for example, scenario studies and wider sustainability impact assessments (SIA) of entire national transportation plans (e.g., France, see Meunier 2012) or provide ex post reporting on fulfillment of national sustainable transportation goals (e.g., Sweden, see SIKA 2006). Such efforts are, however, less widespread than the standard methods and few countries report systematically on the sustainability of their transport projects, plans, and policies.

We may conclude that the project-oriented welfare economics-based framework is dominant in the area of transportation appraisal in many parts of the world, although other frameworks and methods are also applied to assess projects, plans, or policies. Indicators in such frameworks are dominated by aggregated economic and, to a lesser degree, disaggregated environmental variables to support review and decision-making applications. The typical framework has some limitations with regard to sustainability assessment, of which only a few, such as discounting the future, have been noted here (for a deeper critical review, see Gasparatos et al. 2008). It is possible to modify or supplement the core economic framework to address wider aspects of concern for transportation and sustainability, but the extent to which this is actually done varies in practice. Some countries do measure, appraise, or report a broader range of issues covering more sustainability territory, but there appears to be no widespread consensus on what a comprehensive sustainability framework for national, state, or regional transportation assessment and appraisal should look like.

### 7.4.2 Environmental Policy Review and Reporting

Environmental policy review and reporting is a process that addresses the regional, national, or even global level of decision-making rather than individual projects or plans. The reviewing and reporting provides a comprehensive overview of environmental problems and the progress toward green goals or lack thereof. These reviews tend to be retrospective rather than ex ante appraisals. National state-of-the-environment reports prepared by environmental agencies or similar bodies were initiated in Europe and North America in the early 1980s and continue around the world today. Global environmental assessments have been undertaken occasionally since the 1990s. To facilitate the assessment, the reporting is often based on sets of indicators organized in a framework that seek to capture key aspects of the environmental policy. The reports may address the total set of environmental issues or focus on a particular sector such as transportation. International bodies like the Organization for Economic Co-operation and Development (OECD), the EEA, and several divisions of the United Nations have been pioneering the conceptual work and actual reporting in this area.

The OECD conducts regular reviews of environmental policies of its member states. One of the tools used is an indicator-based assessment. The OECD's purpose for using indicators in this effort includes (1) to simplify the measurement of environmental progress and performance of member countries and (2) to undertake effective international comparisons (OECD 2003). For that purpose, the organization adapted the "pressure-state-response" (P-S-R) indicator framework originally developed by Canadian scholars. This framework is applied in the OECD's environmental policy reviews.

In the general P-S-R framework, a set of indicators are defined for a range of 14 key environmental issues, including climate change, acidification, toxic contamination, biodiversity, urban environmental quality, and others. For each issue, three types of indicators are applied, namely:

- "Pressure" indicators, meaning indicators of human activities that exert an indirect stress on an environmental issue, such as measures of production and consumption in general, or a more direct pressure, such as emissions, resource consumption, etc.;
- "State" indicators, meaning indicators measuring the current condition and quality of the environment, such as the concentration of pollutants in environmental media, remaining stocks of renewable or nonrenewable resources, or the status of wildlife and ecosystems; and

- "Response" indicators, meaning indicators that measure how society reacts to environmental problems, for example, via investments in clean technologies, use of environmental taxes, market shares of organic products, waste recycling, etc.

The P-S-R framework supports an integrated assessment, since it not only measures the conditions of the environment but seeks to track developments in those factors that influence and stress the environment (pressures), as well as the results of various initiatives to improve it (the responses). Ideally, this can convey a comprehensive view of a country's environmental performance over time using a limited number of indicators (around 50 in the so-called "core set"). It does not consider social and economic factors as impacts in their own right, only as drivers and responses in an environmental context. Hence it can be said to provide only partial information on sustainable development as formulated in Chap. 2.

To apply the P-S-R approach to the assessment of transport policy, the OECD modified the original framework. Figure 7.3 shows the indicators used for this reporting.

The OECD's other aim is to make international comparisons. The same framework was applied to compare OECD member states with regard to their integration of environmental issues in transportation policies (OECD 1999). The report provides an interesting set of indicators for comparison. In Japan, for example,

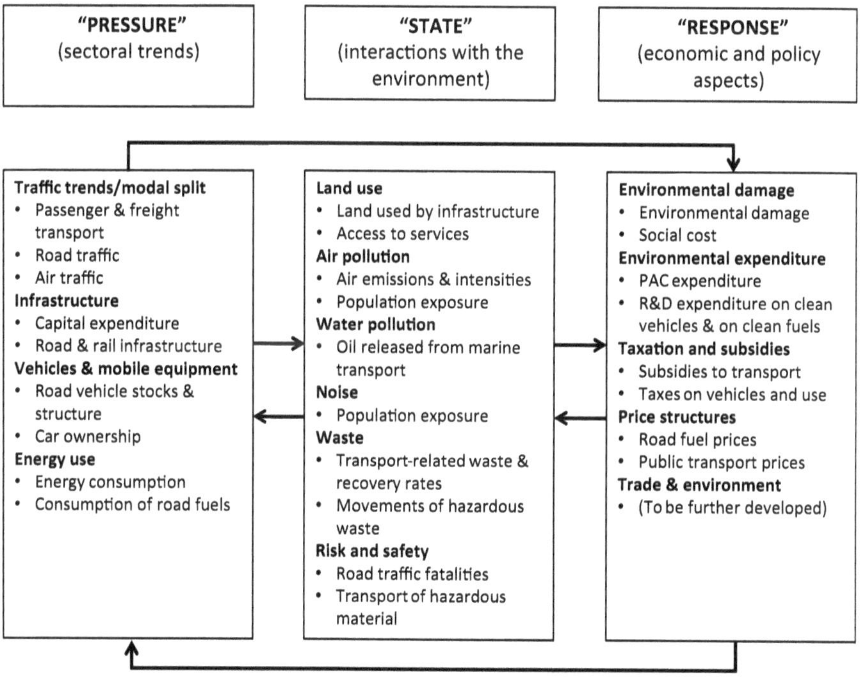


**Fig. 7.3** OECD P-R-S framework adapted to review environmental integration in transportation policies (OECD 1999)

36 % of all passenger transportation is by rail, while the figure is less than 1 % in the USA and Canada. In Australia, transportation energy consumption per capita is nearly double that of Austria. However, comparable data could only be collected for a small number of the indicators in the framework. For example, only one indicator on the response side (road fuel prices and taxes) was included with sufficient data. This means that more diagnostic or decision-relevant indicator applications are difficult to perform. While elaborate frameworks like P-S-R may be needed to provide a comprehensive overview of transportation impacts on the environment, significant data gaps mean that a sophisticated analysis cannot be completed and the framework, in a way, collapses. Moreover, transportation policy data are often specified within a national or local context. This is a notorious challenge to any reliable cross-border comparison. As Litman (2009, p. 5) notes, "The problem researchers face is not an absolute lack of data collection, but a lack of consistency, transparency, and availability of the data that are collected by various organizations and jurisdictions."

The P-S-R system is the original form of a "linkages-based" framework, which generally divides indicators according to logical steps in a chain of events that can be seen as linked in environmental policy making (Jeon and Amekudzi 2005). Figure 7.4 highlights the differences between P-S-R and two well-known extensions,

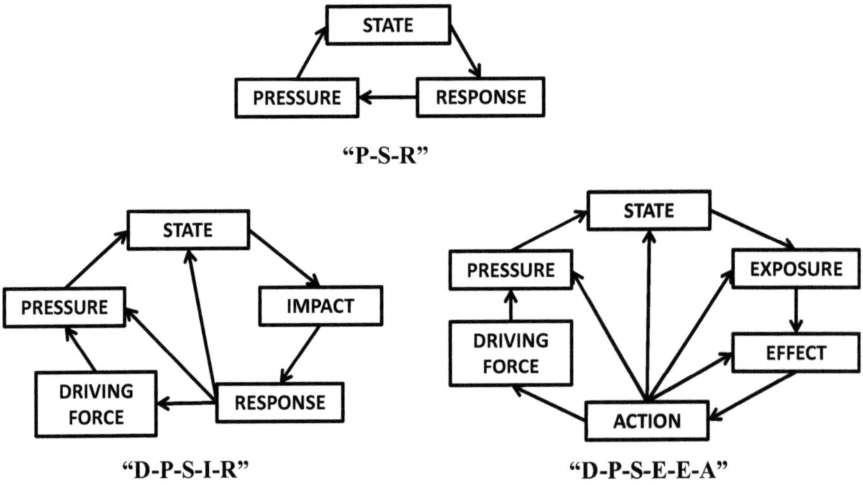


The **D-P-S-I-R** framework includes the analysis of "Driving forces" behind the pressures. These are factors in society like motorization that "drive" environmental problems. Also "Impacts" are added. Impact indicators measure the final outcomes such as cost of pollution.

The **D-P-S-E-E-A** framework focuses on health. It therefore adds indicators for the stage of "Exposure" of human beings to environmental health risk factors. "Effect" is similar to "Impact" in D-P-S-I-R and "Action" is similar to "Response" but the wording suggest that policy actions are more proactive and not simply a reaction to human health problems.

**Fig. 7.4** Examples of linkages-based frameworks

D-P-S-I-R and D-P-S-E-E-A, that have also been applied to transportation and environment policy analysis.

Using indicators for each step allows linkages frameworks to help manage problems during various stages of their development, and eventually learn how a change in a response (e.g., lowering speed limits) reduces a certain pressure (e.g., noise emissions from traffic) and eventually improves the state of the environment (e.g., ambient noise levels). If other sources of noise meanwhile increase, the pressure may have been reduced, which is good, without the state improving, which points to a need for additional measures. In general, the more developed the frameworks are, the more comprehensive and detailed are the assessments that can be supported. However, the more developed the frameworks are, the more demanding they are to populate with indicators and data. All "linkages" frameworks represent simplifications compared to actual causal chains. Waheed et al. (2009) point to a number of weaknesses with linkages-based frameworks for measuring sustainability:

- They do not work effectively if the evidence for causal linkages is missing or vague;
- They lead to oversimplification of spatial and temporal interactions; and
- There can be multiple pressures for most states, and multiple states arising from most pressures, creating difficulties for selecting the best indicators.

### 7.4.3 Sustainability Assessment Frameworks

The area of *sustainability assessment* offers another, more heterogeneous, array of frameworks that are somewhat less mainstreamed than official frameworks for transportation appraisal and environmental policy review. Over the last two decades, multiple sustainability assessment tools and methods have become available, or previous environmental ones have been revised or extended to address sustainability. There is limited agreement about the general logic and even the basic terminology to use in this area.

The OECD (2010, p. 4) defines SIA broadly as "an approach for exploring the combined economic, social and environmental impacts of a range of proposed policies, programs, strategies and action plans." This notion has roots in EIA and SEA, as well as in other forms of project, plan, or policy assessment. In the USA, the EIS required under the National Environmental Policy Act has long included the assessment of other sustainability dimensions. Similarly, in Europe, the European Commission has been instrumental in setting up mandatory Impact Assessment procedures that cover impacts across sustainability pillars for its own policy proposals as well as applications for EU support (Adelle and Weiland 2012). Countries such as the UK, Canada, and South Africa and some multilateral development banks have also taken early initiatives in this area (Pope and Dalal-Clayton 2011). Another approach to sustainability assessment that is not rooted in project or policy assessment includes broad sustainable development indicator (SDI) systems,

**Table 7.2** Tools for sustainability assessment (de Ridder et al. 2007)

| |
|---|
| 1. Assessment frameworks |
| 2. Participatory tools |
| 3. Scenario analysis tools |
| 4. Multi-criteria analysis tools |
| 5. Cost–benefit/cost-effectiveness analysis |
| 6. Accounting tools, physical analysis tools, and indicator sets |
| 7. Modeling tools |

such as the SDI program that was initiated by United Nations after Rio 1992 (UN 2007), the "Genuine Progress" indictor run by the NGO Redefine Progress, and the "Environmental Sustainability Index" (ESI) published by the World Economic Forum (Stiglitz et al. 2009).

Gasparatos et al. (2008) divide sustainability assessment approaches into three groups: "Monetary tools," "Biophysical models," and "Composite indices." Each type can provide an aggregate assessment of sustainability. However, they use completely different methodologies. Twelve specific tools within these groups are described and reviewed.[2] An example of a biophysical model is the "Ecological Footprint," which is a method to calculate the aggregate ecological impact of human activity, using directly and indirectly consumed land area as an indicator for total impacts. This methodology has been applied in academic analyses of transport systems and plans. Martin-Cejas and Sanchez (2010) calculated the footprint of passenger and tourist travel on the island of Lanzarote, Spain. The fossil fuel-based energy use of cars is converted to the corresponding area required to grow equivalent amounts of renewable fuels. This allows comparing current developments with the available bio-capacity to indicate sustainability. The study found that the tourism footprint is large and likely to become critical in the future. According to Gasparatos et al. (2008), all of the aggregation methods in the typology are found to suffer various methodological deficits compared with an ideal holistic assessment of sustainability. The biophysical methods like Ecological Footprint are critiqued for their inability to convey information on the social pillar of sustainability. The monetary tools are critiqued for ignoring the precautionary principle (see Sect. 2.2.6). Most of the tools are, however, found to be able to help predict outcomes for future generations and also to allow some form of stakeholder participation in the assessment process, for example, by including stakeholder preferences and values when impacts are assessed.

De Ridder et al. (2007) groups approaches to SIA into seven categories (see Table 7.2). As in the former typology, the categories refer to frameworks and tools

[2] Singh et al. (2009) identify no less than 70 such indices of relevance for sustainability assessment but do not provide a critical analysis similar to Gasparatos et al. (2008).

with common characteristics, but the grouping also reflects the different roles that each tool can have in an integrated assessment process.

De Ridder et al. (2007) map a tool's functionality to different phases in the planning process. These phases are labeled "Problem analysis," "Finding options," "Analysis," and "Follow-up." This succession of phases can be said to roughly correspond to the distinction of Assmuth and Hilden (2008) between "loosely structured concept-oriented frameworks" (the first two types) and "clearly structured frameworks with specific procedures" (the latter two).

While most tools appear to have uses in different phases, they play a different role in each one. A clear distinction is observed between roles to be played by softer participatory and scenario tools (to support strategic planning, visioning, etc.) versus harder quantitative tools (supporting analysis of objectives, actions, and results). The limited role for MCDM can be extended, for example, to assist the selection of indicator sets (Castillo and Pitfield 2010) or support participatory processes (Munda 2006). The type of framing of sustainability assessment provided by de Ridder et al. (2007) is different from the previous one, as the tools described are not only focused on "measuring sustainability," but also on supporting an informed process that can promote sustainable solutions via careful problem analysis and an open process to seek out and review possible options to choose from. This contrasts to some extent with the framework assumed in classic transportation project appraisal, where the problem is assumed to be well known and the type of solution is given already.

A final example of the framing of sustainability tools is a typology provided by Ness et al. (2007). They distinguish 32 tools in four categories ("Indicators," "Product-related assessment," "Integrated assessment," and "Monetary evaluation tools"). The "Product-related" category includes Life Cycle Analysis (LCA) that is relevant for sustainability since it adopts a long-term comprehensive perspective on the impacts of a particular product, service, or technology. LCA addresses all environmental impacts from the extraction of minerals to the disposal of wastes associated with a certain product, e.g., a car or a road. The life cycle view is applied in the economic dimension under the name of Life Cycle Costing to ensure that all construction as well as future maintenance and renewal costs are considered in an infrastructure project. There is now even a life cycle methodology for social impact analysis (e.g., Jørgensen et al. 2008) and an integrated "sustainability life cycle framework" covering all three pillars, as outlined by Kloepffer (2008). More recently, the LCA framework has been extended to include the "use" phase of highways, whereby relatively small improvements in the rolling resistance of a pavement surface may significantly reduce vehicle emissions over the life of the infrastructure (Bryce et al. 2014). This finding has important implications for the design and maintenance of roadways.

Another useful contribution from the framework of Ness et al. (2007) is a distinction of tools and methods with regard to the *time direction* of the assessment that each tool can best support, namely "prospective" (or "ex ante") tools, such as integrated assessment, versus "retrospective" (or "ex post") tools such as satisfaction surveys or quality-of-life reports. Prospective assessment is often based on models to forecast future outcomes with regard to sustainability or other goals, or to select the best solution. Indicators can be applied here if they are "forecastable" and logically connected to model outputs. Some frameworks bridge different time spans, for example, by using the same sets of indicators to assess ex ante and ex post interventions. However, this is not always possible in practice. While certain indicators such as fuel consumption can be both measured and modeled, others, such as human well-being, can only be measured, since we currently lack tools to model the indicator. The different time dimensions of evaluation are illustrated in Fig. 7.5.

In summary, sustainability assessment methods and tools can be classified within a number of different frameworks emphasizing different aspects such as the underlying paradigm of sustainability, the different phases of analysis for decision-making, the time orientation, the technical content, or the use or not of different aggregation methods to reach a result. How to combine overarching frameworks and particular methods and tools to provide optimal support for sustainability will depend on the context. The most general understanding is that sustainability assessment must consider all three pillars of sustainability and ensure they are managed in an integrated way (OECD 2010; Singh et al. 2009; Ness et al. 2007; Kates et al. 2001). According to Bond and Morrison-Saunders (2009), the inherent flexibility of sustainability appraisal may facilitate outcomes that will in fact not secure this. Current practice may for example allow the "best alternative" to be good enough even when unsustainable. There is a risk that anything addressing the three pillars can be called a "sustainability assessment." Scholars such as Gasparatos et al. (2008) and Meunier (2012) remind us that ethical principles of sustainable development should be drawn upon to guide a sustainability assessment. For example, an assessment framework must address

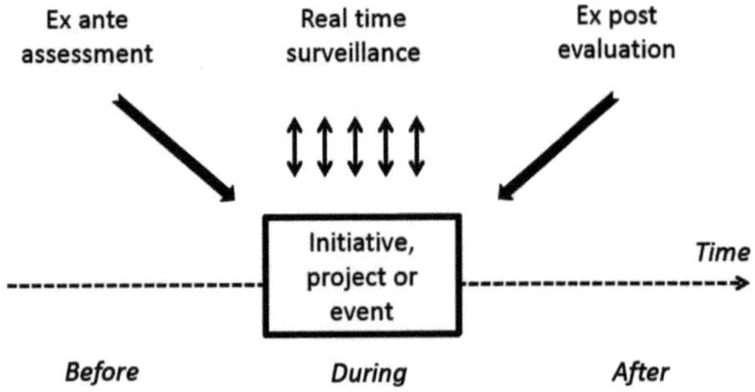


**Fig. 7.5** Simplified assessment types with regard to the time dimension

needs of both future and present generations, a precautionary position should be adopted when relevant, and affected stakeholder interests should be represented in problem identification and decision-making. How to specify such criteria further seems to be an important task. Major distinctions in the field concern whether to assume a strong or a weak interpretation of sustainability, which implies a choice between economic or multidisciplinary indicators, and whether to pursue an aggregated or disaggregate level of assessment, implying a choice between communicating through few or many indicators. More recently, Holden et al. (2013) call for a return to Brundtland's view when evaluating sustainable passenger transportation to ensure that the global environmental and intra-/inter-generational equity challenges society faces are not overlooked by the performance measurement framework (see Sect. 4.2.6).

Sustainability assessment is a broad and heterogeneous area that draws on indicators of all types, from specific product data to all-encompassing aggregated indices. The types of indicator applications supported include mostly raising awareness, description, forecasting, and to some extent review, and learning, with fewer examples of diagnosis, decision-making, and accountability. Most of the encountered methods could well be applicable to frame assessment in the transportation area. However, many of them address issues and apply methods that reach beyond the remit of an individual infrastructure project, and extend the analysis beyond the boundaries of transportation systems.

### 7.4.4 Performance Management Frameworks

Performance management is focused on individual organizations and the results they deliver in relation to organizational goals. Performance *measurement* is an integral part of performance *management* exactly because of the strong emphasis on comparing measured results with goals and targets in order to manage the organization's performance. While there are many performance measurement frameworks available (Ravelomanantsoa et al. 2010), they are generally more homogeneous and focused in their scope than sustainability assessment frameworks. Performance measurement and management can nevertheless be relevant for sustainability since transportation sector agencies through their various mandates and activities can influence how the transportation systems and activities impact the economy, environment, and society (Zietsman et al. 2011). Performance goals and measures can in principle be defined for sustainability-related impacts of transportation systems at the level of an agency.

Transportation is one of the sectors where New Public Management (NPM) reforms have been most widely applied leading to considerable changes in the way transportation systems and markets are governed, for example, via organizational/agency restructuring, privatization, and marketization within subsectors such as road, rail, and air (Sager and Sørensen 2011; Hughes 2011). Among public agencies, mechanisms such as strategic plans, result-based and performance-based contracts, and performance measurement and monitoring systems have been adopted. Additionally, more freedom to manage has been given to senior leadership in several agencies, contingent on fulfilling performance outcome goals.

Performance measurement means measuring the performance of a particular organization (or group of organizations) with regard to strategic goals and corresponding short-term performance targets. The targets are normally set by upper management, potentially with guidance from higher level political bodies and sometimes a wider set of stakeholders. The specification of goals is sometimes connected to the allocation of a budget to the organization or a particular program, so government and taxpayers can see what level of performance they can expect for their money. Goals may also be specified in performance contracts.

In the context of performance management, goals, objectives, and performance measures are hierarchical terms that can be defined as shown in Table 7.3.

According to Falcocchio (2004, p. 220), performance measurement can be used for a number of functions in the organization including to help assess needs, evaluate system performance, set priorities, generate financial resources, allocate funds, and communicate with customers and other stakeholders. Cempel (2010) argues that the most important function is to use performance measurement to drive resource allocation via budgeting and project prioritization. Others emphasize accountability toward the public and stakeholders (Vedung 1997). To be effective, performance measures in any case need to be linked to the goals and objectives that guide transportation decisions.

Most performance measurement frameworks incorporate elements of the standardized production function for measuring organizational performance (see Sect. 6.4.4 in Chap. 6). The organization's performance with regard to fulfilling goals and meeting targets is defined here as the result of a function. Using various examples from the transportation sector, an organization receives an *input* (such as funding, personnel, mandates) to undertake *activities* (such as build, operate, regulate), and produce *outputs* (such as km of paved road, seat kilometers, speeding tickets, studies), which have *outcomes* (such as reduced travel time, perceived accessibility, increased fuel efficiency, less accidents). Indicators for each step are needed for calculating performance, especially for ratios such as *effectiveness* (outcome/goal) and *efficiency* (input/output).

**Table 7.3** Performance hierarchy

| |
|---|
| *Goals* |
| A goal is a general statement of a desired state or ideal function of a transportation system with relevance to a particular organization. An organization may have several goals |
| *Objectives* |
| An objective represents a concrete step toward achieving a goal, stated in measurable terms |
| *Performance measures* |
| A performance measure is a specific indicator (variable) that is used to measure performance with regard to an objective, in a goal-oriented setting |
| *Performance targets* |
| A performance target is the target or standard to be achieved for the performance measure. It specifies a target value using the measurement scale of the performance measure |

**Discussion Topics**

- Consider a city with a diverse range of opportunities (modes) for travel. Assume the city wishes to promote modes like cycling and walking.
- Think of possible indicators that could be used to measure the development in the use of various modes of transport over time. Such indicators could relate to the outcomes or outputs that could explain why mode share might change.
- Select one of those indicators and develop it into a performance measure for the city administration (specifying aspects such as hypothetical goals, targets, etc., that the city can adopt). Discuss why this is a good performance measure and what its limitations might be.

Performance management is found to represent an area where frameworks are generally well developed and indicators have multiple applications. Traditional corporate performance management operates with relatively short-term goals that are focused on the effectiveness and efficiency of an organization and its internal procedures, with limited attention to issues such as environmental impacts. However, using approaches such as the balanced scorecard and especially NPM and New Public Governance (NPG), public agencies have more opportunity to address external, environmental, and politically salient parameters with a performance management-based approach. Often, this is linked to broader national goals that address sustainability. For example, countries like Canada have even mandated the integration of sustainable development strategies in government-wide organizational performance management and reporting. The area of performance measurement has the potential for effectively framing indicators in the interest of sustainability. However, there are still barriers that exist due to inherent weaknesses in the concept and application of performance measurement, the effects of which could include negative reactions, evasion, lack of use in management and decision-making, etc. Additionally, sustainability would still have to "compete for attention" with other short-term concerns of public agencies.

### 7.4.5 Framing Sustainability and Performance Together: An Example

This section presents an example of an overarching framework that has been developed to help transportation agencies in the USA address sustainability comprehensively by incorporating the notion into their strategic planning and management efforts (Zietsman and Ramani 2011; Ramani et al. 2011). It embraces the performance management approach but has a wider scope. The framework is unique in its ambition to integrate principles of sustainability into transportation decision-making by focusing on the specific interconnections between the two. This section describes the conceptual side of the framework.

As noted in Sect. 7.2, generic frameworks provide general directions that can be operationalized through specific practice frameworks. In the same way, an overarching framework can be customized by similar divisions or business units within an organization. These frameworks balance the two types of "framing" discussed in Sect. 7.2—by putting key issues into focus and providing an overall structure for performance measurement and management processes.

These types of overarching, flexible frameworks are useful in places where different organizations with similar goals and functions each need to address a problem in a context-specific manner. This is quite common in the USA, where State Departments of Transportation (DOTs) and Metropolitan Planning Organizations (MPOs) operate autonomously but serve similar goals and functions.

The framework presented by Ramani et al. (2011) supports transportation agencies in adopting a hierarchical approach starting from general overarching sustainability principles moving to transportation specific sustainability goals, and then to agency specific performance measures, across the full range of agency functions. The principles and goals are shown in Table 7.4.

A main feature of the framework is an overarching procedure and guidance for how an agency should combine and apply the different elements to move from principles to practice, as shown in Fig. 7.6. The framework was designed to allow for addressing sustainability over the full spectrum of transportation agency *focus areas*, including planning, programming, project development, construction, operations, and maintenance.

In identifying components to be included in this sustainability framework, the question to be answered is "What does a transportation agency need to be equipped with in order to successfully address sustainability issues through performance measurement?"

As seen in Fig. 7.6, the framework has three types of components:

- ***Fundamental components*** are required for the application of the framework. These include the already described step-by-step elements plus the significance of feedback from performance reporting that can result in revisions being made to components of the framework as needed.
- ***Overarching components*** are elements that need to be considered throughout the framework application process. These include the engagement of stakeholders, partners, and external agencies throughout the process.
- ***Auxiliary components*** are related, but optional, components that can be used to supplement the framework application process. For example, it is optional if the agency will prepare its own definition of sustainability or adopt an existing one.

A core element of the framework is the performance measures. More than 300 examples of actual performance measures are offered in an appendix to the final NCHRP report for agencies to consider (Zietsman and Ramani 2011). They cover all of the 11 sustainability goals shown in Table 7.5, several sub-objectives within each goal, and are differentiated according to agency focus areas such as planning, construction, and maintenance. The framework helps agencies select

**Table 7.4** Principles and goals of sustainability

| |
|---|
| *Principles of sustainability* |
| Sustainability entails meeting human needs for the present and future, while<br>• Preserving and restoring environmental and ecological systems<br>• Fostering community health and vitality<br>• Promoting economic development and prosperity<br>• Ensuring equity between and among population groups and over generations |
| *Goals for sustainability and transportation* |
| • Safety—Provide a safe transportation system for users and the general public<br>• Basic accessibility—Provide a transportation system that offers accessibility that allows people to fulfill at least their basic needs<br>• Equity/equal mobility—Provide options that allow affordable and equitable transportation opportunities for all sections of society<br>• System efficiency—Ensure the transportation system's functionality and efficiency are maintained and enhanced<br>• Security—Ensure the transportation system is secure from, ready for, and resilient to threats from all hazards<br>• Prosperity—Ensure the transportation system's development and operation support economic development and prosperity<br>• Economic viability—Ensure the economic feasibility of transportation investments over time<br>• Ecosystems—Protect and enhance environmental and ecological systems while developing and operating transportation systems<br>• Waste generation—Reduce waste generated by transportation-related activities<br>• Resource consumption—Reduce the use of nonrenewable resources and promote the use of renewable replacements<br>• Emissions and air quality—Reduce transportation-related emissions of air pollutants and greenhouse gases |

indicators according to different cross-cutting applications such as evaluation, decision support, and communication, similar to the indicator application areas discussed in this book. Examples in Table 7.5 illustrate how indicators differ for the same overall objective, depending on which agency function it is to support. For the goal for "economic viability," in the focus area of "planning," the proposed indicator is particularly relevant from a sustainability point of view, as it measures the agency's commitment to include life cycle costs. Many of the other indicators offered in the framework (that are excluded from Table 7.5) are not specific to sustainability and would not necessarily help ensure it. The accompanying guidebook therefore suggests that "it is the application of a collective set of measures, aligned with the objectives and goals and viewed within the context of the sustainability principles, that make them relevant" (Zietsman et al. 2011, p. 8).

In theory, the overarching framework allows transportation agencies to integrate the notion of sustainability into their decision-making by bridging theoretical concepts related to sustainability to the everyday practice of transportation planning and system management. Through a step-by-step process, transportation agencies

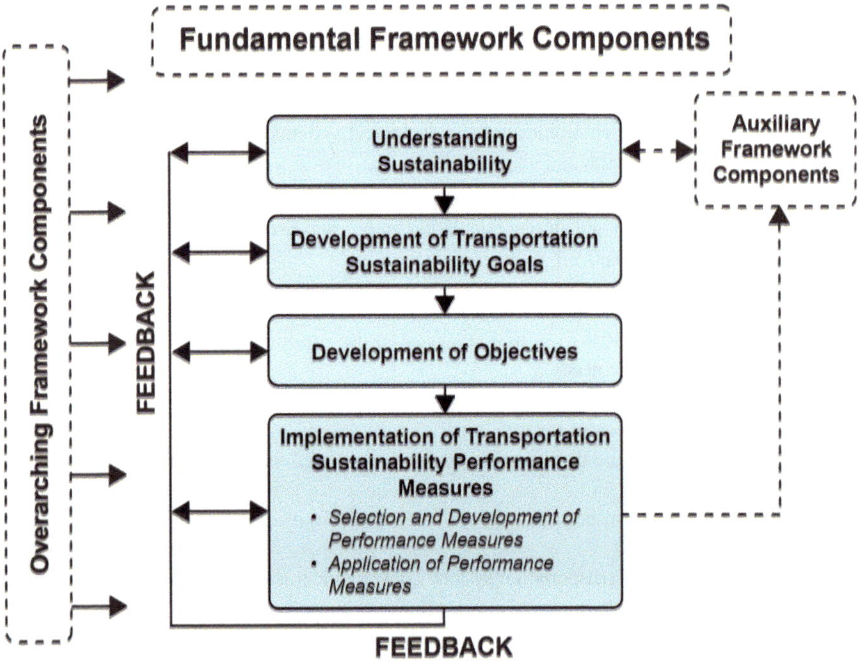


**Fig. 7.6** Overarching framework for transportation and sustainability. *Source*: Zietsman et al. (2011)

**Table 7.5** Examples of performance measures from the compendium in Zietsman et al. (2011)

| | Goal 7: Economic viability | Goal 8: Ecosystems |
|---|---|---|
| Focus area planning | Proportion of projects subjected to life cycle cost analysis (LCCA) | Change in net area of undeveloped land converted to transportation uses (acres) due to project |
| Focus area maintenance | Proportion of projects with maintenance costs within planned budget | Area (in acres) sprayed with herbicides during maintenance |

and practitioners can understand sustainability, develop context-appropriate goals and objectives, and apply performance measures to incorporate sustainability considerations into their activities at various levels, focus areas, and business units.

## 7.5 Toward Sustainable Transportation Frameworks

The review of different types of frameworks in the previous section demonstrates how sustainability in transportation can be "framed" in multiple ways. Framing can deliberately control, or subtly influence, what kind of information is provided and what types of policy, planning, or management decisions are thereby supported, and which ones are not. Framing matters for how sustainability and transportation is measured, be it in purely economic terms as in CBA, with a wide scope as in sustainability assessment, or with agency-oriented systems as in performance management. While frameworks have some flexibility, it is not likely that all indicator applications can be served equally well in any framework. Moreover, frameworks can look good on paper but fail in practice because of economic constraints, lack of data, or limited interest. In other words, there will not be one universal framework for providing indicators of sustainability in transportation, but it is useful to know what type of frameworks could be useful in different situations.

Rather than one framework, what would be of interest is a "framework of frameworks" for characterizing and understanding how various frameworks correspond to different situations. Let us start by considering the generic dimensions of information frameworks that were introduced in Sect. 7.3, namely *intention, concept/ substance*, and *procedure*. We can now place into this typology the factors that were found to be important characteristics for indicator systems in the various fields that were reviewed. Sustainability definitions are, for example, an important aspect of the substance dimension (i.e., what to measure). The intention dimension (i.e., why to measure) is constrained by which indicator applications the framework is designed for, whereas the procedure dimension (i.e., how to measure) may be constrained by the types of variables and data that can be made available. We allocate 11 such aspects to the three dimensions as shown in Fig. 7.7. This can be seen as a basic reference set of categories to consider in the description, analysis, and (possibly) design of suitable frameworks of indicators for sustainability in transportation.

As observed earlier, it is clear that frameworks exist on a gradual scale from broad open concepts to detailed systems with quantified goals and firm procedures. It is not necessarily the case that the most rigorous kind of framework would always be the best one to support sustainability. However, an effective practical framework will most likely need a degree of specification with regard to several categories within the dimensions of substance, intention, and procedure.

*Substance* refers to what is measured. Besides the importance of how sustainability is defined, the delimitation of the transportation system is another key consideration, for example, if a life cycle perspective is adopted or whether walking is counted as transportation or not. How sustainability and transportation are assumed to be related is therefore also an essential aspect.

*Intention* refers to the purpose of the measurement. The purpose may be more or less clearly given by a planning task, a policy decision, or a legal requirement, such as the need to produce an EIS. In general, the governance situation is instrumental to what counts as important, as discussed in Chap. 5. As the review of literature in the four areas has shown, different methods and tools are applied for ex ante and ex post reviews and for assessment at a strategic and project level of decision-making. The case study chapters (Chaps. 8–11) review applications which cover different purposes.

*Procedure* refers to how individual indicators are identified, produced (possibly aggregated), and reported within the framework. Key concerns include how many

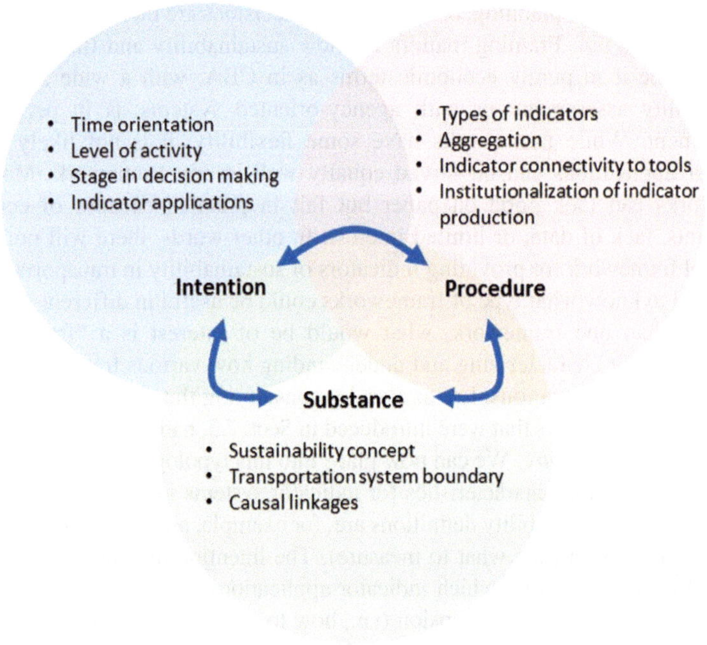


**Fig. 7.7** Dimensions and aspects of frameworks

indicators to include, how many levels of reporting take place before decision makers receive the information, and to what degree a methodology of aggregation of individual indicators is adopted.

As the discussion illustrates, substance, intention, and procedure are partly independent and partly dependent elements of framework building. A high level of detail in one of the dimensions may put demands on the others, whereas a low level of detail overall may disable a framework from providing useful guidance. Either issue may hamper the ability of a framework to foster sustainability in transportation.

A group of researchers at the Georgia Institute of Technology has proposed a set of attributes that they argue any framework should reflect in order to be an effective promoter of sustainability (Pei et al. 2010)—see Table 7.6. These attributes that combine concerns for how to represent sustainability in a measurement framework (substance in our terminology) with more general criteria for effective management (intention and procedure) are discussed below.

**1. Comprehensiveness** The criterion of comprehensiveness is essential to ensure an adequate conceptual and operational representation of sustainability principles (see Chap. 2). Sustainability can be defined in various ways, but according to most of the research literature the integration with a holistic view across the different

**Table 7.6** Attributes of robust performance measurement systems (Pei et al. 2010)

| |
|---|
| 1. A comprehensive and holistic framework encompasses a balanced view of sustainability |
| 2. A framework needs to be connected to the goals and objectives of an agency, thus giving a rationale for the causal relationships between agency action and the desired outcomes |
| 3. A framework should be integrated vertically and horizontally through an agency to allow more effective management |
| 4. A framework should provide the capability of capturing the effects of interactions among variables |
| 5. A framework should reflect stakeholders' perspectives |
| 6. A framework should consider the capabilities and constraints of the agency and those of its stakeholders |
| 7. A framework needs to be flexible and foster self-learning |

pillars and across the dimension of past and future needs is important for measuring whether a development is likely to be sustainable or not.

**2. Connection to goals** Connecting measurement to the goals and objectives of an agency is essential for frameworks that are embedded in a governmental institution. Otherwise there is a risk that sustainability will become sidelined within an organization that is naturally focused on its existing goals and performance indicators, and possibly even formally required to do so by external bodies or owners. This was also strongly emphasized in the US framework presented in Sect. 7.4.5.

**3. Internal integration** A framework should be integrated vertically and horizontally through an agency to allow more effective management. This refers primarily to large public or private organizations charged with the development and/or management of major transportation systems. These entities often have multiple goals, divisions, and layers of management, which creates a diverse and complex performance context. For example, how do the road construction and public transport planning arms of a multimodal agency approach sustainability?

**4. Interactions** A framework should provide the capability of capturing the effects of interactions among variables. Does the framework describe linkages between transportation system impacts, intervention variables, and sustainability outcomes? Does it allow for the identification of synergies among impacts in order to point toward the most critical parameters for sustainability and the most effective measures to enhance it (Pei et al. 2010, p. 77)? A major challenge is obviously the complexity and dynamic character of transportation systems and the multiple ways in which they can generate sustainability impacts, considering all dimensions, long- and short-term consequences, life cycle perspectives, cumulative effects, and system interactions between transportation, land-use, and other sectors (Joumard and Gudmundsson 2010; Himanen et al. 2004; Kahn Ribeiro and Figueroa 2012).

**5. Stakeholder perspectives** Transportation involves and affects all participants in society, some of which are organized as stakeholder groups, others not. The active engagement and involvement of all social groups in decision-making is a fundamental principle and requirement for sustainable development (see Chap. 2). Since the 1992 Rio conference, nine so-called "major groups" have been recognized, including Women, Children and Youth, Local Authorities, Business and Industry, and Nongovernmental Organizations, all of which could maintain autonomous perspectives on transportation goals and decisions.

**6. Agency capabilities and constraints** It is important to put particular emphasis on what an agency can or is allowed to influence (Jeon and Amekudzi 2005). For public road, rail, urban transport, or multimodal agencies, there might be legal documents that describe the scope and purpose of their work, as well as a broader regulatory framework that constrains its activities, for example, with regard to whether or not it can collect revenues or set speed limits.

**7. Flexibility and learning** As already noted, frameworks need to support the continuous development of an organization, program, or policy over time. Due to the complexities and challenges involved in making genuine progress toward sustainability, there is a constant need to evolve and accumulate knowledge and evidence about, for example, causal relations, sustainability impacts, and the effectiveness and efficiency of adopted measures. Agents and stakeholders can learn by being engaged in the definition, application, and assessment of performance in connection with routine operations as well as when facing new tasks such as major construction works, implementation of new regulations, the test of novel information, communication, technology (ICT) equipment, or revisions to management structures and institutional arrangements.

## Discussion Topics

- Recall the different types of frameworks presented in this chapter. For each framework, identify at least one strong feature with regard to promoting indicators that would help sustainable transportation. Are these features compatible with one another?
- Think of your own country or city. What are the remits of the different agencies involved in planning for and delivering transportation services? How important will agency capabilities and constraints be in terms of developing an integrated sustainable transportation performance measurement framework?
- Consider the seven attributes to design a framework for measuring sustainable transportation. Will application of these criteria necessarily deliver a

comprehensive and effective framework? Are there any important aspects missing?

## 7.6 Conclusions

This chapter has emphasized the importance of frameworks. Considerable attention is given to the notion of a framework as a way to organize information according to an overall purpose or practice. A framework generally provides an outer boundary as well as an internal structure for deriving and using information. Frameworks come in different shapes and forms that serve various needs. Broadly speaking they can be grouped into loosely structured or generic frameworks versus clearly structured or contextual frameworks with specific procedures. We have identified some basic features of frameworks with the three dimensions of substance, intention, and procedure, and have suggested 11 aspects within those dimensions that could help to understand and characterize existing frameworks and build new ones in an enlightened way.

The idea of a framework was considered from two perspectives. We first explored a variety of existing frameworks that have been developed by experts or applied in practice by agencies and governments, across the fields of transportation planning, environmental policy, sustainability assessment, and performance management, asking how these frameworks support thinking and action for sustainability in transportation. We then developed an outline for a more ideal framework that could, in theory, inform the creation of various sustainable transportation performance measurement frameworks in practice. Both discussions support the same conclusion, namely that one single framework that works effectively everywhere cannot easily be defined because there are different needs, constraints, and opportunities involved. Nevertheless, the principles of sustainability as laid out in Chap. 2 and further embedded in the transportation context in Chap. 4 apply everywhere and should always be taken into account in a comprehensive manner, either when existing frameworks are reviewed or when new ones are developed.

At a generic level, a framework is more universal and is formulated around principles, as in the approach proposed by Pei et al. (2010) that was discussed in Sect. 7.5, whereas frameworks can also be adapted to a specific context and could look quite different when applied in practice. The framework proposed by Ramani et al. (2011) in Sect. 7.4.5 provides a guide for exactly such a process for US transportation agencies. For all of the examples discussed, what matters most is what happens when frameworks are used and applied. In the worst-case scenario, neatly designed frameworks can end up as window dressing with little impact on actual decisions if they are not firmly embedded in the decision-making entity, or even serve as a smokescreen for selective use of indicators that support unsustainable practices. This is a more obvious risk if the framework just presents a menu of "sustainability" indicators and performance measures to choose from with no strings attached. However, with a sound and explicit sustainability framework in place there will at least be a basis to review and challenge the actions of decision makers or agencies, both with regard to how they measure progress and success and with regard to how they pursue it.

The key message of this chapter is that application of indicators in decision-making processes that move us toward (or away from) more sustainable transportation typically occurs through frameworks. There are many types of frameworks and these have different philosophical backgrounds and different purposes in decision support. This chapter explains the nature of many frameworks in use today and sets out some of the concepts we see as critical to effective framework application. However, as Chap. 5 reminds us, there are many agencies in the public and private sector that are engaged in the transportation planning process. They can have different goals, different resources, and different incentives and think in different timescales. The challenge of transportation governance is to bring agencies together to row in a common direction, toward more sustainable transportation. The guidance set out in this chapter provides some structure to the types of actions that will be more successful in facilitating this objective. However, we stopped short of creating an idealized framework because we recognize the importance and persistence of existing decision-making practice and the need to make sustainable transportation work within existing structures. This is not to say we see the status quo as being likely to lead us down a more sustainable path. It is rather a recognition that we need to start the task now, demonstrate that planning for sustainable transportation leads to better decisions, and work to change the system over time. Shouting from the sidelines seems not to have made a significant difference. To that end, we turn our attention to a series of case studies in the USA, Europe, and Japan in Part II of the book that demonstrate the application of indicators and frameworks in real decision-making around sustainable transportation.

## References


Adelle, C., & Weiland, S. (2012). Policy assessment: The state of the art. *Impact Assessment and Project Appraisal, 30*(1), 25–33.

Arrow, K. J., Cropper, M. L., Gollier, C., Groom, B., Heal, G. M., Newell, R. G., et al. (2012, December). *How should benefits and costs be discounted in an intergenerational context? The views of an expert panel. Resources for the future* (Discussion paper 12–53). Accessed May 2, 2015, from http://www.rff.org/RFF/Documents/RFF-DP-12-53.pdf

Assmuth, T., & Hilden, M. (2008). The significance of information frameworks in integrated risk assessment and management. *Environmental Science and Policy, 11*, 71–86.

Bakker, P., Koopmans, C., & Nijkamp, P. (2010). Appraisal of integrated transport policies. In M. Givoni & D. Banister (Eds.), *Integrated transport: From policy to practice*. Abingdon, Oxon: Routledge.

Becker, J. (2007). How frameworks can help operationalize sustainable development indicators. *World Futures, 63*, 137–150.

Bickel, P., Burgess, A., Hunt, A., Laird, J., Lieb, C., Lindberg, G., et al. (2005). State-of-the-art in project assessment. HEATCO Deliverable 2. HEATCO developing harmonised European approaches for transport costing and project assessment.

Boezeman, D., Leroy, P., Maas, R., & Kruitwagen, S. (2010). The (limited) political influence of ecological economics: A case study on Dutch environmental policies. *Ecological Economics, 69*(9), 1756–1764.

Bond, A. J., & Morrison-Saunders, A. (2009). Sustainability appraisal: Jack of all trades, master of none? *Impact Assessment and Project Appraisal, 27*(4), 321–329.

Bristow, A. L., & Nellthrop, J. (2000). Transport project appraisal in the European Union. *Transport Policy, 7*, 51–60.

Browne, D., & Ryan, L. (2011). Comparative analysis of evaluation techniques for transport policies. *Environmental Impact Assessment Review, 31*(3), 226–233.

Bryce, J., Flintsch, G., & Hall, R. P. (2014). A Multi-Criteria decision analysis technique for sustainable infrastructure management business practices. *Transportation Research Part D: Transport and Environment, 32*, 435–445.

Castillo, H., & Pitfield, D. E. (2010). ELASTIC—A methodological framework for identifying and selecting sustainable transport indicators. *Transportation Research Part D, 15*, 179–188.

Cempel, E. (2010). *A CFO's handbook on performance management*. Washington, DC: American Association of State Highway and Transportation Officials.

Cobb, C., & Rixford, G. (1998, October 30–31). Competing paradigms in the development of social and economic indicators. In *CSLS conference on the state of living and the quality of life in Canada*, Ottawa, ON.

de Ridder, W., Turnpenny, J., Nilsson, M., & von Raggamby, A. (2007). A framework for tool selection and use in integrated assessment for sustainable development. *Journal of Environmental Assessment Policy and Management, 9*(4), 423–441.

DG Regio. (2008). *Structural funds, cohesion fund and instrument for pre-accession*. Brussels: European Commission, Directorate General Regional Policy. Guide to Cost-Benefit Analysis of investment projects.

EEA. (2011). *Laying the foundations for greener transport TERM 2011: Transport indicators tracking progress towards environmental targets in Europe* (EEA Report No 7/2011). Copenhagen: European Environment Agency.

Falcocchio, J. C. (2004). Performance measures for evaluating transportation systems—Stakeholder perspective. *Transportation Research Record: Journal of the Transportation Research Board, 1895*, 220–227.

FHWA. (2003, August). *Economic analysis primer*. Washington, DC: U.S. Department of Transportation, Federal Highway Administration.

Fischer, T. (2006). Strategic environmental assessment and transport planning: Towards a generic framework for evaluating practice and developing guidance. *Impact Assessment and Project Appraisal, 24*(3), 183–197.

Fischer, T., Dalkmann, H., Lowry, M., & Tennøy, A. (2010). The dimensions and context of transport decision making. Chapter 4. In R. Joumard & H. Gudmundsson (Eds.), *Indicators of environmental sustainability in transport: An interdisciplinary approach to methods*. (pp. 79–102). INRETS report, Recherches R282, Bron, France, 422 p. Accessed April 26, 2015, from http://hal.archives-ouvertes.fr/hal-00492823/fr/

Gasparatos, A., El-Haram, M., & Horner, M. (2008). A critical review of reductionist approaches for assessing the progress towards sustainability. *Environmental Impact Assessment Review, 28*, 286–311.

Giorgi, L., & Tandon, A. (2002). Introduction: The theory and practice of evaluation. In L. Giorgi & A. Pearman (Eds.), *Project and policy evaluation in transport* (pp. 1–13). Aldershot: Ashgate.

Hayashi, Y., & Morisugi, H. (2000). International comparison of background concept and methodology of transportation project appraisal. *Transport Policy, 7*, 73–88.

Highways Agency. (2013, January). *Value management requirements, Version 2*. London: Highways Agency, Department for Transport.

Himanen, V., Lee-Gosselin, M., & Perrels, A. (2004). Impacts of transport on sustainability: Towards an integrated transatlantic evidence base. *Transport Reviews, 24*(6), 691–705.

Holden, E., Linnerud, K., & Banister, D. (2013). Sustainable passenger transport: Back to Brundtland. *Transportation Research Part A, 54*, 67–77.

Hughes, B. (2011). *Performance reporting of transport reform outcomes: Measuring the value of what can be managed.* Bentley, WA, Australia: Curtin–Monash Accident Research Centre, School of Public Health, Curtin University.

Jeon, C. M., & Amekudzi, A. M. (2005). Addressing sustainability in transportation systems: Definitions, indicators, and metrics. *Journal of Infrastructure Systems, 11*, 31–50.

Jeon, C. M., Amekudzi, A. M., & Guensler, R. L. (2013). Sustainability assessment at the transportation planning level: Performance measures and indexes. *Transport Policy, 25*, 10–21.

Jørgensen, A., Le Bocq, A., Nazarkina, L., & Hauschild, M. (2008). Methodologies for social life cycle assessment. *The International Journal of Life Cycle Assessment, 13*(2), 96–103.

Joumard, R., & Gudmundsson, H. (Eds.). (2010). Indicators of environmental sustainability in transport: An interdisciplinary approach to methods. *INRETS report*, Recherches R282, Bron, France, 422 p. Accessed April 26, 2015, from http://hal.archives-ouvertes.fr/hal-00492823/fr/

Joumard, R., & Nicolas, J. P. (2010). Transport project evaluation methodology within the framework of sustainable development. *Ecological Indicators, 10*(2), 136–142.

Kahn Ribeiro, S., & Figueroa, M. J. (Eds.). (2012). Chapter 9—Energy end-use: Transport. In *Global energy assessment—Toward a sustainable future* (pp. 1203–1306). Cambridge, UK/Laxenburg, Austria: Cambridge University Press/International Institute for Applied Systems Analysis.

Kates, R. W., Clark, W. C., Corell, R., Hall, M. J., Jaeger, C. C., Lowe, I., McCarthy, J. J., Schellnhuber, H. J., Bolin, B., & Dickson, N. M. (2001). Sustainability science. *Science, 292*, 641–642.

Kloepffer, W. (2008). Life cycle sustainability assessment of products (with Comments by Helias A. Udo de Haes, p. 95). *The International Journal of Life Cycle Assessment, 13*(2), 89–95.

Litman, T. (2009, November 14). *Sustainable transportation indicator data quality and availability.* Victoria, BC: Victoria Transport Policy Institute. Accessed April 26, 2015, from http://www.vtpi.org/

Mackie, P., & Worsley, T. (2013). *International comparison of transport appraisal practice.* Leeds: Institute for Transport Studies, University of Leeds.

Martin-Cejas, R. R., & Sanchez, P. P. R. (2010). Ecological footprint analysis of road transport related to tourism activity: The case for Lanzarote Island. *Tourism Management, 31*(2010), 98–103.

Meunier, D. (2012). Towards a sustainable development approach in transport assessment. *Procedia—Social and Behavioral Sciences, 48*, 3065–3077.

Munda, G. (2006). Social multi-criteria evaluation for urban sustainability policies. *Land Use Policy, 23*, 86–94.

National Roads Authority of Ireland. (2008, November 20). *Environmental impact assessment of national road schemes—A practical guide*. Dublin, Revision 1, 106 p.

Ness, B., Urbel-Piirsalu, E., Anderberg, S., & Olsson, L. (2007). Categorising tools for sustainability assessment. *Ecological Economics, 60*(3), 498–508.

Niemeijer, D. (2002). Developing indicators for environmental policy: Data-driven and theory-driven approaches examined by example. *Environmental Science and Policy, 5*, 91–103.

OECD. (1999). *Indicators for the integration of environmental concerns into transport policies.* ENV/EPOC/SE(98)1/FINAL. Paris: Organisation for Economic Co-operation and Development.

OECD. (2003). *Environmental indicators—Development, measurement and use.* Reference paper. Paris: OECD.

OECD. (2010). *Guidance on sustainability impact assessment.* Paris: OECD.

Pei, Y. L., Amekudzi, A. A., Meyer, M. D., Barrella, E. M., & Ross, C. L. (2010). Performance measurement frameworks and development of effective sustainable transport strategies and indicators. *Transportation Research Record: Journal of the Transportation Research Board, 2163*, 73–80.

Polletta, F., & Ho, M. K. (2006). Frames and their consequences. Chapter 10. In R. E. Goodin & C. Tilly (Eds.), *The oxford handbook of contextual political studies*. Oxford: Oxford Univ. Press.

Pope, J., & Dalal-Clayton, B. (2011). From SEA to sustainability assessment? Chapter 34. In B. Sadler, R. Aschemann, & J. Dusik (Eds.), *Handbook of strategic environmental assessment*. London: Earthscan.

Quinet, E. (2011). Cost–benefit indicators and transport programming. *Fiscal Studies, 32*(1), 145–175.

Ramani, T., Zietsman, J., Hall, R. P., Marsden, G., & Gudmundsson, H. (2011). A generally applicable sustainability assessment framework for transportation agencies. *Transportation Research Record: Journal of the Transportation Research Board, 2242*, 9–18.

Rametsteiner, E., Pulzl, H., Alkan-Olsson, J., & Frederiksen, P. (2011). Sustainability indicators development—Science or political negotiation? *Ecological Indicators, 11*, 61–70.

Ravelomanantsoa, M., Ducq, Y., & Vallespir, B. (2010). Integration of requirements for performance indicator system definition and implementation methods. In: B. Vallespir & T. Alix (Eds.), *APMS 2009, IFIP AICT*, Vol. 338, pp. 530–537.

Sabatier, P. (Ed.). (1997). *Theories of the policy process*. Boulder, CO: Westview Press.

Sager, T., & Ravlum, I.-A. (2005). The political relevance of planners' analysis: The case of a parliamentary standing committee. *Planning Theory, 4*(1), 33–64.

Sager, T., & Sørensen, C. H. (2011). Planning analysis and political steering with new public management. *European Planning Studies, 19*(2), 217–241.

Samberg, S., Bassok, A., & Holman, S. (2011). Method for evaluation of sustainable transportation—Toward a comprehensive approach. *Transportation Research Record: Journal of the Transportation Research Board, 2242*, 1–8.

SIKA. (2006). *Follow-up of the Swedish transport policy objectives*. Östersund: SIKA. Accessed June 9, 2010, from http://www.sika-institute.se/Doclib/2006/sr_2006_4_eng.pdf

Singh, R. K., Murty, H. R., Gupta, S. K., & Dikshit, A. K. (2009). An overview of sustainability assessment methodologies. *Ecological Indicators, 9*, 189–212.

Sørensen, C. H., & Gudmundsson, H. (2010). The impact of governance modes on sustainable transport—The case of bus transport in Greater Manchester, UK. *World Review on Intermodal Transportation Research, 3*(1–2), 8–25.

Stiglitz, J. E., Sen, A., & Fitoussi, J. P. (2009). Report by the Commission on the Measurement of Economic Performance and Social Progress. Accessed April 26, 2015, from http://www.stiglitz-sen-fitoussi.fr/documents/rapport_anglais.pdf

UN. (2007). *Indicators of sustainable development: Guidelines and methodologies* (3rd ed.). New York, NY: United Nations. October 2007.

Vedung, E. (1997). *Public policy and program evaluation*. New Brunswick, NJ: Transaction. 336 p.

Waheed, B., Khan, F., & Veitch, B. (2009). Linkage-based frameworks for sustainability assessment: Making a case for driving force-pressure-state-exposure-effect-action (DPSEEA) frameworks. *Sustainability, 1*, 441–463.

WSDOT. (2011). *The gray notebook*. GNB 43, Quarter ending September 30, 2011 published November 18, 2011. Office of the Washington State Department of Transportation. Accessed April 26, 2015, from www.wsdot.wa.gov/accountability

Zietsman, J., & Ramani, T. (2011). *Sustainability performance measures for state DOTs and other transportation agencies—Contractors final report* (NCHRP Report 708). Washington, DC: Transportation Research Board of the National Academies. Accessed April 26, 2015, from http://onlinepubs.trb.org/onlinepubs/nchrp/docs/NCHRP08-74_FR.pdf

Zietsman, J., Ramani, T., Potter, J., Reeder, V., & DeFlorio, J. (2011). *A guidebook for sustainability performance measurement for transportation agencies* (NCHRP Report 708). Washington, DC: Transportation Research Board.

## Additional Readings

Bell, S., & Morse, S. (2003). *Measuring sustainability: Learning from doing*. London: Earthscan.

Dalal-Clayton, B., & Sadler, B. (Eds.). (2014). *Sustainability appraisal: A sourcebook and reference guide to international experience*. Abingdon: Routledge.

ITF. (2011). *Improving the practice of transport project appraisal. ITF Round Tables No 149. International Transport Forum*. Paris: OECD.

Nilsson, M., Jordan, A., Turnpenny, J., Hertin, J., Nykvist, B., & Russel, D. (2008). The use and non-use of policy appraisal tools in public policy making: An analysis of three European countries and the European Union. *Policy Sciences, 41*, 335–355.

# Part II
# Case Studies

## Preamble

What do concepts and measures of sustainability in transportation mean in practice? This is arguably the most important question to address, and certainly one of the most frequently asked when the idea of sustainability is debated among policy makers and practitioners.

The second part of the book explores this question via a set of case studies drawn from today's front line of sustainability applications. Here the focus shifts from theory and history to ongoing efforts by existing organizations to incorporate the measurement of sustainability into practice at different decision-making levels in various parts of the world. The cases were selected to represent different contexts and applications, while at the same time covering relatively advanced and ambitious programs and initiatives.

It should already be clear from the first part of the book that measuring sustainability needs to be embedded in meaningful frameworks in order to make a difference. Even when agencies or decision makers share the same overarching commitment to sustainability, the ways in which they choose to measure it will be somewhat different depending on the requirements and opportunities of the situation. The combined questions of "what," "how," and "why" to measure should guide the development of frameworks that are tailored to the geographic and political context. The following four cases demonstrate more precisely how context matters for engaging in sustainability measurement in transportation.

Yet, as also laid out in the first seven chapters, sustainability is not an infinitely flexible notion. There are basic dimensions and principles that need to be respected if claims to measure progress toward or away from sustainability are to be justified, and there are issues such as economic development, climate change, and biodiversity that may be particularly important to address. Moreover, variations in measurement and management approaches should reflect a systematic consideration of indicator applications in order to meet the communication and decision-making needs they are intended to support. If sustainability measurement is not framed appropriately, there is a risk that it will be ignored or misapplied. Thus, "context" is not everything. There are principles and criteria that can enable sustainability to be taken into account in a systematic way.

The crux of the matter is that framing and applying sustainability to transportation practice involves a delicate balance between universal values and day-to-day pressures and opportunities that face agents and agencies; transformative aspirations and existing mechanisms of delivery; and science and politics. A key message of this book is that there is no universal formula for how to administer this balance in practice due to the complexity, diversity, and constantly evolving situation of transportation planning and decision-making today. However, this does not imply that lessons cannot be learned from existing practices or that these practices cannot be improved. On the contrary, there is a direct need to ask questions that help reveal the trade-offs that agencies must inevitably make. What do factors like different policy ambitions, how problems are perceived, preexisting assessment procedures, and knowledge of sustainability mean for the ways in which real decision-making bodies and agencies address sustainability? Do these factors enable or inhibit the adoption of a holistic approach to measuring sustainability, as advocated in this book? Phrased more generally: How does context matter in practice, and to what extent should it be allowed to do so? These are issues on the horizon of research on sustainability measurement in transportation that the final part of this book aims to explore.

The following chapters address these questions using key concepts and distinctions from the first part of the book to describe the way sustainability is approached in the case studies. The chapters analyze and explain differences among the cases with a view to highlighting strengths and weaknesses in each case, and the prospects for more influential and potentially transformative future measurement actions. The chapters are not comparative in the sense of looking at identical decision points in different locations or different decisions in one location. The state of practice in measuring sustainability in transportation is not yet so evolved as to allow such an approach, although we have selected case studies which capture elements of best practice. The aim is to present different types of policy and planning situations in order to illustrate different challenges and constraints in the real world, and also to demonstrate the importance of sustainability concerns across contexts. By choosing case studies with elements of leading edge practice, we hope to highlight the potential for improvement and advance the discussion of key limitations and weaknesses.

As laid out in Chap. 2, sustainable development is concerned with the present and future generations, while simultaneously considering economic, social, and environmental dimensions of development. Any approach to measure progress should be anchored in these overarching dimensions. The extent to which each dimension is addressed will be influenced by theoretical paradigms, such as weak or strong sustainability; indirect political goals formulated by international, national, or local bodies; or by considerations that may be difficult to ascertain such as the influence of organizational culture. The more holistic and explicit the approach, the better the chance that it will be consistent with the overall sustainable development agenda, although this approach may also potentially be at odds with more pragmatic or problem-driven decision-making.

Chapters 3 and 4 help to place transportation within the context of sustainable development by adopting a systems perspective that not only emphasizes the interconnections between the components of the transportation system such as vehicles, energy sources, and infrastructure, but the close dependencies between transportation and other social, economic, and environmental systems. The sustainability of transportation is thus not primarily determined by the successful perpetuation of movement itself, but by its positive and negative impacts in these domains. The system view also draws attention to the delimitation of system boundaries. Setting boundaries (technical, geographical, or modal) may be necessary for certain planning and decision-making functions, but may also install unfortunate "blind spots" in the measurement of eventual sustainability impacts. An important takeaway from the first part of the book is that a conventional "transportation-centered" planning approach to sustainability may be too limited in some cases and that the indicators used need to go beyond a traditional "silo" approach to transportation performance measurement.

Chapter 5 explores the importance of these institutional and operational boundaries. Given the holistic nature of the issues at play and the preexisting configuration of transportation agencies and responsibilities within the sector, Chap. 5 points to the importance of transportation governance through networks of interconnected governmental and nongovernmental agencies. The extent to which decision-support frameworks bring together or divide transportation and wider agendas, such as land-use and energy policy, plays out in day-to-day decision-making. Similarly, the extent to which public and commercial interests are aligned and fully incorporate sustainability concerns emerges from this interplay. As institutional structures change slowly, our case study analysis looks for ways to improve coordination in the here and now as well as identify the critical gaps and issues which create the case for longer-term reforms. Chapter 5 also points to the different types of decisions that are made from long-term planning to short-run operational decisions. Our case studies capture some initial insights into the extent to which different issues will be important to different types of decisions.

The principal components in the measurement of sustainability in transportation are the indicators that are used. As discussed in Chap. 6, indicators are variables or a combination of variables selected to represent wider issues or characteristics of interest. There are multiple types of indicators available to report on sustainability in transportation, and they need to be selected and designed in a way that reflects the intended indicator application(s). Hence, indicator selection should consider criteria of relevance for representation, practice, and context. In particular, two aspects of indicators are important to review when looking at practice cases: How sustainability is represented (what to measure?), and how the indicators fit with the purpose or application (how and why to measure?). The case chapters will review both the particular indicators used and which applications they serve drawing from the typology in Chap. 6.

Finally, as discussed in Chap. 7, indicators need frameworks to connect the individual variables to each other and to the governance and communication context. The indicator selection aspects of representation, practice, and context need to be reflected at an organizing level, sometimes embodied in formalized framework paradigms like "D-P-S-I-R," Cost–Benefit analysis, or the "Balanced Scorecard" for organizational performance, but also in less explicit or ad hoc ways. It is highly relevant to review how actual organizations handle the challenges involved in framing indicators in a way that is both loyal to notions of sustainability and sustainable transportation and to the management and reporting regimes their practice may otherwise be inscribed in. As we reflected in Chap. 5, it is also important to see the use of information as part of the political process and the extent to which the frameworks are intended to shift practice to a significantly more sustainable position is considered.

Each of the following case chapters applies key elements from the previous chapters as summarized above. The following section structure is used in each chapter:

*Background* This section provides the general context for the case study with a focus on the national, state, or regional policy ambitions in regard to sustainability and the way that transportation decision-making is generally organized.

*Framework* This section conveys how sustainability measurements are framed in each case. The framing is partly derived from the general governance and planning context, and the stage of decision-making that is considered in the case (e.g., planning or delivery; ex ante or ex post assessment). In addition, this section is where any explicit considerations for how to operationalize sustainability for performance measurement will be included, if such considerations are observed in the case.

*Indicators used* This section identifies and reviews specific key indicators used in the case. Specific attention is paid to how indicators are selected to represent certain system features or impacts considered to be important in each case. This analysis of indicators allows us to discuss strong and weak points in how sustainability concerns are operationalized.

*Indicator applications* This section addresses the types of indicator applications that the case represents, among the generic types defined in Chap. 6: Describe; Forecast; Review; Diagnose; Decide; Account; Learn; and Communicate. These applications are not always clearly specified by the measuring body and it is the aim of the analysis to uncover the *implied* applications, and how these may enable or constrain the endeavors of the measuring agency.

*Discussion* The discussion summarizes the observations of each case and reflects on the lessons learned from the application of indicators and frameworks in practice.

The four cases are scaled from the continental, to the national, to the state, to the local level, and involve indicators framed in different ways from ex ante assessment, to ex post evaluation, to comprehensive learning cycles. These cases reflect examples where empirical material was available to the authors or could be collected with reasonable effort, and where the case studies contained elements of what we consider to be advanced practice. A summary of the four case studies and how they relate to the key concepts set out in Part I of the book is provided in the table below.

The case studies exemplify different interpretations of sustainability, governance issues, frameworks, and applications of indicators. The diversity of the cases should not be seen as an artifact of incomplete analysis, but as a fully intended result of the approach when looking into the particular context and evolution of each case with the knowledge we have today. In future editions of the book, we hope to expand the scope of cases beyond the examples included here to cover additional parts of the world, a greater variety of applications, and even more advanced practices.

## Comparison of Case Studies

| Case study | Scale | Approach to sustainability | Domain/ functional area Chapter 5 | Indicator applications Chapter 6 | Framework Chapter 7 |
|---|---|---|---|---|---|
| 8—European Transport White Paper | International | Sectoral but multimodal (air, maritime, and surface)<br>Holistic, covering three pillars<br>Tending to stronger sustainability with some reference to targets and constraints | Strategic Planning at European Commission | Describe<br>Review<br>Diagnose<br>Forecast<br>Decide | Multi-criteria framework organized under the three pillars |
| 9—High-Speed Rail in England | National—Route | Sectoral but project specific (includes impacts on other modes)<br>Holistic, covering three pillars<br>Tending to strong sustainability in comprehensive framework but weaker in trade-off analysis application | Programming by the UK Department for Transport | Describe<br>Review<br>Diagnose<br>Forecast<br>Decide | Multi-criteria framework organized under the three pillars |

(continued)

| Case study | Scale | Approach to sustainability | Domain/ functional area Chapter 5 | Indicator applications Chapter 6 | Framework Chapter 7 |
|---|---|---|---|---|---|
| 10—New York's GreenLITES Rating Systems | State—Projects under NYSDOT | Sectoral. Extensive coverage of environment with limited social metrics Tending to weak sustainability as aspects of the application are focused on implementation which may not be consistent with a broader strong sustainability framing | Developing, constructing, operating and maintaining | Learn Decide Forecast Communicate | Checkbox compliance frameworks operating at multiple levels in organization |
| 11—Japan's "Eco-Model City" Program | City | Cross-sectoral. Focus on greenhouse gas emissions Weak sustainability framing, focusing on a narrow set of indicators | Strategic planning and programming | Review Diagnose Learn Communicate | Performance reporting framework |

# European Union Transport White Paper

# 8

This case study examines the European Transport White Paper of 2011 as an example of a planning domain application. The European Union is an economic and political union of some 28 states in Europe. Some of the key premises of the European Union have been to open up markets to free trade within the region and to seek to harmonize technical standards, laws, and processes to facilitate the free movement of people and goods. It is interesting that sustainability is also emphasized as an overarching goal for all areas of European policy making. The European Union employs a wide range of assessment and monitoring mechanisms to track progress on the implementation and performance of policies in the transport area, which is seen as a keystone for the so-called Single European Market. These mechanisms clearly reflect tensions between the different goals and pose challenges for the interpretation of indicators on transportation trends. Also the sheer size and complexity of the Union raise challenges for the effective application of indicator-based evidence. The chapter will place the indicator- and application-rich example of the ex ante assessment of the European Transport Policy White Paper in the wider context European Union policy making.

## 8.1 Background

The European Union is a complex entity comprising several different institutions operating a unique set of competencies and procedures. It is a form of confederation, in some regards resembling a federal state, and in others an international organization. The basic functions are laid out in a number of Treaties adopted by Member States from 1951 and onwards. Figure 8.1 provides a vastly simplified overview of some of the key bodies and their roles in the current configuration. Notice in particular the European Commission which has key responsibilities with regard to proposing, implementing, and overseeing common European policies. Policies proposed by the Commission have to be adopted by Council and Parliament jointly in the form of Directives to carry any legal status. This is referred to as



H. Gudmundsson et al., *Sustainable Transportation*, Springer Texts in Business and Economics, DOI 10.1007/978-3-662-46924-8_8

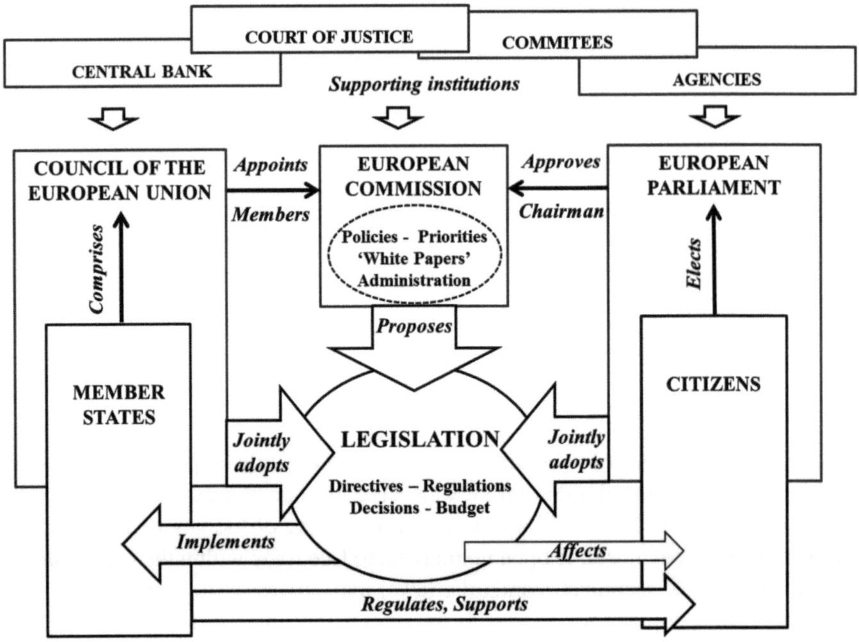


**Fig. 8.1** Simplified overview of key European institutions

the "ordinary legislative procedure." However the European Union is also aligning policies in areas where goals and measures are not formally adopted but coordinated in a more open way with stronger reliance on political agreements, funding, and indicators of progress (Trubek and Trubek 2005). A key characteristic of the EU is the amount of steering (governance) of the system rather than direct action (Piattoni 2009). Policy "White Papers" are one of the tools used by the Commission to chart out proposals for a policy area over a longer time horizon than the annual Work Program, inviting the Council (representing Member countries) and the Parliament (representing European citizens) to debate on the need for subsequent legislations. White Papers are not legally binding, but can nevertheless exert a strong influence on policy directions.

The Treaty on the Functioning of the European Union sets out the aims and extent of the competencies of the European Union with respect to transportation. These we suggest are of fundamental importance to the types of decisions that can be taken, and therefore also the types of information and scale at which information is seen to be required. The Treaty fundamentally sets out the goal of the common transport policy as the removal of obstacles at the borders between Member States so as to facilitate the free movement of goods and people, thereby fulfilling promises of a more integrated Europe. Transport is seen as a cornerstone of the general European integration process closely linked to the creation and completion of the so-called internal market, which has been created to stimulate jobs and economic growth. Transportation policy is thus focused on overcoming barriers between Member States and creating a single "European transport" area with level competition conditions for and between the different forms of transport. Key aims have been to "open up" national markets in air and rail transport to enhance competition and to remove "unnecessary" differences in technical and administrative standards (EC 2013). It is easy to see that such purely "transportation-centered" goals could potentially conflict with provisions for more sustainable transportation as discussed in Chap. 4.

The European Union is also influential in major infrastructure construction through the development of the Trans European Transport Network (TEN-T) that connects Europe across the borders of member states. The European Commission does not have the power to insist on new infrastructure but clearly acts as an important lever for implementation through supporting funds. The Commission has a shared responsibility with national governments on road safety, but urban mobility is largely the responsibility of the member states under the principle of subsidiarity, meaning that decisions are to be taken at the level of authority that is closest to citizens. The main means of influencing urban mobility have therefore been through the use of structural funds to support regeneration and the development of voluntary programs of innovation such as the CIVITAS initiative providing a platform for exchange of knowledge and sharing of good practice.[1]

The transportation policy scope of the European Community extends into the ways in which the operators are allowed to develop, operate, and maintain the infrastructure, vehicles, and staffing (aligned with Sussman's internal transport system components approach—Fig. 3.3) and to moderate the inputs to the system such as fuel mix and usage, materials, and labor and the environmental impacts (as with the input–output model—Fig. 3.5). The role of the transport system as a key driver in the economic system is also at the fore (Fig. 3.4).

In another vein of policy, it is noteworthy that the goal of Sustainable Development since 1997 has been incorporated as a part of the founding Treaties of the European Union, not only for the Union itself but also for its relations to the rest of the world (EU Treaty 2012). This has also had significant implications for how transportation policy is reviewed and monitored. Specific principles and provisions are defined in the European Union's Sustainable Development Strategy (SDS) adopted by the Council in its current form in 2006. Here "Sustainable Transport" is directly stated as one of seven key challenges along with others such as Climate Change, Protection of Natural Resources, and the promotion of Good Public Health. The overall objective for "Sustainable Transport" is to "ensure that our transport systems meet society's economic, social and environmental needs while minimising their undesirable impacts on the economy, society and the environment" (Council of The European Union 2006, p. 27). The SDS also comes with a set of more operational objectives (see Table 8.1) and a set of indicators, reported biannually by EUROSTAT, the statistical branch of the European Commission.

[1] http://www.civitas.eu/

**Table 8.1** Objectives for "sustainable transport" in the EU SDS (modified from Council of The European Union 2006)

| **Overall Objective** |
|---|
| To ensure that our transport systems meet society's economic, social, and environmental needs whilst minimizing their undesirable impacts on the economy, society, and the environment |
| **Operational objectives and targets** |
| • Decoupling economic growth and the demand for transport with the aim of reducing environmental impacts |
| • Achieving sustainable levels of transport energy use and reducing transport greenhouse gas emissions |
| • Reducing pollutant emissions from transport to levels that minimize effects on human health and/or the environment |
| • Achieving a balanced shift towards environment friendly transport modes to bring about a sustainable transport and mobility system |
| • Reducing transport noise both at source and through mitigation measures to ensure overall exposure levels minimize impacts on health |
| • Modernizing the EU framework for public passenger transport services to encourage better efficiency and performance |
| • Realizing specific targets for $CO_2$ emissions from light duty vehicles |
| • Halving road transport deaths by 2010 compared to 2000 |

In the most recent report, it was observed that little progress could be observed with regard to sustainable transport indicators such as transportation energy consumption relative to GDP ("decoupling"), or shift from road and air to rail and maritime transport. Significant improvement on traffic safety is noted, although the goal of halving deaths by 2010 was not quite met (EUROSTAT 2013).

Transportation and sustainability is also addressed within a number of other major policies and frameworks covering Economic growth and employment, Environmental protection, Climate action, Enterprise, etc. Several of these even come with their own goals, targets, and monitoring schemes. The most elaborate and persistent system for the tracking of transportation specifically is undoubtedly the so-called "Transport and Environment Reporting Mechanism" (TERM) that the European Environment Agency has published annually for over 10 years, reporting more than 30 indicators, and showing a mixture of positive and negative trends regarding the integration of environmental protection in the transportation sector (see EEA 2013). The mechanism is environment focused but, by being based in the "D-P-S-I-R" approach, it conveys a broader scope (see Chap. 7).

All in all, it could be noted that the use of goals, indicators, and reporting schemes by the European Union is extensive and even somewhat confusing, due to the increasing scope and scale of European policy making. It is occasionally asked, even by the Commission itself (EC 2014, p 14ff.), if such schemes really make any difference to policy making and subsequent outcomes, given the diversity and complexity of the policy agendas. In studies on earlier uses of European transport policy indicators, Gudmundsson (2003) and Gudmundsson and Sørensen (2013) found that several indicators were used in policy processes, while their actual influence seemed more limited, sometimes serving more to "rationalize" than to generate decisions, or even as mere symbolic gestures.

## 8.2 Framework

This chapter focuses on the current most comprehensive expression of the Common integrated transportation strategy, namely the 2011 European Transport Policy White Paper (ETPWP) called "Roadmap to a Single Transport Area—Towards a competitive and resource efficient transport system" (EC 2011a). The preparation of this document is an example of the mandatory framework for Impact Assessment (see Chap. 7) adopted by the European Commission (EC 2009), which includes an evaluation of the potential economic, social, and environmental consequences of a proposed policy. This section will introduce the ETPWP before moving into how it was assessed drawing from the paper itself as well as supporting documents.

The first attempt at an integrated transport strategy for Europe which sought to realize the transport goal of the Treaty while also embracing sustainability was published in 1992 (EC 1992). Some ten years later, in 2001 the next, more comprehensive White Paper "Time to decide" was issued (EC 2001). The implementation of that strategy was subsequently evaluated (EC 2006) serving in 2011 as preparation for the current operating strategy, the ETPWP (EC 2011a). The scale and focus of these documents have evolved with the changing economic and political circumstances as well as with (uneven) progress in fulfilling the original vision of a "frictionless" market (Faludi 2004).

The overarching aims of the current ETPWP are to "build a competitive transport system that will increase mobility, remove major barriers in key areas and fuel growth and employment. At the same time, the proposals will dramatically reduce Europe's dependence on imported oil and cut carbon emissions in transport by 60 % by 2050" (EC 2011b). The policy also aims to help "define a long-term strategy that would transform the EU transport system into a sustainable system by 2050" (EC 2011c, p. 27). In the White Paper itself it is stated that "Transport is fundamental to our economy and society. Mobility is vital for the internal market and for the quality of life of citizens as they enjoy their freedom to travel. Transport enables economic growth and job creation: it must be sustainable in the light of the new challenges we face" (EC 2011a, p. 3). Tensions between the promotion of economic growth and free movements on one side and the broader systemic consideration of sustainable development embodied in the SDS on the other are clear and seem further exacerbated by the adoption of the numerical 60 % emission reduction target. However, neither the definition of how transport contributes to the broader definition of sustainable development nor what is meant by sustainable transport is made clear. This raises obvious challenges for the measurement and interpretation of progress.

The ETPWP acknowledges some of the potential conflicts with sustainable development goals explicitly (EC 2011c, para 106). In particular, it acknowledges that fossil fuels retain a significant attractiveness for low-cost mobility and this could be in tension with greenhouse gas emissions targets. In addition, the ETPWP aims to lower congestion which may bring demands for greater infrastructure and in turn further sprawl in land use at the cost of accessibility for those without cars. While some potential conflicts are acknowledged others remain hidden. The ETPWP for example states that "Curbing Mobility is not an option" (EC 2011a, p. 5), which presupposes that these tensions can all be resolved without significant demand management. This position is highly contested in the sustainable transportation debate (Banister 2008) and is also somewhat in contradiction to the strong focus on "decoupling" of transportation from economic growth and shifts in the modal split assumed by the EU's SDS (Council of The European Union 2006; EUROSTAT 2013).

A document like the ETPWP serves several purposes and, as such, it draws on indicators in different ways. On one level it aims to act as a lodestar, providing a signal of the role of transport policy within the broader European agenda. As is highlighted above, the need to support competitiveness and employment is paramount as is tackling climate change emissions and issues of fuel security and prices. Indicators on key trends in these areas would be needed to support this function of the policy. Underneath these broad objectives, the ETPWP sets out a series of more than 130 specific measures, which the EU will adopt to help facilitate change across different sectors and modes of transport. The list includes measures such as "to revise Slot Regulation to favor more efficient use of airport capacity," to "define new rules on Air Cargo screening to enhance security," and to set "Appropriate standards for $CO_2$ emissions of vehicles in all modes" (EC 2011a). To support these policy areas, performance type indicators to track the actual adoption, implementation, and effectiveness of the individual measures would be relevant.

First and foremost, the ETPWP is a forward looking document, proposing policies for the next decade. In making the case for the policy agenda, there is a need to set out a clear analysis of the current trends and the reasons why taking no action or the continuation of business as usual would be insufficient as a means to meet the stated policy objectives. The Impact Assessment Guidelines of the European Commission specifically require an impact analysis to consider the nature and scale of the problem: how is it evolving, what objectives should be set to address the problems, what the main policy options are, how monitoring could be organized, and whether the European Union should be involved at all (EC 2009). To fulfill these requirements, the ETPWP and its accompanying "Impact Assessment" report look across a wide range of indicator application areas that *describe* the current position, *review* how things are progressing, and *diagnose* how things have worked to date. It looks ahead a decade and beyond informed by a model-based forecasting exercise (EC 2011c) which is described in more detail below. At least some of the indicators used therefore have to have the capability to be ***forecast***. As the Impact Assessment Guidelines stresses, the assessment aims to support and does not replace decision-making. The adoption of a policy proposal is always a political decision (EC 2009, p. 4). It is not possible to derive simply from reading the policy document the exact extent to which the forecasted indicators have been influential in ***deciding*** on specific courses of action in an instrumental way, or if their role has been more symbolic or rationalizing. However, it seems likely that the indicators and modeling exercises have been used to define and test out different intensities and combinations of action that have been considered, similar to the role of indicators in a comparable exercise in 2006 (Gudmundsson and Sørensen 2013). We review the key indicators and their application below.

## 8.3 Indicators Used

This section presents a summary of the full range of indicators described in the supporting analysis document which accompanies the White Paper (EC 2011c) before describing a sample of those that appear to have seen widest application in the subsequent justification of the ETPWP. Table 8.2 summarizes the range of indicators considered in the forward looking assessment.

The table highlights a mixed picture in terms of coverage and clarity in indicators. Fourteen of the 24 indicators are quantified. It is, of course, not possible to quantify everything. However, nine of the ten indicators that are qualitative are also not clearly defined which would make it difficult for other analysts to repeat the analysis and look for consistency.

**Discussion Topics**

- Examine Table 8.2 and identify three strengths and weaknesses of the range of indicators selected.
- What in your view is missing from the list and how easy would it be to include such an indicator for a pan-European level analysis?
- Using existing literature look for potential indicators that can be used to fill one of the undefined indicators for the ETS and the main White Paper document.

We move on now to a more detailed analysis of four of the indicators and then demonstrate how these have been used through the different indicator application areas noted above. The selection of four is to provide an indication of the broader context for the indicators and to explore their scientific qualities.

### Congestion

Different measures of congestion are referred to throughout the White Paper. These include delays on roads, reliability of travel time, overcrowding on public transport,

**Table 8.2** Indicators used in the ETPWP supporting analysis (EC 2011c, p. 56ff)

| Impact areas | Indicators |
|---|---|
| **Economic Impacts** | |
| Transport Activity | Passenger kilometers and tonne kilometers by all modes |
| Modal Shift | Share of passenger or tonne kilometers by mode |
| Transport costs to users | Unit cost per passenger or tonne transported (including capital costs, fixed operation costs, and variable fuel and non-fuel costs) |
| *Economic growth* | *Qualitative inference of policy impacts on GDP only* |
| *Efficiency of the transport system* | *Not defined but incorporates notions of smart pricing, efficient networks, fuel efficiency, and vehicle purchase costs* |
| **Congestion** | **Average speed and use of available road capacity** |
| Household costs | The share of passenger transport costs within the household income of the average EU household |
| *Transport related sectors* | *Not defined but qualitative inference of the potential of the strategies to support the European vehicle manufacturing industry* |
| *Innovation and research* | *Not defined but qualitative inference about the impact of the strategy on research spent on green innovation* |
| *Reduction of administrative burden* | *Not defined but qualitative inference about overall levels of administration* |
| *EU budget* | *Not defined as will be assessed on a case-by-case basis* |
| *International relations* | *Not defined but qualitative inference about the potential synergies and conflicts with international organizations* |
| **Social Impacts** | |
| Degree of mobility | Refers to the % change in total transport activity for passengers with an overall reduction being negative |
| *Choice* | *Not defined but qualitative inferences about rail investments improving choice* |
| Accessibility | Potential accessibility is a generalized cost-based measure. Larger areas are more attractive and cost, time, and distance are negative separation factors |
| *Distributional Impacts* | *Not clearly defined although the analysis refers in part to the distribution of household costs by income band* |
| Employment level and conditions | Number of jobs in the transport sector. Skills and working conditions are not defined clearly |
| **Safety** | **External costs of accidents and total number of accidents** |
| **Environmental Impacts** | |
| **Climate Change** | **Total $CO_2$ emissions from transport. Both transport and well to wheel analyses are presented** |
| Air pollution | Emissions of $NO_x$ and $PM_{10}$ and external costs of these pollutants |
| Noise pollution | External costs of noise pollution |
| Energy use/energy efficiency | Total energy demand from transport Millions of Tonnes of Oil Equivalent (reviewed in detail below). Energy intensity is an efficiency indicator that uses total energy demand and transport activity to create a ratio for passenger and freight |
| **Renewable energy use** | **Total energy demand split by fossil fuels, biofuels, and electricity** |
| *Biodiversity* | *Not defined but qualitatively refers to fragmentation, land-take, loss of biodiversity, and damage to ecosystem services* |

*Note*: Impacts in italics had no quantified indicators. Indicators in bold are reviewed in this text queues at airports, and economic costs of congestion. Congestion is noted to have impacts on fuel consumption, lowering efficiency by 30 % in heavy congestion, impacts on productivity, competitiveness of the economy, and quality of life. It is also said to cause dissatisfaction to people and businesses.

There is no reliable monitored data on congestion levels at a European level in part due to the difficulties in achieving a common definition and in part due to inconsistencies in data collection. The data used in the ETS therefore come from modeling exercises. The Impact Assessment document states that "Estimating the costs of congestion is not straightforward, because it occurs mostly during certain times of the day, often caused by specific bottlenecks in the network" (EC 2011c, p. 144). So, while reliability and overcrowding feature in the narrative, fairly aggregate demand–capacity relationships are used to estimate changes in speed and from this to calculate the costs of congestion. A further modeling exercise also uses "use of available traffic capacity" on the road network, but the units are not provided.

The two indicators of congestion (speed and use of available traffic capacity) are both quantitative, modeled outcome variables. Speed is a descriptive statistic, whereas use of available traffic capacity refers to a normative threshold.

Both indicators hold good scientific properties of *reliability* and *sensitivity*. *Validity* is more debatable as it is not clear that either fully capture the notions of congestion as experienced by the user. For example, it is understood that infrequent significant delays are more problematic to users than small ongoing delays. This relates more to variability of travel time which is not fully captured. Low *data availability* at the European level has already been noted. While both constructs have a clear basis in traffic flow modeling literature as being indicators of congestion, it is worth reflecting on the extent to which they are *interpretable* and *actionable* at a European Scale. What level of investment or management measures would make a measurable difference to speeds and available capacity at a Europe wide scale given the aggregation required? We suggest that these indicators are useful but imperfect proxies for congestion. They require further interpretation at a national or local scale to inform detailed investment decisions. This is consistent with the partnership approach to TEN-T investments.

We note that while multi-modal congestion concerns are raised, the metrics are dominated by the road network and little insight is given to the treatment and understanding of congestion elsewhere in the networks.

**Safety and Accidents**

The White Paper frequently makes reference to safety and security, although these refer to quite separate concepts (the former covering death and injury in the course of traveling and the latter including perceived personal risks and matters such as terrorist threats). While a multi-modal perspective is adopted which covers vehicle safety and certification as well as use, there is a clear quantified target set for road traffic casualties updating the previous goal of "halving" between 2001 and 2011.

> By 2050, move close to zero fatalities in road transport. In line with this goal, the EU aims at halving road casualties by 2020. Make sure that the EU is a world leader in safety and security of transport in all modes of transport (EC 2011a, p. 10).

The safety impact therefore allows applying a quantified normative indicator with clear targets set for 2020. This indicator has sound scientific properties in so far as it is a valid measure of a critical safety outcome (although only one of several) and is reliable and sensitive (for example, to changes in in-vehicle safety standards). Safety statistics are gathered for all forms of transport by national governments and have, for some time, been the subject of effective target-setting approaches in the road transport sector. The policy and target relevance is therefore obvious, and actionability is also present at European, national, and local levels. Both rail and aviation are safety critical systems with very low accident rates. These both feature in the policy initiatives but are not covered in the background analysis.

**Greenhouse Gas Emissions**

The European Union has been pushing for global climate policy solutions since the early days of international negotiations. Overall, the EU aims to reduce emissions by 80–95 % below 1990 levels by 2050, albeit depending on efforts undertaken by other developed countries, in order to allow global warming to stay below 2 °C. Climate change features prominently in European strategies for economic growth, sustainable development, energy, R&D, and transportation. The White Paper provides a very clear statement on greenhouse gas emissions from the transport sector:

> Commission analysis shows that while deeper cuts can be achieved in other sectors of the economy, a reduction of at least 60 % of GHGs by 2050 with respect to 1990 is required from the transport sector, which is a significant and still growing source of GHGs. By 2030, the goal for transport will be to reduce GHG emissions to around 20 % below their 2008 level. Given the substantial increase in transport emissions over the past two decades, this would still put them 8 % above the 1990 level.

This provides clear quantified targets for the analysis of transportation sector $CO_2$ emissions up to 2030 and 2050. Carbon emission accounting is conducted in a standardized way across the EU in accordance with guidelines developed by the IPCC. For the purposes of forecasting, the modeling exercise makes assumptions about average emission rates, fuel mixes, and activity patterns to arrive at estimates for the change in emissions from different policy packages. The analysis shows that the goals can be reached with different combinations of policies (EC 2011c, p 72ff.).

The indicator has good scientific properties being valid, reliable, and sensitive to transport policy interventions, although there remains debate about the extent to which any level of atmospheric $CO_2$ concentration will lead to a particular level of climate change. Data availability, transparency, and interpretability are well supported by the established regimes for calculations and modeling. While there are still reasons to question if effective measures will in fact be deployed, allowing the move from a position *above* the 1990 emissions levels in 2030 to far *below* it in only 20 years' time, the ETPWP has at least developed a clear set of normative goals for transport to work towards.

**Energy Use and Oil Dependence**

The EWPTP states that "Oil will become scarcer in future decades, sourced increasingly from uncertain supplies. As the IEA has recently pointed out, the less successful the world is in decarbonizing, the greater will be the oil price increase. In 2010, the oil import bill was around € 210 billion for the EU" (EC 2011a, p. 3). Oil dependence as an indicator is therefore different, although related to, greenhouse gas emissions. This is further underscored by an emphasis on fuel security and relying on imports from "proven reserves in politically less stable regions" (EC 2011c, p. 134). Although not a direct feature of the supporting analysis, oil security is clearly an important political theme in the transportation policy.

Oil dependence is measured as the percentage of the energy supply for transport that comes from oil (and this is captured by the renewable energy use indicator). This is both monitored and also forecast on the basis of various demand, price, and technological development scenarios. Oil is an input variable (in terms of energy used), but oil dependence is an outcome of various policy choices including efficiency, mode share, and fuel and vehicle technology. Oil dependence has some sound scientific properties (even if the aim is politically rather than scientifically defined, and there are concerns also with some alternative forms of energy (biofuels, coal, nuclear). The data to measure the current level of oil use and dependence are accessible in the public domain in aggregate form, while forecastability is limited due to difficulties in assessing future oil prices and developments in markets for energy technologies (Hamilton 2008). Still the oil dependency measure is actionable in regard to changes in policies on biofuels, electromobility, etc., and it has high policy target relevance.

**Discussion Topics**

- Which of the indicators in Table 8.2 exist at a national or local level in your country? How consistent are the definitions? If they are inconsistent does it matter?
- The supporting analysis for the ETS includes total energy demand as an indicator as well as the split by different sources. The ETS itself focuses more on oil dependence as a key measure. What is the distinction and why is that important?
- Conduct an analysis of one of the other indicators from the list in Table 8.2 using the supporting analysis for the ETS and the main White Paper document.

## 8.4 Indicator Applications

### 8.4.1 Describe

The ETPWP begins with a description of the current problems facing the transport system and its users in the EU. This is supported by a variety of current statistics and indicators some of which are used in other frameworks mentioned in Sect. 8.2. A few examples are shown in Table 8.3.

The examples include single-point as well as time series indicators showing how trends have changed (or not) in the recent past. The aim is to set the scene and thereby justify the need for policy actions, as requested in the Impact Assessment Guidelines. The first one seeks to make the point that the transportation sector is very important, and the second one that it is even growing in significance over time. The two final ones establish the proposition that the current state of the transport system is not sufficient to meet sustainable policy aspirations.

### 8.4.2 Review

The review function asks "how are we doing," often ex post. It differs from the description function as it moves towards establishing whether the existing course of action has promoted the type of changes that are required to achieve the stated goals. This function is served by an ex post analysis of the 2001–2010 White Paper, included as Appendix 2 of the Impact Assessment (EC 2011c). The Commission's analysis of its 2001 policy to decouple transport growth from economic growth provides good insight into the review function. The context of the decoupling target is revisited, with a reminder that in 2001 the EU had economic term growth forecasts of 3 % per annum (higher in Eastern European states). It was not deemed possible to accommodate levels of traffic growth at or above 3 % and also to meet congestion and environmental concerns. The focus was therefore on achieving growth with less travel (Fig. 8.2). The Commission's analysis shows that freight traffic was rising faster than GDP over the period right up to the economic downturn where it fell sharply. By contrast, passenger traffic grew more slowly than GDP

**Table 8.3** Selected indicators describing transport conditions (EC 2011c, pp. 10–12)

| |
|---|
| • The transport services sector accounted for 4.6 % of total EU gross value added in 2008 |
| • Average mobility per person in the EU increased by 7 % between 2000 and 2008, mainly through higher motorization levels as well as more high-speed rail and air travel |
| • Today transport accounts for one-quarter of EU $CO_2$ emissions; $CO_2$ emissions from transport have been growing over the last 20 years |
| • Transport continues to rely nearly entirely on oil and oil-based products: for more than 95 % of its needs worldwide and 96 % in the EU-27 |

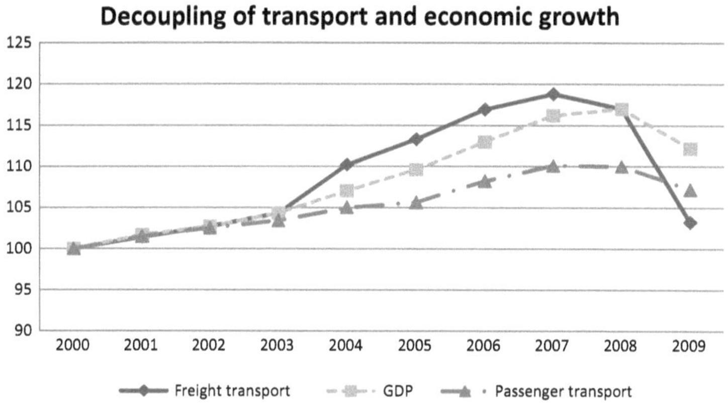


**Fig. 8.2** Evolution of GDP, passenger and freight transport in the EU27 between 2000 and 2009 (Index 2000 = 100) (adapted from EC 2011c)

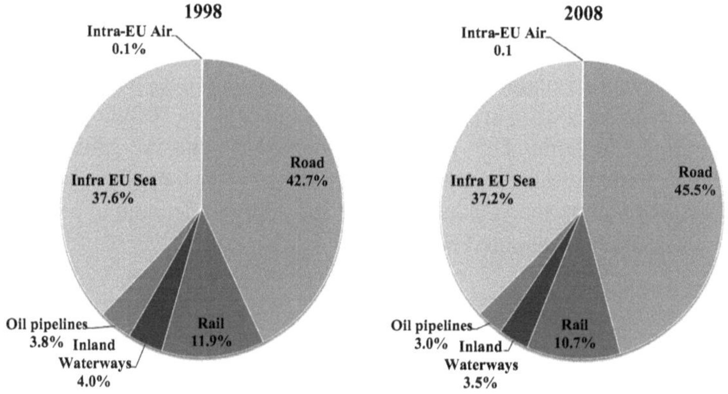


**Fig. 8.3** Modal split of European freight transport (adapted from EC 2011c)

over the whole period. Decoupling occurred for passenger but not for freight transport, although the economic recession from 2008 creates some confusion.

Moreover, slower economic growth than hoped for meant that decoupling became less of a policy concern than it was in 2001.

The decoupling strategy was revisited in 2006 shifting the emphasis more directly to reducing the impact of transport on the environment, by decoupling transport from emissions rather than from the economy (as also expressed, for example, by the 60 % greenhouse gas reduction target). Hence, the review did not only support a revision of the planned policy measures but also of the policy objective itself.

Figure 8.3 shows another ex post indicator that perhaps shows a lack of progress. We see the modal split of European freight transport in 2 years. The nearly identical diagrams fail to provoke the mind much, albeit they do in fact constitute a normative indicator. It bears evidence of significant failure regarding another strong policy ambition in the previous policy: to obtain a shift from road transportation to other modes.

### 8.4.3 Diagnose

The diagnosis function breaks down overall developments to individual components and driving forces. This works hand in hand with the review and as we shall see shortly also with the forecasting exercises in the ETPWP assessment process. So, for example, in the discussion on environmental decoupling, the analysis suggests that insufficient progress has been made because while "new vehicles have become more fuel efficient and emit less $CO_2$ per km than earlier model[s] did ... these efficiency gains have been more than compensated for by rising vehicle numbers and increasing traffic volumes" (EC 2011c, p. 99). The diagnosis explains why such decoupling has been harder to achieve than anticipated.

In contrast, another diagnosis explains success with a more than 30 % reduction in particulate matter ($PM_{10}$) and a halving of ozone forming pollutants as a result of cleaner engine standards. This has been achieved despite the growth in overall traffic levels. Nonetheless, the analysis also points out that there remain many urban areas where the pollutant concentrations are above levels deemed to be healthy and there remains an uncoordinated approach to resolving this (EC 2011c, p. 100).

### 8.4.4 Forecast

Forecasting means an assumption or prediction of future trends and outcomes; this was a key element in the ETPWP assessment. Forecasting can contribute to the identification of a need for new measures, transport investments, or other initiatives, or to decisions that such choices are unnecessary. Forecasting is also applied in policy analysis and decision-making to support the setting of realistic targets and the selection of appropriate measures. It is often based on numerical or simulation models, to handle complex interactions in a systematic way (although we note this is not the same as presuming they are correct).

The development of the ETPWP was supported by a range of large-scale model-based forecasting exercises which included GEM-E3 (a structural economic model), TRANSTOOLS (a European transport networks model), TREMOVE (a model to assess emissions and pollution impacts of policies), and PRIMES-TREMOVE (a demand forecasting model). To be capable of performing European wide analyses, such models have, by necessity, to be relatively aggregate in their

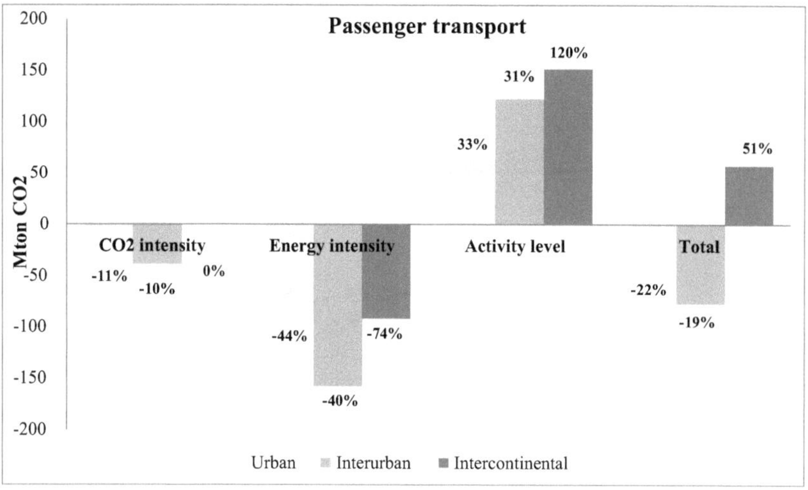


**Fig. 8.4** Diagnosis of the factors influencing $CO_2$ emissions for the reference scenario—passenger transport case (2005–2050). *Note*: *Bar* heights refer to absolute emission levels and are not compatible with the percentage changes (adapted from EC 2011c)

composition. TRANSTOOLS for example works at the NUTS3 area resolution which equates to just over 1,300 "regions" within the EU27.

The forecasting involved first the definition of a reference or do-minimum case and then a series of alternative policy packages which were to be examined and compared with the reference and each other. The time horizon was 2005–2050 and each policy package was designed to achieve the 60 % reduction target for $CO_2$ but through different policy mixes.

The reference forecasting itself included a diagnostic component as illustrated in Fig. 8.4.

It is shown that, even without new policies, technical improvements would overcome increased transport activity and lead to reduced emissions in the period towards 2050, except for intercontinental passenger transport (air). Yet these trends would be far from sufficient to reach the 60 % reduction target. Hence new policies are needed.

Three policy packages that could reach the target were developed. Option 2 has a high level of demand management policies and relatively slow predicted change in technology. Option 3 adopts a firmer approach on technology standards and a more rapid switch in engine powertrain technology to non-fossil fuels. Option 4 was a compromise between these two more "extreme" scenarios. Table 8.4 shows an example output which compares the levels of transport activity that each of the three scenarios generates relative to the reference case (Option 1). Obviously Option 2 shows the strongest effects with regard to changes in mobility. The "compromise" Option 4 is much less constraining for road transport, but does imply sizable increases in rail and Waterways. As noted earlier in the case study, only 14 of the 24 indicators can be quantified through the modeling exercise.

**Table 8.4** Change in passenger and freight transport activity (EC 2011c, p. 54)

| | Policy option 2 | | | Policy option 3 | | | Policy option 4 | | |
|---|---|---|---|---|---|---|---|---|---|
| *Compared to Reference case (%)* | 2020 | 2030 | 2050 | 2020 | 2030 | 2050 | 2020 | 2030 | 2050 |
| *Passenger transport* | −3 % | −9 % | −18 % | 0 % | 0 % | −2 % | −1 % | −2 % | −7 % |
| Road | −3 % | −12 % | −23 % | −1 % | 0 % | −2 % | −1 % | −3 % | −9 % |
| Rail | 9 % | 19 % | 35 % | 0 % | 0 % | 5 % | 8 % | 17 % | 27 % |
| Aviation | −12 % | −12 % | −22 % | 1 % | 3 % | −8 % | −11 % | −9 % | −17 % |
| *Freight Transport* | −1 % | 0 % | 1 % | 0 % | 2 % | 5 % | 0 % | 2 % | 5 % |
| Road | −11 % | −27 % | −47 % | −3 % | −1 % | 2 % | −5 % | −5 % | −3 % |
| Rail | 9 % | 14 % | 58 % | 2 % | 3 % | 3 % | 7 % | 15 % | 18 % |
| Inland Waterways | 10 % | 20 % | 35 % | 3 % | 3 % | −1 % | 11 % | 22 % | 21 % |
| Maritime | 1 % | 3 % | 5 % | 1 % | 3 % | 6 % | 1 % | 3 % | 6 % |

Discussion on other indicators is provided but is quite difficult to interpret relative to the apparent detail that the models provide. For example, on the EU budget the analysis states: "In principle, all policy options envisaged in this Impact Assessment report have a direct impact on the EU budget. However, the impact of individual measures on the EU budget will be assessed in the context of individual impact assessments" (EC 2011c, p. 65).

### 8.4.5 Decide

It is not possible through a documentary analysis alone to say to what extent the impact assessment has informed the decisions to be pursued in the final version of the White Paper itself or in subsequent proposals. Nonetheless, the supporting analysis shows an overall comparative analysis framework for the various policy packages and a process for determining the preferred policy package upon which the ETS is based.

Table 8.5 shows the assessment of the impacts of each of the policy packages again relative to the reference case. It is worth noting that the assessment is reduced to a discrete seven-point scale from triple negative to triple positive. Quantitative and qualitative indicators are all converted to this seven-point scale for comparative purposes. The three dimensions of sustainable development are all well covered in principle, while there is no attempt to summarize indicators at the level of each dimension, or in total.

The selection of the European Commission's preferred policy package is based on three criteria. The first is *effectiveness* which compares the ability of the packages to cut $CO_2$ emissions and reduce congestion as well as the degree of associated technological risk. This clearly places the technology-oriented package at a disadvantage as it may even stimulate some traffic growth and is reliant on significant technological risk. The second criterion is *efficiency* which is defined as the total cost of the policy packages relative to policy Option 1. This is a full social cost approach including external costs as well as infrastructure, vehicle, operation, and maintenance costs. Mitigation cost per tonne of $CO_2$ is also included. The final criteria is *coherence* which is described as being a measure of the extent to which the package achieves the objectives with the lowest possible trade-offs between the economic, social, and environmental criteria (as presented in Table 8.5 above). Table 8.6 shows the result of this analysis, which reduces the data still further to a qualitative assessment of each package in the three criteria. The supporting analysis goes on to rule out policy Option 3 as, despite being the least cost means of achieving $CO_2$ reductions, it comes with high technological risk and fails to address pricing problems and other key goals. There is relatively little to choose between policy packages two and four and there are many common policy elements. Unsurprisingly, Option 4 was identified as the preferred one as it "offers the advantage of greater balance between system improvement and technological development" (EC 2011c, p. 86).

**Table 8.5** Comparative analysis of policy packages relative to reference

| | Policy Option 2 | Policy Option 3 | Policy Option 4 |
|---|---|---|---|
| **Economic Impacts** | | | |
| Transport Activity | −− | = | − |
| Modal Shift | ++ | = | + |
| Transport costs to users | −−− | = | −− |
| Economic growth | ++ | + | +++ |
| Efficiency of the transport system | ++ | + | +++ |
| Congestion | ++ | = | + |
| Household costs | −− | − | −− |
| Transport-related sectors | + | +++ | +++ |
| Innovation and research | + | +++ | ++ |
| Reduction of administrative burden | + | = | + |
| EU budget | = | = | = |
| International relations | −− | − | − |
| **Social Impacts** | | | |
| Degree of mobility | −−− | = | − |
| Choice | ++ | = | ++ |
| Accessibility | ++ | = | ++ |
| Distributional Impacts | = | − | + |
| Employment level and conditions | ++ | ++ | +++ |
| Safety | ++ | = | + |
| **Environmental Impacts** | | | |
| Climate Change | +++ | +++ | +++ |
| Air pollution | +++ | ++ | ++ |
| Noise pollution | +++ | ++ | + |
| Energy use/energy efficiency | +++ | ++ | +++ |
| Renewable energy use | +++ | +++ | ++ |
| Biodiversity | + | − | = |

+ improvement (from + to +++) = no change - worsening (from - to ---)

**Table 8.6** Selecting preferred policy package

| | Effectiveness | Efficiency | Coherence |
|---|---|---|---|
| Policy Option 1 | No | No | No |
| Policy Option 2 | High | Low | Medium |
| Policy Option 3 | Low | High | Low |
| Policy Option 4 | Medium | Medium | Medium |

## 8.5 Discussion

The European White Papers and the accompanying processes are perhaps the largest scale attempt to provide a guiding framework for sustainable transportation that exists. The discussion of the strengths and weaknesses of such an attempt needs to be conducted with a keen eye on the context of European transport policy. There is a significant diversity within the EU and tensions exist between the needs of the so-called Accession countries where there are significant infrastructure deficits and stronger economic growth projections and the original member states where connectivity is stronger and a major focus is on actions to stimulate an economic recovery while pursuing ambitious environmental and social goals.

In Europe, as elsewhere in the world, there remains a challenging (and possibly inconsistent) narrative surrounding the extent to which "green growth" will be possible and the resulting balance between technological innovation in the vehicle market and demand management measures (Marsden et al. 2014). The ETPWP notes for example that moving away from fossil fuel dependence could have short-run additional costs which are economically difficult. The analysis also points to the need for consistency with other policy areas such as innovation policy, energy policy, and policies to open up markets, acknowledging the cross-sectoral nature of the problem which we drew out in Chaps. 3 and 4. The politics of target setting also matter and any sustainability framework will likely be influenced by those politics (Flinders et al., 2014).

The ETPWP identifies itself as sitting within a broader sustainable development agenda for the EU. While it offers a broad understanding of sustainable transportation, this is clearly set within this wider context and transportation is seen as contributing to sustainability rather than being sustainable in and of itself. That said, while reference is given to definitions of sustainable transportation, there is not a particularly clear rationale given to the selection of indicators that are applied in the supporting analysis for the ETS. Over and above this, a read of the policy document without the supporting analysis gives weight to some of the indicators more than others (e.g., oil dependence compared with total energy use) and also some concepts (such as security and market openness) that do not feature in the supporting assessment. We can therefore observe that the range of indicators used to assess the direction of travel of European Transport Policy and to steer the future course is deficient in some respects compared with the most comprehensive definitions of sustainable transportation that we identified in Chap. 4. Further, a close read of the objectives in Table 8.1 reveals the emphasis given to "reducing" or "minimizing" impacts, with no reference to any limits within which the transportation sector needs to operate. The transportation-centered framing also deemphasizes the need for cross-sectoral collaboration to advance the EU's sustainable development agenda. Given these observations, the EU approach to sustainable transportation can be clearly positioned within the "weak" sustainability model.

The spatial scale and diversity of Europe and the subsequent challenges this provides to data collection and the development of robust forecasting tools should also not be overlooked. May et al. (2008) reviewed some of the inconsistencies in data collection which occur at a national and local scale. Such inconsistencies clearly limit the extent to which any aggregation of data can work. When we consider the modeling tools that can be developed to work at this type of scale, they necessarily vastly simplify the networks and origin-destination matrices that can be considered and the ability to infer impacts (for example air quality or safety).

This is also true in the extent to which any of the proposed policy interventions can be deemed to be effective. Research on policy transfer is clear that there will be a highly differentiated set of outcomes from any given policy depending on local context (Marsden et al. 2011), yet an analysis at this scale has to work with a fairly coarse set of assumptions.

The ETPWP is, to its credit, quite transparent in recognizing some of its shortcomings (EC 2011c, p. 29). That said, other issues are not thoroughly discussed such as equity and intergenerational assessment where the coverage is weak. The indicators for example focus on the distribution of household mobility costs by income band and for the average EU household. There is so much diversity within the EU-28 that one might question the worth of these measures. It is certainly the case that the modeling tools are coarse and therefore insensitive to the detail of local travel opportunities that are so critical in defining availability, accessibility, and inclusion (Lucas 2004). On the other hand, high spatial resolution may be less essential for committing to common technical standards, overarching priorities for infrastructure investment levels, innovation programs, or moving towards the adoption of frameworks that allows the internalization of external costs. Other aspects of detail and operational relevance of the supporting knowledge may however also be required in such areas.

How much should one expect the policy analysis of large-scale strategies to resolve these issues? The ETPWP process sets an overall framework and direction of travel for the Common transportation policy in the EU and it guides the development of some policies in which the EU has competence over delivery. It can however be seen as an umbrella strategy underneath which sit national, regional, local, and even community-based strategies, not to mention industry, citizen, and stakeholder driven ones. At each of these scales, the issues of context can be brought more to the fore and the level of detail in terms of data, range of indicators, and, to some degree, modeling tools can become richer. The coarseness of the ETPWP is not necessarily problematic, provided these other strategies cascade the principles and direction of policy in a reasonably consistent manner and with similar levels of ambition and timing of action. There are no strong mechanisms to ensure that this does happen. It should be recognized however that when these policies come to be implemented, they could be inconsistent with local preferences and interpretations of sustainable development and transportation. The limited detail in many aspects of the indicator set could lead to the promotion of policies that are not as consistent or favorable as shown in the ETS supporting analysis for particular contexts. Whatever these tensions might be, it is difficult to argue with the contention that a clear European transportation strategy, underpinned by sustainable development principles, at least provides a position against which other interpretations can be argued. In the absence of such a strategy, it would surely be more difficult for member states to agree on a shared future direction for transportation policy.[2]

[2] By contrast, the lack of a national sustainable transportation policy in the USA means that states and regions are left to formulate their own approach, which could result in conflicting outcomes if

The series of White Papers shows several different and increasingly reflected indicator applications. Strategy development typically follows a rational objectives-led approach with description, problem analysis, and alternative evaluation as key features in choosing the future strategy. This requires description, review, and diagnosis in the first instance. In reality, these application areas are interconnected in the ETS, with description and review feeding into discussions regarding diagnosis.

There is more of a divide between the diagnosis and the forecast and decide function. Within the ETPWP, this is where the gap between indicators which can be measured and reported on and those which can be forecast comes to the fore. The modeling exercise used to underpin the strategy development focuses largely on transportation network investments, new fuel, and engine technologies, prices, and mode choice at a fairly coarse scale. Since modeling tools are still very much focused on transport outcomes, they inevitably underplay the broader sustainability outcomes. Such an exercise is, for example, insensitive to some of the broader aspects of the strategy such as labor market conditions, stimulation of innovation, international maritime regulation, and consumer rights. These are important parts of the strategy but not major features of the analysis and justification within the supporting analysis. Space precludes a more thorough review of each of these policies, but it is important to acknowledge that none of the policies evaluated are new or exist in a vacuum. They all represent a continuation or reframing of existing directions of travel (e.g., the narrative over continuing to internalize the external costs of transport).

The decision function is shown to be applied in a logical sequence with the indicators informing a comparative evaluation of the different policy options. While this suggests that indicators could have been a major driver in the decision-making process, it is important to acknowledge that it is not uncommon for specific policies to be selected and justified subsequently through the analysis or as part of a broader package (where inconsistencies in policy choice are hidden or diminished), rather than being the outcome of such analysis as discussed in Chap. 5. Gudmundsson and Sørensen (2013), as well as indicator scholars, more generally (Astleithner et al. 2004; Innes and Booher 2000) demonstrate that it is hard to find evidence for a notion of "instrumental" indicator use. The very selection of the policy packages is itself a highly politicized process as packages that do not fit with the ethos of the organization or which challenge existing logics may very well be ignored at an early stage (Marsh and Sharman 2009). Whatever the process for selecting the preferred policy packages, the presentation of an indicator-based analysis provides a transparent justification for the final selection which is connected to the problem analysis and diagnosis.

We can conclude from the review of the case study that indicators have been an important feature in the development of the European Transport policy at least

---

coordination among agencies is poor (see the related discussion in Sect. 4.2.3 and the case study on how New York State Department of Transportation (NYSDOT) has approached sustainable transportation in Chap. 10).

since the 2001 White Paper and increasingly so. There has also been a significant attempt to connect these to the principles of sustainable development. There are shortcomings in the approach which result from factors including the politics of collective decision-making in the EU, limitations in data collection and sharing, and the size of the challenge and capability of assessment tools to provide all the information that is required. There are important choices to be made to simplify and make the task manageable. The extent to which these compromise an effective understanding of sustainable development is a matter of discussion and interpretation. Nonetheless, there is a general rule to consider that the larger the spatial scale, the coarser the assumptions will be. Some of this can only be fully understood by seeing how these issues are cascaded further down the governance chain. While this set of issues has been pulled out from an analysis of the European Transport Strategy, similar issues are likely to exist in any context where higher tiers of government seek to set out a direction for implementation at a lower tier. It seems, from our experience, that this European level strategy is being interpreted quite differently in the different member states and so, as part of a governance framework, it is important but only partially effective. The flip side to that discussion is that smaller spatial scales for the strategy allow more refined assumptions. The risks here are associated with the weaker coverage of inter-spatial issues and policies being selected that have an important bearing on the smaller spatial unit. The tensions between scale and coverage cannot be "resolved," but they should be recognized and discussed (and to an extent the ETPWP does this) as part of understanding the system boundaries for the assessment.

**Discussion Topics**

- The analysis above has not covered issues of evaluation and learning. Review the ex post evaluation of the 2001 White Paper and also the ongoing TERM indicator reports from the European Environment Agency. How are indicators being used in the evaluation and learning process?
- The analysis has highlighted a range of challenges to developing a sustainable transportation strategy at this scale. What improvement would you prioritize for the sustainability analysis of the ETS and why?

## References

Astleithner, F., Hamedinger, A., Holman, N., & Rydin, Y. (2004). Institutions and indicators—The discourse about indicators in the context of sustainability. *Journal of Housing and the Built Environment, 19*, 7–24.

Banister, D. (2008). The sustainable mobility paradigm. *Transport Policy, 15*, 73–80.

Council of the European Union. (2006, June 9). Renewed EU sustainable development strategy. 10117/06. Brussels.

EC. (1992). The future development of the common transport Policy—A global approach to the construction of a community framework for sustainable mobility. Commission of the European Communities, Brussels.
EC. (2001). European transport policy for 2010: Time to decide. White paper. COM (2001) 370. Commission of the European Communities. Brussels, 12/09/2001.
EC. (2006). Keep Europe moving—Sustainable mobility for our continent. Mid-term review of the European Commission's 2001 Transport White Paper. European Commission, Brussels. http://ec.europa.eu/transport/themes/strategies/2006_keep_europe_moving_en.htm (accessed on April 24, 2015).
EC. (2009). Impact assessment guidelines. SEC (2009) 92. European Commission, Brussels. http://ec.europa.eu/smart-regulation/impact/index_en.htm (accessed on April 24, 2015).
EC. (2011a, March 28). Roadmap to a single European transport area—Towards a competitive and resource efficient transport system. European Commission COM (2011) 144 Final, Brussels.
EC. (2011b). European strategies. White Paper 2011. http://ec.europa.eu/transport/themes/strategies/2011_white_paper_en.htm (accessed on April 24, 2015).
EC. (2011c, March 28). Impact assessment, accompanying paper to white paper: Roadmap to a single European transport area—Towards a competitive and resource efficient transport system. Commission Staff Working Paper, SEC(2011) 358 Final, Brussels.
EC. (2013). The European Union explained: Transport. European Commission, Directorate-General for Communication, Brussels.
EC. (2014). Taking stock of the Europe 2020 strategy for smart, sustainable and inclusive growth. COM(2014) 130 final. European Commission, Brussels, 5.3.2014.
EEA. (2013). *A closer look at urban transport TERM 2013: Transport indicators tracking progress towards environmental targets in Europe* (EEA Report No 11/2013). Copenhagen: European Environment Agency.
EU Treaty. (2012). *Consolidated version of the treaty on European Union*. http://eur-lex.europa.eu/legal-content/EN/ALL/?uri=CELEX:12012M/TXT (accessed April 24, 2015).
EUROSTAT. (2013). *Sustainable development in the European Union. 2013 monitoring report of the EU sustainable development strategy*. Luxembourg: Publications Office of the European Union, Eurostat Books.
Faludi, A. (2004). Spatial planning traditions in Europe: Their role in the ESDP process. *International Planning Studies, 9*(2–3), 155–172.
Flinders, M., Bache, I., Bartle, I., & Marsden, G. (2014). Blame games and climate change: Accountability, multi-level governance and carbon management. *British Journal of Politics and International Relations*. doi:10.1111/1467-856X.12040.
Gudmundsson, H. (2003). The policy use of environmental indicators. Learning from evaluation research. *The Journal of Transdisciplinary Environmental Studies, 2*(2), 1–12.
Gudmundsson, H., & Sørensen, C. H. (2013). Some use—Little influence? On the roles of indicators in European sustainable transport policy. *Ecological Indicators, 35*, 43–51.
Hamilton, J. D. (2008, November). *Understanding crude oil prices* (Working Paper 14492). Cambridge, MA: National Bureau of Economic Research. http://www.nber.org/papers/w14492 (accessed on April 24, 2015).
Innes, J. E., & Booher, D. E. (2000). Indicators for sustainable communities: A strategy building on complexity theory and distributed intelligence. *Planning Theory & Practice, 1*(2), 173–186.
Lucas, K. (2004). *Running on empty: Transport, social exclusion and environmental justice*. Bristol: Policy Press.
Marsden, G., Frick, K. T., May, A. D., & Deakin, E. (2011). How do cities approach policy innovation and policy learning? A study of 30 policies in Northern Europe and North America. *Transport Policy, 18*, 501–512.
Marsden, G., Mullen, C., Bache, I., Bartle, I., & Flinders, M. (2014). Carbon reduction and travel behaviour: Discourses, disputes and contradictions in governance. *Transport Policy, 35*, 71–78.

Marsh, D., & Sharman, J. C. (2009). Policy diffusion and policy transfer. *Policy Studies, 30*(3), 269–288.

May, A. D., Grant-Muller, S., Marsden, G., & Thanos, S. (2008). *Improving the collection and monitoring of urban travel data: An international review.* Presented at 87th annual Transportation Research Board meeting, Washington, DC.

Piattoni, S. (2009). Multi-level governance. *Journal of European Integration, 31*(2), 162–180.

Trubek, D. M., & Trubek, L. G. (2005). Hard and soft law in the construction of social Europe: The role of the open method of co-ordination. *European Law Journal, 11*(3), 343–364.

# High-Speed Rail in England 9

This case study refers to the ex ante assessment of the case for the development of a High-Speed Rail network in England for domestic travel connecting London and cities to the north including Birmingham, Manchester, and Leeds (see Fig. 9.1). While interpretations of what constitutes High-Speed Rail vary, the UK government sees this to be trains operating above 250 km/h.

The case study is unique in that the estimated costs of making the scheme operational are £15.4 billion–£17.3 billion (US$24 billion–US$27 billion) for the first phase between London and Birmingham alone. This therefore represents the second largest investment decision in the English transport system for many decades (after London CrossRail). Although the cost of the scheme is distinctive, the decision is being taken within the same framework which governs all major transport infrastructure investments in the UK. As such, the additional scrutiny which has been brought to bear provides insight into the way in which information is used in decision-making processes. The scheme is so large and significant to the UK that it has led to the development of a bespoke sustainability appraisal, the design and application of which are fundamental to the themes of this book.

A feature of major infrastructure schemes is the amount of time it takes for them to be brought forward, assessed, and constructed. Such schemes can have many iterations before they are finally constructed. Schemes need to be justifiable under different political leadership and with differing policy priorities. The case of High Speed 2 (HS2) in England is a good example of this. In order to understand the role of the sustainability appraisal in the decision-making process, it is necessary to look at both the national sustainable development strategy and the transportation policies that were in place at the time the decision to consider High-Speed Rail was taken.



H. Gudmundsson et al., *Sustainable Transportation*, Springer Texts in Business and Economics, DOI 10.1007/978-3-662-46924-8_9

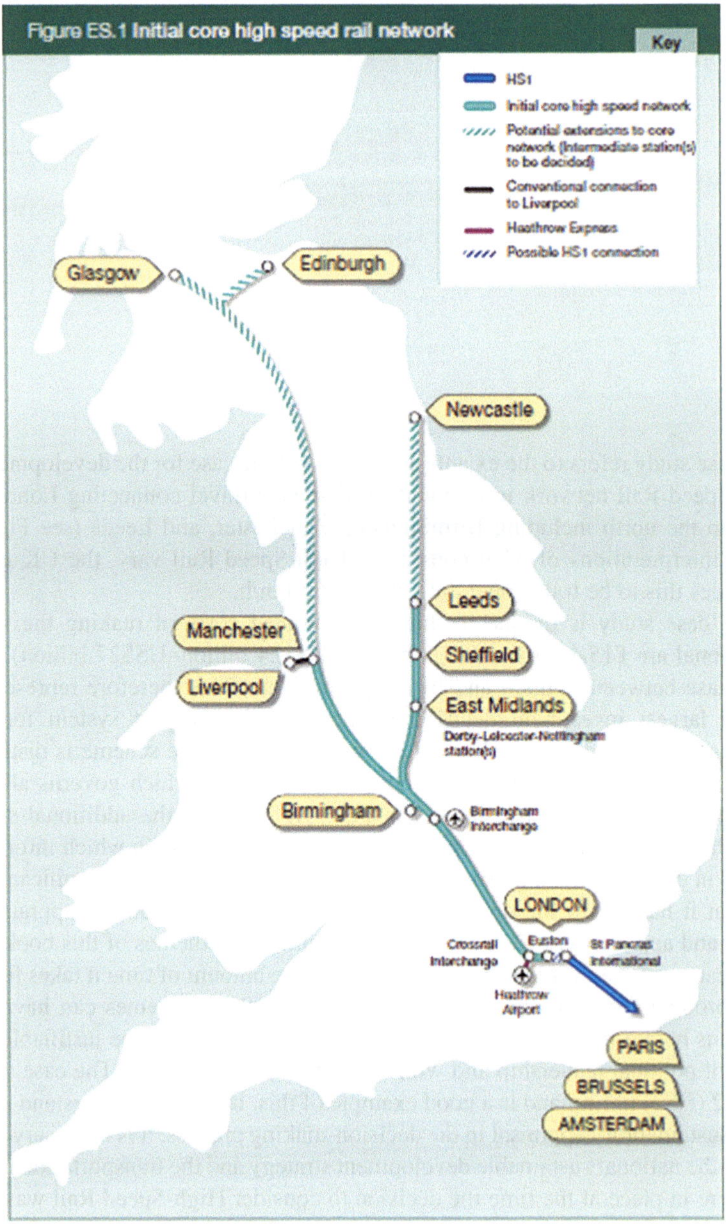


**Fig. 9.1** High-speed 2 route concept map. Reproduced with permission from HS2 Ltd

## 9.1 Background

In 2005, the UK published a revised Sustainable Development Strategy (DEFRA 2005). This strategy set out the overarching goals for sustainable development and provided a set of indicators which could be used to track progress. The key goals of the strategy were:

- Living within environmental limits;
- Ensuring a strong, healthy, and just society;
- Achieving a sustainable economy;
- Promoting good governance; and
- Using sound science responsibly.

The strategy was set in a period when new public management ideologies were at the fore of the UK government so each government department's contribution to these indicators was established through a series of "Public Service Agreements" that the departments were supposed to work towards. The key transportation-related indicators are shown in Table 9.1.

In 2007 and 2008, a white paper and implementation plan were developed that set the national context for transport strategy development. In 2007, the report "Delivering a Sustainable Transport System" laid out five key goals for sustainable transportation as follows (DfT 2007a):

- Goal 1—maximize the competitiveness and productivity of the economy;
- Goal 2—address climate change, by cutting emissions of carbon dioxide ($CO_2$) and other greenhouse gases;
- Goal 3—protect people's safety, security, and health;
- Goal 4—improve quality of life, including through a healthy natural environment; and
- Goal 5—promote greater equality of opportunity.

A recently commissioned report on economic competitiveness had concluded that the UK was well connected and accessible and that the main focus of investment attention should be on pinch points and major international gateways rather than the construction of "grand projects" (Eddington 2006).

In 2008, the report "Towards a Sustainable Transport System" provided information about implementation priorities and how decisions would be taken forward. It signaled the development of a Manchester–Birmingham–London corridor study. In doing this, the method was supposed to ensure that the problems were tackled (rather than solutions being proposed in search of a problem). The study was supposed to "generate a broad range of options. This might include widening of motorways, active traffic management, road pricing, or the construction of new rail capacity either through a conventional (c. 125 mph) or a high-speed (c. 200 mph) line. Equally, the right solution might be a combination of two or more of these. Some radical options (double-deck motorways, Maglevs, and dedicated freight

**Table 9.1** Sustainable development strategy indicators linked directly to transport

| Indicator | Commentary |
|---|---|
| Greenhouse gas emissions | A range of different indicators are put forward relating to the UK's commitments on climate change emission reduction. These run from the macro-level cross-governmental commitments, to sectoral commitments to a breakdown across different modes of transport |
| $CO_2$ emissions by end user (industry, transport, domestic, other) | |
| Aviation and shipping emissions | |
| Private vehicles: $CO_2$ emissions, car-km, and final household expenditure | |
| Road freight: $CO_2$ emissions, tonne-km, tonnes, and GDP | |
| Road transport emissions $NO_x$, $PM_{10}$, $CO_2$ emissions, and GDP | Total toxic emissions from road transport are also identified as important as are measures of the impacts on the natural environment (e.g., eutrophication) and public health as assessed against published air quality standards |
| Emissions of air pollutants | |
| Air quality and health | |
| Ecological impacts of air pollution | |
| Mobility | This metric appears to relate to choice and captures the number of trips by mode and the distance traveled by journey purpose. Success is linked to targets to increase public transport use, walking, and cycling |
| Getting to school | This could be seen as a subset of the mobility objective, but relates specifically to interventions on school travel plans and safe routes to school |
| Accessibility | This relates to access to key services such as health, education, and employment. The strategy specifically identifies rural accessibility although urban access and social exclusion was also a major feature of policy at the time |
| Road accidents | Connected to the 2000 road safety strategy which had targets for fatality and casualty reduction for 2010 |

links) have been considered and rejected as inappropriate or unaffordable, but others may emerge in the option generation process. Value for money will be a key consideration" (DfT 2008, p. 66).

In 2007, a rail White Paper was also published which stated that "it would not be prudent to commit now 'all-or-nothing' projects, such as network-wide electrification or a high-speed line, for which the longer-term benefits are currently uncertain and which do not reflect today's priorities" (DfT 2007b, p. 8). Further to this, it stated that "Higher speed is not the only or best way of cutting journey times. Nor is it without cost. Increasing the maximum speed of a train from 200 km/h to 350 km/h means a 90 per cent increase in energy consumption. In exchange, it cuts station-to-station journey time by less than 25 per cent and door-to-door journey-time by even less. . . . The argument that high-speed rail travel is a "green option" does not stand up to close inspection on the basis of the present electricity generation mix. The Government estimates that carbon emissions per passenger for a journey between London and

Edinburgh will be approximately 7 kg of carbon for conventional-speed rail, 14 kg for high-speed rail, and 26 kg for aviation" (DfT 2007b, p. 62).

It is not clear what the outcome of the corridor study was, although it is worth noting that previous corridor studies have failed to achieve an integrated set of policy outcomes (Marsden 2005 and Chap. 5). However, in 2009 the Secretary of State established a new company (HS2 Ltd) to explore options for a line from London to Birmingham via Heathrow. This came as part of a statement on a possible third runway at Heathrow. HS2 Ltd subsequently clarified its remit as focusing on, in order of priority:

- Passenger capacity: "this is the driving consideration, including capacity released on classic lines";
- Speed;
- Land use and development objectives and the support of new housing development; and
- Developing the line to be capable of handling freight for greater network resilience (Rowlands 2009).

HS2 Ltd also clarified that modal shift from air to rail was "not expected to be a key objective for HS2" (Rowlands 2009).

In 2009, HS2 submitted options to the Secretary of State and in 2010 a White Paper was published setting out a commitment to a preferred route concept for Birmingham, Manchester, and Leeds (see Fig. 9.1).

In 2010, there was a change of government but a continued commitment to High Speed Rail across all of the political parties. More detailed route alignments and business case development continued over the period to 2013. In November 2013, a Hybrid Bill was submitted to the UK Parliament to grant the powers necessary to construct and operate Phase One of HS2 between London and the West Midlands. Importantly for the context of this book, a major Appraisal of Sustainability (AoS) was developed and issued in 2011. This is an example of the general "sustainability appraisal" category identified in Chap. 7 (Sect. 7.4.3). In the final discussion of this chapter, a reflection is given on the ongoing debate around the benefits and rationale for the HS2 project. However, the key question we address here is what the sustainability appraisal comprised, how consistent it was with the sustainable development strategy, and, crucially, how it influenced the decision-making process.

## 9.2 Framework

The HS2 AoS report describes its purpose as both providing an assessment of the extent to which the proposed scheme between London and the West Midlands supports sustainable development objectives ("through the integration of environmental, social and economic considerations") and informing the design of the scheme (Geisler et al. 2011, p. 2)—see Fig. 9.2.

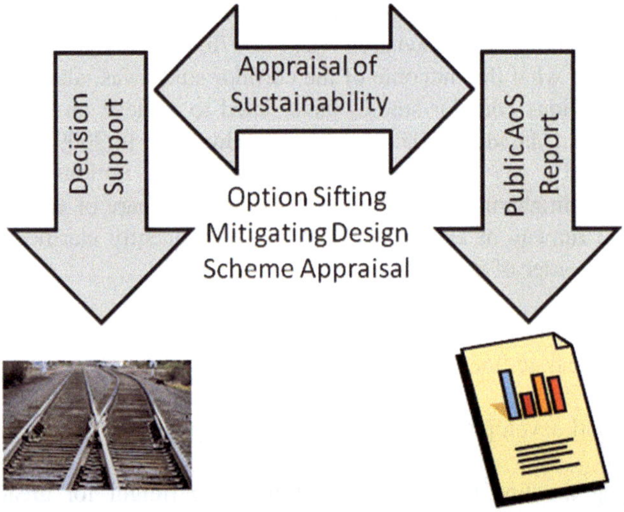


**Fig. 9.2** Overview of role of the appraisal of sustainability. *Source*: Adapted from Geisler et al. (2011)

In setting the context within which the AoS was developed, the document refers directly to the overarching sustainable development goals from the UK's 2006 strategy and the interpretation of these presented in "Delivering a Sustainable Transport Strategy" as set out above. The goals for the AoS were also defined by drawing on existing transport appraisal guidance[1] and on the specific objectives of the HS2 scheme which are set out as:

- To enhance passenger capacity;
- To create faster journeys;
- To encourage modal shift;
- To improve connectivity; and
- To support regeneration and growth (Geisler et al. 2011, p. 1).

This led to the development of 18 sustainability issues and 33 objectives as shown in Table 9.2. These 33 objectives in turn led to the deployment of 66 evaluation criteria (or indicators), a sample of which are reviewed in the next section. Overall, this is a set of technically derived ex ante indicators which cover the three pillars of sustainable development. There is no scope for citizen participation within this stage of the AoS.

The AoS was applied to varying levels of detail throughout the decision-making process. An overview of the process is shown in Fig. 9.3. The process shows that the initial options were produced by consideration of all of the different components

[1] Guidance on the transport modeling and appraisal framework is available at https://www.gov.uk/transport-analysis-guidance-webtag (accessed 1/1/2015).

**Table 9.2** Sustainability issues and objectives

| Key sustainability issue | Objective |
|---|---|
| *Reducing greenhouse gas emissions and combating climate change* | |
| Climatic factors and adaptability | • Improve resilience of the rail network against extreme weather events |
| Greenhouse gases | • Contribute to the reduction of greenhouse gas emissions by facilitating modal shift from road and air to rail<br>• Reduce relative contribution made by rail to greenhouse gas emissions by promoting energy efficient technologies |
| *Natural and cultural resource protection and environmental enhancement* | |
| Landscape and townscape | • Maintain and enhance existing landscape character<br>• Maintain and enhance existing townscape character |
| Cultural heritage | • Preserve and protect archeological assets<br>• Preserve and protect historic buildings<br>• Preserve and protect historic landscapes |
| Biodiversity | • Maintain and enhance biodiversity |
| Water resources | • Protect surface water resources<br>• Protect groundwater resources |
| Flood risk | • Conserve and enhance the capacity of flood plains |
| *Creating sustainable communities* | |
| Air quality | • Maintain and enhance local air quality |
| Noise and vibration | • Maintain and enhance the local noise environment<br>• Maintain the local vibration environment |
| Community integrity | • Maintain and enhance community integrity |
| Accessibility | • Maintain and enhance pedestrian access<br>• Maintain and enhance access to public transport<br>• Maintain and enhance public transport interchange |
| Health and well-being | • Maintain and improve mental well-being<br>• Maintain and improve physical health<br>• Reduce health inequalities |
| Safety and security | • Contribute to the reduction of road traffic accidents<br>• Protect against crime and fear of crime |
| Economic prosperity | • Support economic competitiveness and make efficient use of public funds<br>• Support wider economic growth and maintain and enhance employment opportunities |
| Economic welfare | • Support wider economic growth<br>• Support planned developments<br>• Maintain and enhance regeneration |
| *Sustainable consumption and production* | |
| Soil and land resources | • Maintain and enhance land resources<br>• Encourage the use of brownfield sites |
| Waste generation | • Prevent and minimize waste protection |
| Resource use | • Conserve and protect primary material resources |

*Source*: Geisler et al. (2011, p. 29)

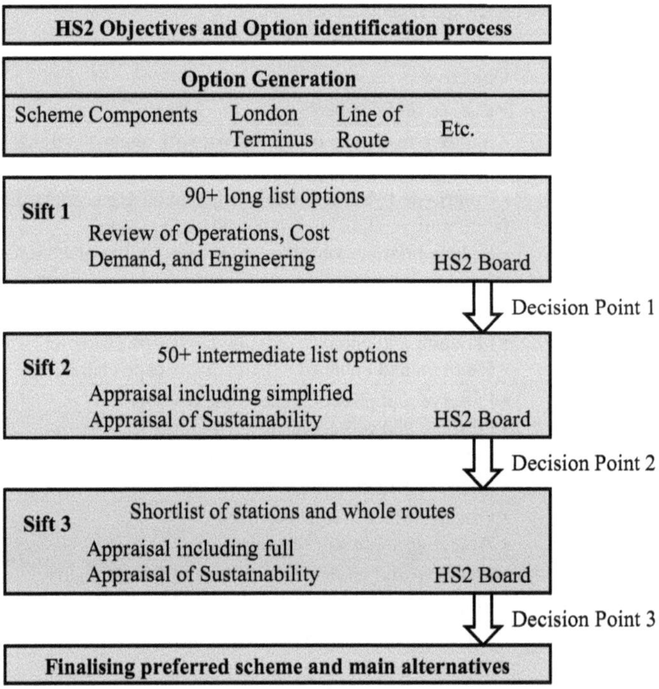


**Fig. 9.3** Option sift and appraisal process for HS2. *Source*: Adapted from Geisler et al. (2011, p. 32)

that could go together to make up the HS2 route options. This included different station designs, different spur points on the network, and alternative alignments for the route. Taken together these created a potential long list of more than 90 options to examine. Here, as in any option generation exercise, there needs to be a proportionate assessment made of options as the costs of detailed analysis of unlikely options would be prohibitive compared to the benefits from such an assessment. The HS2 process had an expert-led review team look at issues such as operational constraints (e.g., at stations and interchanges), capacity relative to demand, engineering issues, and costs. At this stage, there are clearly lots of indicators at play (e.g., cost and utilized capacity), but these are not derived from or directly feeding into the AoS.

Sift 2 had 50+ options and began to study in more detail the impacts of specific route alignments. Here, issues such as number of Sites of Special Scientific Interest that would be affected and other protected natural habitats began to be identified. However, the process again had to be streamlined for affordability purposes so this was used to sift out the most environmentally challenging proposals or to identify the need for greater mitigation where these options performed well on other grounds. A limited number of stations and routes were taken through for a full AoS set out in Geisler et al. (2011, p. 32).

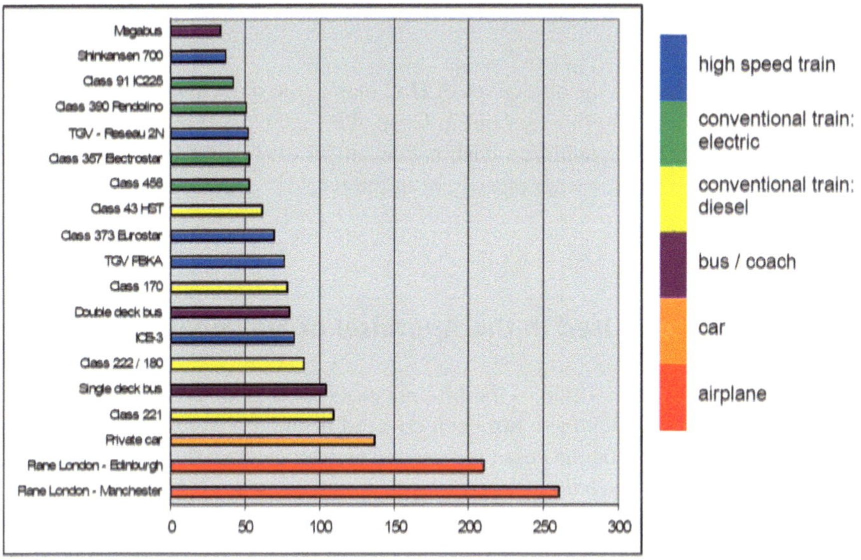


**Fig. 9.4** Comparative emissions of different forms of transportation—grams $CO_2$/passenger km. *Source*: Geisler et al. (2011, p. 56)

The AoS had to take account of the fact that the proposed HS2 line would not open for at least a further 15 years and therefore the baseline for with and without HS2 would need to be extrapolated from existing conditions. The conditions against which the impacts of the HS2 line would then be compared also had to be extrapolated further over the appraisal assessment period. There is clearly considerable uncertainty in projecting so far into the future, but some account must be taken of likely changes to the baseline conditions (e.g., from technology, population growth, and land take for other purposes and flood risk). Figure 9.4 shows the baseline assumptions for the relative emissions of each mode. The report also discusses the likely magnitude of change given anticipated improvements in the efficiency of vehicles and the greening of electricity generation.

The baseline assessment concluded that the main existing sustainability issues included $CO_2$ emissions by air, road transport, and power generation; development of housing and employment growth areas along the route; poor condition of some water courses and sites of special scientific interest; poor air quality, and noise issues on some parts of the route. The baseline assessment also separates out issues such as carbon emissions and economic growth that affect the whole HS2 proposition, from those that affect urban areas (such as specific air quality management areas and known areas of deprivation) and rural areas (which are typically more related to landscape and biodiversity).

**Discussion Topics**

- To what extent do the objectives of HS2 map to the objectives of a sustainable transportation system as set out in Chap. 4?
- What objectives would you wish to see set for evaluating High-Speed Rail? Are there examples from elsewhere in the world?

## 9.3 Indicators Used in the Appraisal of Sustainability

This section reviews a sample of the 66 indicators used in the AoS, highlighting a number of features which have not been covered in the other case study chapters. The text draws heavily on the detail reported in Geisler et al. (2011) and appropriate section numbers are attributed.

**Flood Risk**

Flood risk is assessed by estimating the length of the route that will pass through areas that are predicted to flood more frequently than once in every 100 years and land which is predicted to flood less frequently than once every 100 years but more frequently than once in every 1000 years (Sect. 8.2.5). This is an indicator that is based on a combination of historical observation and forecasts of rainfall intensity provided by the UK Environment Agency. There are some uncertainties regarding flood risk given the predictions of more frequent and more intensive rainfall events that may occur as a result of climate change. However, it is the Environment Agency that has the statutory responsibility for reporting on flood risk so the AoS has to work with the values currently agreed upon.

**Tranquility**

The HS2 AoS uses the definition of tranquility developed by the Campaign for the Protection of Rural England (CPRE) (a nongovernmental organization). It is "a complex concept that can best be described as 'getting away from it all'" (Sect. 7.5.25, p. 70). CPRE developed maps of tranquility for every 500 $m^2$ of the country on the basis of a large sample survey of factors which people said did and did not correspond to tranquility. In total, 44 factors go together to make up the tranquility index including the presence of wildlife, light pollution, the presence of large numbers of people, and car traffic (some of these factors adding to and some detracting from the tranquility index). While tranquility is clearly a difficult concept to define, this provides some evidence base to allow a quantification of the extent to which tranquil areas will be affected.

**Land Resource**

This indicator describes the area of land of different categories whose productive use will be damaged or enhanced by the introduction of HS2 (Sect. 7.6.2).

In particular, the report refers to the impacts on Grade 1 and Grade 2 agricultural land. These definitions are based on advice from the Ministry with responsibility for farming and agriculture (MAFF 1988). For example, Grade 1 land is defined as excellent quality "with no or very minor limitations to agricultural use. A very wide range of agricultural and horticultural crops can be grown and commonly includes top fruit, soft fruit, salad crops and winter harvested vegetables" (MAFF 1988, p. 9). So, while this is a quantitative indicator (of the areas affected), the classification of the land remains somewhat subjective, although the classification is independent of the assessment team.

**Noise**

There are well–established noise assessment thresholds in the UK and a noise mapping procedure that is deployed at a national level. The specific appraisal criteria used in the assessment of noise along the route were:

- Dwellings potentially exposed to "high average noise" levels, i.e., greater than or equal to 73 dB $L_{Aeq,18hr}$;
- Dwellings that could qualify for noise insulation, based on the Noise insulation (Railways and Other Guided Transport Systems, Regulations 1996); and
- Dwellings that could have a noticeable (although not necessarily significant) increase in "average" daytime railway noise levels, defined as having a rail noise level of 50 dB $L_{Aeq,18hr}$ or more with an increase in existing noise levels greater than or equal to 3 dB $L_{Aeq,18hr}$ (the level of change perceptible to humans) (Sect. 10.2.6, p. 97).

Assessments of the impact of HS2 require an estimate of the noise generated by mechanical noise, rolling noise, and aerodynamic noise, with the latter dominating at high speed. Calculations of noise impacts based on the assumed frequency and operational speed of the services and using measurements taken from other comparable high-speed rail systems enable the estimates of the number of homes affected to be identified.

**Economic Prosperity**

The impacts of High-Speed Rail on the UK economy have become a major source of debate for both those for and against HS2. Some argue that HS2 will be transformational for the UK economy, closing the North–South productivity divide. Others claim that the benefits will all fall to wealthy travelers and business travel and will do little to aid productivity given the advancement of mobile technology and working as we travel. Economic prosperity is therefore an important indicator to the case. The AoS describes it as relating to competitiveness across the UK as it reduces business costs and brings about higher business output or GDP. It also captures potential benefits realized by bringing labor markets closer together.

The AoS draws on an analysis that suggested a potential £11 billion (US$16.7 billion) of benefits to business over the 60-year appraisal period. The benefits are accounted for by the journey time savings from large numbers of users switching from the current standard rail line as well as some newly generated patronage. The time saved is monetized by multiplying by agreed values of time for different classes of travel (business having a higher value than commute or leisure travel). The report provides some distributional analysis of where the benefits will fall, with the largest beneficiaries being trips originating in London (34 %) and then the West Midlands (24 %). The report acknowledges that "the way that benefits actually flow through the economy are difficult to predict" (Sect. 8.15.4, p. 111).

## 9.4 Indicator Applications

This section discusses the indicator applications outlined in Chap. 6 providing practical examples that show how the indicators have been put to use in fulfilling the functions.

### 9.4.1 Describe

The AoS baseline is primarily a describing function, presenting data on existing conditions along the route. Some examples of the described application are given above in the indicator descriptions. A further example is under the sustainable consumption and production heading as mineral deposits. Here, the report describes where river sands and gravel are, where quarried deposits are accessed, and where limestone deposits and igneous rock exist which may be commercially viable. This allows the HS2 route assessment to then understand whether particular route alignments will affect access to these mineral deposits.

### 9.4.2 Forecast

As discussed above, the AoS has a baseline and this baseline has to be established for decades into the future. Against this forecast baseline, the potential impacts of the different HS2 alignments can then be compared (discussed further in the decide application). One of the more complex descriptions of forecasting in the AoS relates to carbon dioxide emissions. Here, the AoS looks at the broader policy context where The Climate Change Act 2008 has set a legally binding target to reduce greenhouse gas emissions in the UK by at least 80 % by 2050. Within this, interim 5-year carbon budgets (for 2008–2012, 2013–2017, and 2018–2022) have been set by the Government at levels leading to a 34 % reduction (on 1990 levels) in greenhouse gas emissions by 2020. Each sector has a plan to contribute towards the reduction in $CO_2$ emissions.

Within the transport sector, the assumptions on total $CO_2$ emissions depend on the uptake of electric or hydrogen vehicles coupled with the extent to which the power sector decarbonizes. In addition, the mix of biofuels and environmental benefits which these deliver will be a factor. There is a central range estimate of the reduction of $CO_2$ emissions provided in the AoS as a basis for looking at the contribution of HS2 in the future. Rail accounts for less than 2 % of the overall UK $CO_2$ emissions from transport and so, while the emissions from HS2 are not negligible, they are also only a small piece of a much bigger jigsaw of carbon policy.

### 9.4.3 Review

The air quality baseline provides an example of the review application. The AoS states that "Air quality in the UK as a whole has improved over the last 20 years, and most of the area between London and the West Midlands is currently achieving national air quality standards. However, air quality is a problem in London, with 2010 European limit values for $NO_2$ and particulates not being achieved in many parts. In Hillingdon the council designated an air quality management area (AQMA) for $NO_2$ from its southern borough boundary up to the Chiltern-Marylebone railway line. In addition Camden Council has declared its entire borough an AQMA for $NO_2$ and $PM_{10}$. Euston Station is located within an area of this borough that is of particularly poor air quality" (Geisler et al. 2011, p. 67). Other problem areas are also identified. This shows an element of review as it looks at the levels of air quality relative to an established benchmark and provides an assessment as to what this level of performance signifies. The report also looks ahead in the forecast function and suggests that technological improvements may solve the problems of air quality alongside much of the proposed route but not at the London Terminus in Euston.

### 9.4.4 Diagnose

The AoS contains very little by way of diagnosis. This is a feature of the nature of the scheme and the purpose of the report. This is not a general analysis of strategy options but is instead a document designed to assess the potential for different route options for a scheme that is already planned. The key issues relate therefore to impacts relative to the baseline and so describe, forecast, and review applications come to the fore. There are elements of diagnosis within the AoS (such as that shown in Fig. 9.4), but these are typically for contextual reasons rather than to suggest alternative solutions. Figure 9.4, for example, will inform the $CO_2$ impact of HS2 relative to not building the scheme.

### 9.4.5 Decide

Figure 9.3 shows how the AoS has been used as part of the route selection process. A short form AoS was used in the second sift of around 50 options and a full AoS was used in determining routing and design features of the final proposed alignment.[2] The final AoS results and descriptions of the decide application can be found in Section 8 of the report. For example, in response to the flood indicators where "About 16 km of the surface route (some 7 % of its total length) would cross Flood Zone 3, all of which is susceptible to increased incidents of flooding due to climate change. The outline design of the proposed scheme assumes that all of this would be on viaduct to ensure its protection from flood events" (Geisler et al. 2011, pp. 77–78). The landscape impacts have been particularly influential on route selection and design as it crosses an Area of Outstanding Natural Beauty (AONB). "Given the sensitivity of the AONB in this area the development of the proposed route has sought to integrate the line within the landscape as far as practicable. Some 6.5 km of the route would be in tunnel and in the northern part of the Chilterns over 5 km of the surface route would be within an existing transport corridor, running alongside the A413 arterial valley. In total, around 9 km of the surface route would be in cutting and therefore fully or partially hidden from many views" (Geisler et al. 2011, p. 84). These excerpts indicate how the AoS has been a key part of the design process.

### 9.4.6 Other Applications

Neither the account nor learn applications are features of the AoS given its focus on outlining options for a potential new transportation scheme. The AoS is directly addressing the role of the scheme promoter in understanding the impacts of the scheme and, while design lessons were incorporated as the AoS progressed, it is too early to see the scheme process as having a learning loop function. However, the AoS has been subject to public consultation, receiving over 30,000 responses. In that sense, indicators have been used to communicate key information and have played a role in the accountability process for this major infrastructure proposal.

**Discussion Topics**

- Has the AoS of HS2 led to the development of a scheme consistent with the UK Sustainable Development Strategy?
- Can you identify examples where a package of transport schemes has emerged from a sustainable development strategy? What is different about such a process compared to the HS2 case described here?
- The AoS for HS2 suggests that where large numbers of options or option combinations are to be considered, the level of detail of the assessment needs

[2] It is important to recognize that design is an on-going iterative process and further work continues to be done in response to investigations, cost information, and public consultation.

to be less. What would the core components of a short form AoS be and what are the limitations of such an approach?
- How should we treat scheme proposals which result from identified operational priorities rather than strategic exercises?

## 9.5 Discussion

HS2 is a very large infrastructure scheme proposal. It is not however unique and also not atypical of a range of major infrastructure schemes. They typically have very long gestation periods before a commitment to proceed is reached and then long lead times to construction. Inevitably, therefore, they need to have the backing of multiple political administrations, sometimes at multiple levels.

This chapter began by showing that the rationale for HS2, while developed in a period where the UK had both a sustainable development and sustainable transport strategy, was never couched in terms of its sustainability credentials. As the Director of HS2 wrote to the Secretary of State, the key priorities for the scheme were rail capacity, speed, and unlocking areas of economic growth. On paper, the UK sustainable development strategy leans towards strong sustainability with its goal of "living within environmental limits." Had the initial idea for HS2 been conceived within this framework, the environmental and social impacts might have received a more balanced consideration alongside the emphasis given to economic growth. This broader analysis framework could also have revealed opportunities to address negative impacts through non-transportation investments. For example, the higher energy consumption and carbon emissions per passenger from high-speed rail when compared with conventional rail could be offset by strategic investments in renewable energy. Given the scale of the infrastructure project, adopting a holistic and cross-sectoral approach might have led to a package of solutions that better aligned with the goals stated in the UK sustainable development strategy. In the absence of such an approach, the HS2 project was conceived via the more traditional approach to infrastructure development.

The AoS that was developed is comprehensive and, in many senses, an example of good practice. It provides an open and transparent account of the baseline, future conditions, and likely impacts of HS2. It was also used not just to assess but also to shape the design of the preferred route. Nonetheless, it is important to acknowledge that this has been applied to the assessment of what the preferred scheme should be, not whether a scheme should be built. Given that this is one of the largest infrastructure projects the UK will build in the coming decades, the secondary importance of broader sustainability goals to the decision to proceed is a significant concern.

The overall outcomes of the AoS bear reflection as part of the discussion. Table 9.3 shows the summary of the final assessment for HS2.

**Table 9.3** HS2 Sustainability appraisal summary table

| | Likely impact of proposed HS2 | Likely change between the current baseline and future baseline | Cumulative impacts |
|---|---|---|---|
| *Reducing greenhouse gas emissions and combating climate change* | | | |
| Resilience of the rail network | + | 0 | + |
| Greenhouse gas emissions | +/– | + | + |
| *Natural and cultural resource protection and environmental enhancement* | | | |
| Landscape character | –– | – | –– |
| Townscape character | 0 | 0 | 0 |
| Archeological assets | – | 0 | – |
| Historic buildings | – | – | – |
| Historic landscapes | – | – | – |
| Biodiversity | – | – | – |
| Surface water resources | – | 0 | – |
| Groundwater resources | – | – | – |
| Capacity of flood plains | – | – | –– |
| *Creating sustainable communities* | | | |
| Local air quality | U | + | + |
| Local noise environment | –– | – | –– |
| Local vibration environment | – | 0 | 0 |
| Community integrity | 0 | – | 0 |
| Pedestrian access | 0 | + | + |
| Access to public transport | + | + | ++ |
| Public transport interchange | + | + | ++ |
| Mental well-being | 0 | 0 | 0 |
| Physical health | 0 | + | + |
| Health inequalities | 0 | 0 | 0 |
| Road traffic accidents | 0 | 0 | 0 |
| Crime and fear of crime | 0 | 0 | 0 |
| Economic competitiveness | ++ | U | ++ |
| Wider economic growth and employment | ++ | U | ++ |
| Employment | ++ | U | ++ |
| Support planned development | – | ++ | ++ |

(continued)

**Table 9.3** (continued)

| | Likely impact of proposed HS2 | Likely change between the current baseline and future baseline | Cumulative impacts |
|---|---|---|---|
| Regeneration | + | ++ | ++ |
| *Sustainable consumption and production* | | | |
| Land resources | – | – | – |
| Brownfield sites | + | + | + |
| Waste protection | – | + | + |
| Primary material resources | – | – | – |

*Source*: Adapted from Geisler et al. (2011, pp. 127–128)
– –, highly unsupportive of objective; –, unsupportive of objective; 0, neutral; +, supportive of objective; ++, highly supportive of objective; U, unclassified

The identified benefits are primarily economic with welfare gains, employment, and regeneration benefits. The scheme also provides some additional network resilience by means of opening up alternative routes. It will improve public transport access and interchange. By contrast, it scores negative or neutral on most of the environmental and social indicators as a scheme, although some of these impacts (e.g., climate change emissions and physical health) are set within a context of a general improvement in baseline conditions. The most important negative impacts are, unsurprisingly, landscape impacts. The AoS therefore highlights the much discussed tension between the three pillars of sustainability. It does not resolve them. This is a matter of politics and probably underlines why so much scrutiny has been placed on the Government's assessment of the economic worth of the scheme. The National Audit Office in the UK has traced the evolution of the business case and assessment procedures noting that the benefit-to-cost ratio for Phase 1 to the West Midlands has fallen from 2.6 to 1.6 (excluding wider economic benefits) while costs have risen and it anticipates that this may fall still further (NAO 2013).[3]

Such a large infrastructure scheme will inevitably be controversial, not least because of the residents that will be affected by major new infrastructure in a fairly densely populated country. A contributing factor to the inability of Government to build consensus of the need for the scheme is, it is suggested, the lack of a clear strategic case for it. It has been put forward as a scheme which will close the North–South productivity divide, as a scheme necessary for capacity on the network and as an engine of economic growth. The National Audit Office concluded that "The Department's strategic reasons for developing High Speed 2 are not presented well" and "the Department has focused on developing the economic case" (NAO 2013, p. 6). It appears that despite the completion of a comprehensive

[3] The report does acknowledge that the economic case for the Y shaped network which includes Leeds and Manchester is greater, although it suggests that there are also more cost uncertainties at this stage with these further extensions.

AoS, it is not clear whether this investment is indeed a key plank of the UK's sustainable development strategy.

The positive developments we identify from the assessment are that it has had a tangible impact on the choice of route alignment and design. In theory, the assessment could have identified some environmental or social conditions which outweighed the economic benefits and challenged the scheme's viability. However, as noted in Chap. 5, the decision about what information is allowed to be presented and how it is couched (e.g., relative to background trends) is all highly political and it seems unlikely that the AoS was ever commissioned to challenge the decision to proceed.

## References

DEFRA, Department of Environment, Food and Rural Affairs. (2005). *Securing the future: Delivering the UK sustainable development strategy*. London: TSO.

DfT. (2007a). *Towards a sustainable transport system*. Department for Transport, London, Cm 7226. London: TSO.

DfT. (2007b). *Delivering a sustainable railway*. Department for Transport, London, Cm 7176. London: TSO.

DfT. (2008). *Delivering a sustainable transport system*. Department for Transport, London, Cm 7226. London: TSO.

Eddington, R. (2006). *The case for action: Sir Rod Eddington's advice to government*. HM treasury. Accessed December 20, 2013, from http://webarchive.nationalarchives.gov.uk/

Geisler, N., Bryant, A., Mayes, A., Pownall, A., Sheenagh, M., Turner, S., et al. (2011). *HS2 London to the West Midlands appraisal of sustainability: Main report volume 1*. Report to HS2 limited. Accessed December 20, 2013, from http://hs2.org.uk/news-resources/publications

MAFF. (1988). *Agricultural land classifications of England and Wales*. London: The Ministry of Agriculture, Fisheries and Foods.

Marsden, G. (2005). The multi modal study transport investment plans. *Proceedings of Institution of Civil Engineers: Transport, 158*(2), 75–87.

NAO. (2013, May). *High Speed 2: A review of early programme preparation*. Report by the Comptroller Auditor General, HC124. London: National Audit Office.

Rowlands, D. (2009, February 13). Letter to Lord Adonis (Secretary of State for Transport) from the Chairman of HS2 Limited.

# New York's GreenLITES Rating Systems 10

New York State Department of Transportation's (NYSDOT's) GreenLITES (Leadership In Transportation and Environmental Sustainability) programs are collectively one of the leading sustainable transportation performance assessment systems in the USA. This case study begins by examining the historical context that enabled the GreenLITES programs to be created, after which each of the programs is discussed. A unique aspect explored in the case is how the GreenLITES certification (or rating) programs are promoting change within NYSDOT. Particular attention is paid to how the data (or indicators) from the programs are used within a multi-actor decision-making context (Holden 2013). The case study concludes by examining how NYSDOT's four asset management teams (specifically the Sustainability team) and the Comprehensive Program Team (CPT) build on the GreenLITES programs to promote a culture of sustainability across the agency. Attention is paid to how the GreenLITES certification data and sustainability concepts are used in the context of the Learn, Decide, Forecast, and Communicate indicator application types (described in Chap. 6).

## 10.1 Background

NYSDOT is a large transportation agency with 8300 public employees serving New York State (NYSDOT 2014). The agency is responsible for the creation of comprehensive transportation plans and policy for the State and for the development and safe operation of transportation facilities and services for highways, railroads, mass transit systems, ports, waterways, and aviation facilities.[1]

New York State has around 115,000 lane miles of roadway of which NYDOT is responsible for 15,000 lane miles (13 %) (BTS 2012) that carry 52 % of the State's

[1] Source: NYSDOT, Responsibilities and Functions, https://www.dot.ny.gov/about-nysdot/responsibilities-and-functions (accessed on May 24, 2015).



H. Gudmundsson et al., *Sustainable Transportation*, Springer Texts in Business and Economics, DOI 10.1007/978-3-662-46924-8_10

total VMT (vehicle miles traveled) (NYSDOT 2012c).[2] The majority of the remaining lane miles are managed by counties (18 %) or municipalities (67 %), with the federal government managing less than 1 % (BTS 2012). NYSDOT is also responsible for operating and maintaining 7860 road bridges, over 100,000 small and large culverts, and hundreds of thousands of acres of right of way.[3]

Of the 19.6 million people living in New York State, around 8.3 million live in New York City (NYC). Given the limited availability, cost, and discouragement of parking in the city, only 30 % of NYC residents have a driver's license, compared to 87 % in the USA (NYSDOT 2012c). In fact, around one-fifth of all households in the USA that do not own a car can be found in NYC (*ibid.*). The constraints placed on vehicle use mean that public transportation—consisting of buses, a subway, commuter rail, and ferries—is the dominant mode of travel in NYC. Each weekday, around 70 % of all workers who arrive in Manhattan commute using public transportation. The State also has a comparatively high number of people who walk or bike to work (e.g., 11 % in NYC and 7 % in New York State versus 3 % for the USA) (*ibid.*). Statewide, public transportation systems move 2.75 billion passengers annually, with the vast majority of these passengers (2.45 billion) being located in NYC's metropolitan region (*ibid.*). Overall, the State accounts for around one-third of the nation's transit riders.

The scale and use of public transportation in NYC results in New York State having the most energy efficient transportation sector in the USA when considering per capita measures such as gallons of fuel consumed and VMT (NYSDOT 2012c).

From an organizational perspective, NYSDOT has 11 regional offices that serve 64 counties. The State has 13 Metropolitan Planning Organizations (MPOs) that are responsible for developing regional Transportation Improvement Programs (TIPs) that are generally adopted every 2 years. The TIPs form the basis for the Statewide Transportation Improvement Program (STIP) that is updated a minimum of every 4 years. The STIP contains a comprehensive list of all highway and transit projects that plan to use Federal funds.

NYSDOT is tasked with coordinating the development of a comprehensive transportation policy for the state and supporting the development and operation of transportation facilities and services. It also develops and maintains the statewide long-range transportation master plan. Consistent with most US departments of transportation, the administration and advancement of public safety programs is viewed as a priority throughout all of the agency's activities.

Given the extent of the transportation infrastructure in New York State, it should be no surprise that NYSDOT's approach to transportation planning has evolved from building systems to investing in maintaining and enhancing these systems in the face of growing transportation demands (McVoy et al. 2010). The agency also

---

[2] Note: US metrics are used in this case study to present the information in its original form.

[3] Source: NYSDOT, Operations Certification Program, https://www.dot.ny.gov/programs/greenlites/operations-cert (accessed on May 24, 2015).

**Fig. 10.1** NYSDOT's forward four principles. *Source*: Adapted from NYSDOT (2014)

needed to act to address the legacy of deferred maintenance and underfunding that left the physical infrastructure in a state of decline (NYSDOT 2007).

New York State's Transportation Master Plan for 2030 provides insight into how the existing transportation systems could be enhanced in response to future demands. The plan emphasizes the need to create a more seamless multimodal system for travelers and businesses (NYSDOT 2006). In addition to the more technical aspects of managing critical transportation corridors and improving the efficiency and reliability of services, the master plan "strongly supports increased cooperation and coordination by all of New York's transportation providers" to promote "a far more *collaborative approach* to planning and investment decision making among local governments, MPOs [metropolitan planning organizations], transportation operators and the State" (NYSDOT 2006, p. 1, emphasis added). This case study pays particular attention to how NYSDOT is taking steps to realize the need for institutional change, and how data and sustainability are used to help realize this goal.

Given the need to maintain and enhance existing systems and promote cooperation among the actors providing transportation services, and the growing importance given to protecting and improving the environment, NYSDOT recently established four principles to guide the agency's decisions and investments (Nelson et al. 2011). The "Forward Four" principles, captured in Fig. 10.1, have their roots in NYSDOT's 1998 Environmental Initiative, Context Sensitive Solutions, and Sustainability. The following section takes a closer look at the Environmental Initiative, which laid the foundation for the agency's sustainability policy and the GreenLITES programs that now form a central pillar of NYSDOT's efforts to promote sustainable transportation.

### 10.1.1 New York State's Environmental Initiative

In April 1998, NYSDOT's Environmental Analysis Bureau released the Environmental Initiative Statement that set a new direction for their organization (McVoy et al. 2000, 2010; Nelson et al. 2002). The strategy behind the Environmental Initiative was to move beyond simply "greening" projects or "streamlining" the agency's environment-related activities. Instead, NYSDOT sought "a new paradigm," whereby the agency leveraged its organizational capacity to promote a new culture of environmental stewardship (McVoy et al. 2000, p. 92). The idea was to build partnerships between environmental agencies (such as the New York State Department of Environmental Conservation) and groups, who might have previously adopted an adversarial stance to transportation projects and programs. As the largest public works agency in New York State, NYSDOT considered itself to have an obligation and responsibility to protect the State's environmental resources (NYSDOT 1999), and in doing so, provide an approach that other public works agencies could emulate (Nelson et al. 2002). Box 10.1 provides a summary of five major objectives of the Environmental Initiative.

**Box 10.1: Five Major Objectives of NYSDOT's 1998 Environmental Initiative**

1. Promote and strengthen an environmental ethic throughout the department. Staff should feel a responsibility to leave project sites in better condition than they found them and look for opportunities to enhance New York's environment.
2. Advance state environmental policies and objectives with NYSDOT resources. Advance environmental policies as part of the department's normal work. Fund environmental benefit projects, including stormwater retrofits, wetland restorations, habitat enhancements, recreational access, informational signs, landscaping, and environmental research.
3. Partner with others to construct environmental enhancements. Pursue opportunities for joint development. Incorporate environmental elements or facilities funded by other agencies, municipalities, or environmental groups into NYSDOT construction and maintenance projects. NYSDOT provides design and construction engineering support.
4. Pilot new environmental protection and enhancement methods. Cooperatively research and pilot new methods to, for example, reduce environmental toxins, improve air quality, and increase the use of recycled materials.
5. Strengthen relationships with environmental agencies, organizations and local municipalities. Improve communications, streamline permitting, share program information, and conduct joint training. Gain their confidence in NYSDOT's ability to self-regulate (McVoy et al. 2000, p. 93).

The importance given to creating a new collaborative approach to transportation planning and decision-making is evident in the objectives listed in Box 10.1. The spirit of collaboration is also captured in NYSDOT's guidelines and procedures for implementing the Environmental Initiative (NYSDOT 1999). The guidelines emphasize the importance of context-sensitive design—i.e., that each project should consider the unique social/cultural and environmental characteristics of the area in which it is located—and the inclusion of "environmental betterments" that enhance the environment. However, the most interesting aspect of the document is the detail given to how NYSDOT directors and managers should "coordinate and communicate" within the agency and with other agencies and groups. The *institutionalization of collaboration* within the agency could be seen as one of the most transformative aspects of the Environmental Initiative.

Soon after its implementation, NYSDOT's perception of the Environmental Initiative was that it was leading to noticeable improvements in the morale of public servants. As McVoy et al. (2000, p. 96) comment, "Because it's "OK to be green," designers enjoy more freedom and flexibility in their work, and NYSDOT has stronger, more positive working relationships with external agencies, local municipalities, and other environmental groups. These improved relationships result in avoided costs by reducing delay, litigation, frustrating do-overs, and effort wasted on arguing contentious issues. By working together at the start, projects are accomplished in a more timely and productive manner for all concerned."

The Environmental Initiative established an "environmental ethic" in NYSDOT that began to permeate all functional areas of the agency—e.g., planning, design, construction, maintenance, and operations (Venner Consulting and Parsons Brinckerhoff 2004). This organizational cultural change can be linked to the core objectives of the initiative and to the codification of how NYSDOT leadership should coordinate and communicate. The environmental ethic combined with a focus on collaboration laid the foundation for NYSDOT's efforts to promote a more sustainable transportation system.

**Discussion Topics**

- The above discussion highlights how NYSDOT pursued a "collaborative" approach to advancing an environmental agenda. What are the pros and cons of pursuing such an approach?
- How important is the *process* of creating a performance assessment system relative to the final system that is implemented?

### 10.1.2 From Environmental Stewardship to Sustainability

In 2008, NYSDOT submitted to the New York Legislature the *Multimodal Transportation Program Submission: 2009–2014* (NYSDOT 2008). While this capital program was tailored to the statewide priorities for economic development, energy efficiency, and smart growth, the $25.6 billion of investment focused primarily on stabilizing the condition of infrastructure along with some investment for system expansion. Further, the emphasis the program placed on energy efficiency (to reduce costs and greenhouse gas emissions), environmental protection, and multimodal planning was not reinforced by a clear set of targets that could steer and advance progress toward these priorities.

While the Multimodal Transportation Program falls short from a sustainable transportation perspective, it should be viewed in the context through which the planning and delivery of capital projects occurs. In parallel with the development of the capital program, NYSDOT established a comprehensive sustainability policy and GreenLITES, a transportation and environmental sustainability rating program, discussed in the following section. These two initiatives have a direct impact on how capital projects are approached and provide a mechanism through which sustainability principles can be incorporated into agency investments.

With a vision to "exemplify how transportation can support a sustainable society," NYSDOT established a sustainability mission "to fully integrate sustainability into the Department's decisions and practices in planning, designing, constructing, maintaining and operating New York State's transportation system. NYSDOT will also model and advance sustainability in managing its internal resources."[4] The mission effectively calls on the agency to emulate the approach it used to integrate environmental stewardship into its daily routines, practices, and decisions.

NYSDOT defines a sustainable society in the spirit of the Brundtland definition of sustainable development.

> A sustainable society manages resources in a way that fulfills the social (community), economic and environmental needs of the present without compromising the needs and opportunities of future generations.
>
> A transportation system which supports a sustainable society is one that:
>
> 1. Allows individual and societal transportation needs to be met in a manner consistent with human and ecosystem health with equity within and between generations.
> 2. Is safe, affordable, accessible, operates efficiently, offers choice of transport mode, and supports a vibrant economy.
> 3. Protects and preserves the environment by limiting transportation emissions and wastes, minimizes the consumption of resources and enhances the existing environment as practicable.

Two observations can be made in relation to how NYSDOT approaches sustainability. First, the agency does not talk about sustainable transportation per se; rather it focuses on the role that transportation plays in supporting a sustainable society. Second, NYSDOT's definition sidesteps the problem we identified with the European Council's (EC's) definition discussed in Chap. 4. Whereas the EC's definition states that a "sustainable transportation" system "limits emissions and waste within the planet's ability to absorb them" (EC 2001, pp. 15–16), NYSDOT's definition focuses on the need to protect and preserve the environment. The former framing highlights the need to explicitly acknowledge the role of other economic sectors that impact the environment. By sidestepping the issue of "environmental limits," NYSDOT's definition moves away from the "strong" model of sustainability toward "weak" sustainability. Another way to view the definition is

[4] Source: NYSDOT Sustainability Policy, https://www.dot.ny.gov/programs/greenlites/sustainability (accessed on May 24, 2015).

that NYSDOT has taken care to frame the scope of what it is realistically able to address.

Notwithstanding the potential limitations with the definition, the sustainability leadership role that NYSDOT has adopted in New York State provides a platform from which it could work with other public agencies to identify the most cost-effective way to limit statewide pollutants to agreeable levels. Thus, NYSDOT provides an example of how a large public works agency can take a lead role in promoting decision-making for sustainable development at a statewide level. Of course, the challenge with this approach is that it requires strong agency leadership and a sustained focus on implementing the principles of sustainable development to address problems such as conflicting agendas, persistent policy paradigms and mind-sets, and politics. The significance of these political economic considerations will vary by region, which means a transportation agency should gauge the broader political landscape to determine whether it could successfully adopt an approach similar to that followed by NYSDOT.

## 10.2 Frameworks and Indicators

Since the GreenLITES initiative consists of multiple frameworks that contain numerous indicators, the discussion of these two aspects is combined in the following sections.

### 10.2.1 The GreenLITES Initiative

In an effort to integrate sustainability principles into the multimodal capital program, NYSDOT launched the GreenLITES (Leadership in Transportation Environmental Sustainability) self-certification program in 2008 to distinguish projects that incorporate environmental sustainability into their design. The program was modeled after the University of Washington's Greenroads initiative,[5] as well as the Leadership in Energy and Environmental Design (LEED)[6] program adopted by the building industry (NYSDOT 2010a).

GreenLITES uses a points system to determine if new projects are environmentally certifiable. The rating system has four certification levels: 0–14, no rating; 15–29, *Certified*; 30–44, *Silver*; 45–59, *Gold*; 60 and up, *Evergreen* (Table 10.1). The range of points for each certification level were divided into thirds to represent low (no-rating), medium (*Certified*), and high (*Silver*, *Gold*, and *Evergreen*) levels of environmental performance (Fig. 10.2). The highest rating, *Evergreen*, is given to designs that are considered to significantly advance the state of sustainable transportation solutions (NYSDOT 2010a). In 2008, NYSDOT rated 26 projects

[5] See Greenroads, https://www.greenroads.org/ (accessed on May 24, 2015).

[6] See LEED, http://www.usgbc.org/leed/rating-systems (accessed on May 24, 2015).

**Table 10.1** GreenLITES certification levels

| Level | Description | Points |
|---|---|---|
| None | *Non-certified*: A project design has not incorporated a sufficient number of sustainability choices to be certified | 0–14 |
| GreenLITES Certified | *Certified*: This certification highlights a project design that has incorporated a number of sustainable choices | 15–29 |
| GreenLITES Silver | *Silver*: Silver certification highlights a project design that has incorporated a number of sustainable choices with several of these choices having a high level of impact, or having advanced the state of practice | 30–44 |
| GreenLITES Gold | *Gold*: Gold certification highlights a project design that has incorporated a substantial number of sustainable choices with many of these choices having a high level of impact, or having advanced the state of practice | 45–59 |
| GreenLITES Evergreen | *Evergreen*: Evergreen certification highlights a project design that has incorporated the highest number of sustainable choices with many of these choices having an extremely high level of impact. Additionally, these projects may advance the state of practice or are innovative in the way environmental sustainability is approached on the project | 60 and up |

*Source*: Adapted from NYSDOT (2010a, p. 6)

that had recently been completed to create realistic point ranges for each certification level. The expectation was that as the program developed over time, the bell curve shown in Fig. 10.2 would move to the right as the quality of project designs improves.

New transportation project designs are rated against five categories of criteria that cover sustainable sites, water quality, materials and resources, energy and atmosphere, and innovation (or unlisted categories/items). These five categories cover over 175 sustainable practices that could be deployed in a project. Table 10.2 summarizes the 21 subcategories under the five main categories. Figure 10.3 provides a snapshot of part of the scorecard for the sustainable sites category. The figure shows how points are awarded if a certain action has been taken to enhance a project. Any unique characteristic of a certified project is incorporated into the relevant category of the rating tool to advance the state of practice. This flexibility enables the GreenLITES program to evolve as new ideas emerge and provides project designers with an incentive to try new innovations. The subjective allocation of points for each best practice does present a potential challenge in terms of consistency, but this is addressed by ensuring that each evaluation is reviewed by multiple experienced personnel.

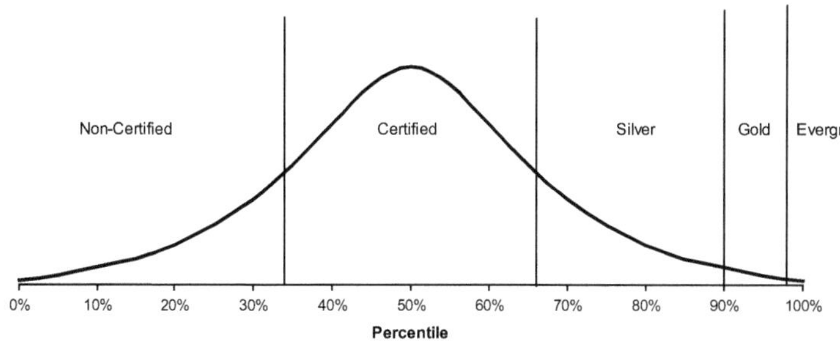


**Fig. 10.2** Initial GreenLITES award distribution. *Source*: NYSDOT (2010a, p. 7)

**Table 10.2** Category and subcategories of the GreenLITES scorecard

| GreenLITES category | Subcategories (number of best practices listed in subcategory) |
|---|---|
| Sustainable sites (S) | S1—Alignment selection (10)<br>S2—Context sensitive solutions (13)<br>S3—Land use/Community planning (12)<br>S4—Protect, enhance, or restore wildlife habitat (14)<br>S5—Protect, plant, or mitigate for removal of trees and plant communities (10) |
| Water quality (W) | W1—Stormwater management (Volume and Quality) (7)<br>W2—Best management practices (BMPs) (6) |
| Materials and resources (M) | M1—Reuse of materials (21)<br>M2—Recycled content (8)<br>M3—Local materials (2)<br>M4—Bioengineering techniques (5)<br>M5—Hazardous material minimization (3) |
| Energy and atmosphere (E) | E1—Improve traffic flow (16)<br>E2—Reduce electrical consumption (6)<br>E3—Reduce petroleum consumption (11)<br>E4—Improve bicycle and pedestrian facilities (27)<br>E5—Noise abatement (8)<br>E6—Stray light reduction (2) |
| Innovation (I) | I1—Innovation (general) (1)<br>I2—Innovation (general) (1)<br>I3—NYSDOT street design manual (1) |

As should be expected, the subcategories and best practices align well with the environmental domain of sustainable transportation (see Table 4.3) and provide some limited coverage in the social domain through, for example, improved access to pedestrian and bicycle facilities. Further, the best practices listed in the scorecard consist of a mixture of "outcome" and "process" indicators (see Sect. 6.4.4). The outcome indicators measure whether a specific design objective has been realized (such as minimizing the use of land which forms part of significant contiguous wildlife areas). The process indicators measure, for example, whether the design of a project followed a best practice in terms of project planning (such as incorporating aesthetic design guidance on bridge building). The majority of indicators in the

**GreenLITES Project Environmental Sustainability Rating System Scorecard** v 2.1.0 — POINTS — Project: — PIN: — Type: — Contact Name:

*Please fill in all* yellow highlighted cells *and follow all instructions in* red text.

| | CATEGORY | ID | DESCRIPTION | Available | Scored | INSTRUCTIONS |
|---|---|---|---|---|---|---|
| Sustainable Sites (S) | S-1 *Alignment Selection* | S-1a | Avoidance of previously undeveloped lands (open spaces or "greenfields"). | 2 | | <= Please enter 0 or 2 |
| | | S-1b | Selecting an alignment that establishes a minimum 100-foot buffer zone between the edge of pavement and a natural watercourse or significantly sized natural wetland to serve the purpose of stormwater filtration. | 2 | | <= Please enter 0 or 2 |
| | | S-1c | Alignments which minimize overall construction "footprint." Examples: use of retaining walls, selecting design option with minimal footprint. | 2 | | <= Please enter 0 or 2 |
| | | S-1d | Design vertical alignments which minimize total earthwork. (Applicable only for projects modifying existing vertical alignments.) | 1 | | <= Please enter 0 or 1 |
| | | S-1e | Adjust alignment to avoid or minimize impacts to social/environmental resources (avoidance of parklands, wetlands, historic sites, farmlands, residential and commercial buildings, etc.). | 1 | | <= Please enter 0 or 1 |
| | | S-1f | Alignments that optimize benefits among competing constraints. (The goal is not always the minimum-length alignment, but the one with the best benefit overall.) | 1 | | <= Please enter 0 or 1 |
| | | S-1g | Micro-adjustments that do not compromise safety or operation but make the difference in providing sufficient clear area for tree planting. | 1 | | <= Please enter 0 or 1 |
| | | S-1h | Clear zones seeded with seed mixtures that help to reduce maintenance needs and increase carbon sequestration. | 1 | | <= Please enter 0 or 1 |
| | | S-1i | Provide a depressed roadway alignment. | 1 | | <= Please enter 0 or 1 |
| | | S-1j | Use of launched soil nails as a more cost effective option to stabilize a slope rather than, for example, closing a road to construct a retaining wall which may negatively affect traffic flow and neighboring properties. | 1 | | <= Please enter 0 or 1 |
| | S-2 *Context Sensitive Solutions* | S-2a | Adjust or incorporate highway features to respond to the unique character or sense of place (both natural and built) of the area ("Unique character" means whatever identifiable elements make a place distinctive, memorable, important to the community, etc. - landmarks, views, historic bridges & buildings, parkways, characteristic use of materials, a notable stand of trees, etc.). | 2 | | <= Please enter 0 or 2 |

**Fig. 10.3** GreenLITES project scorecard. *Source*: GreenLITES scorecard in Excel (version 2.1.0)

Land Use/Community Planning category can be classified as process indicators since they focus on activities undertaken.

The GreenLITES project design certification program is primarily an internal management program for NYSDOT that enables the agency to measure the performance of its projects over time and identify where improvements to system designs are needed. Data from the project design scorecard are not used to prioritize which projects are selected; rather it helps project designers make decisions that can enhance the sustainability of a project. An example of a best practice improvement is the use of the more sustainable "piano key" crosswalk design rather than a "ladder" design. Likewise, the use of living snow fences is becoming more common largely due to being highlighted as a best practice in the "Operations" GreenLITES awards.

As best practices become standardized, the GreenLITES project design scorecard is updated. For example, the initial scorecard listed LED traffic control lights as a one-point item. Since LED traffic lights are now mandatory, this item was removed from the scorecard and replaced with LED street lights.

In addition to evaluating all new projects proposed from within NYSDOT, the certification program can evaluate projects proposed by local governments, nongovernmental organizations, and other New York government agencies and authorities that use federal funding. While there is no mandatory requirement for assessment outside of NYSDOT, agencies and groups can use the tool to demonstrate their commitment to environmental sustainability. In this regard, the GreenLITES program establishes a shared standard across multiple actors, who operate in one of the 11 regional divisions of NYSDOT or act alongside the agency in these regions.

### 10.2.2 Expanding the GreenLITES Program

Following its implementation, the value of the GreenLITES program in advancing other areas of transportation service delivery became apparent. A year after the GreenLITES project design program was launched, the GreenLITES maintenance/operations certification program was piloted. This self-rating program used a similar certification process as the project design program, with the main difference being its application to the operation and maintenance of assets such as bridges, pavements,[7] drainage, signals, lighting, signs, the roadside environment, facilities, and rest areas (NYSDOT 2012a). It also covers snow and ice operations. Whereas the project design program certifies a project, the GreenLITES operations certification applies to Residencies (i.e., the facilities that maintain highways), Regional Bridge Maintenance Groups, and where applicable, Main Office or Regional Operations Program Areas. A GreenLITES operations scorecard is developed for each region and included in a Comprehensive Maintenance Operations Program (MOP) Summary that provides NYSDOT managers with data on how various groups are performing (NYSDOT 2012a).

In 2013, the GreenLITES operations program was modified. To better transfer best practices, each operational group across the state is required to submit their top three sustainability practices to the Main Office. These best practices are then reviewed and the top five/six submissions are candidates to receive the Evergreen certification. All of the submitted sustainability practices are shared with all the operational groups for implementation consideration, recognizing that not all ideas can be applied by each group. Those operational groups that are able to implement the most innovative practices from the previous year become candidates to receive an additional award for implementing best practices.

The GreenLITES initiative has also been expanded into the planning domain (NYSDOT 2011). In contrast to the project design and operations "certification" programs, the GreenLITES sustainable planning program was created to bring a more balanced approach to transportation decision-making at the local level. The program offers a GreenLITES project solicitation tool that highlights a series of sustainable transportation planning practices that MPOs should consider when assessing projects. The tool asks seven questions that are each followed by a series of sub-questions (Table 10.3). Each time the answer to a sub-question is "yes" a point is awarded. A total of 26 points can be granted. Interestingly, a higher score is not necessarily equivalent to a more sustainable project. The purpose of the tool is to help ensure that projects are vetted using a comprehensive planning process that pays particular attention to environmental, social, and economic factors. It is meant

[7] In the USA, the term pavement refers to a road surface.

**Table 10.3** GreenLITES project solicitation tool for MPOs

| Goal | Criteria (*1 point is awarded for each "yes" answer*) |
|---|---|
| Is the project consistent with current local comprehensive plan? If the community does not have a plan, answer "no" to the questions | Has the Plan been developed within the last 10 years? |
| | Does the Plan provide a vision of community objectives and priorities? |
| | Does the Plan incorporate "walkable communities" and/or "complete streets" concepts? |
| | Has the Plan been developed through an enhanced public outreach effort? This would involve reaching out to all members of the community |
| | Does the Plan promote population and development densities that are sufficient to warrant public transit? |
| | Is the project consistent with the objectives of the Plan? |
| Does the project support many of the "livability principles"? | Does the project provide for more transportation choices (modes) that are safe, reliable, and affordable? |
| | Does the project enhance economic competitiveness through reliable and timely access to employment centers, housing, educational opportunities, and expanded business access to markets? |
| | Does the project contribute toward the revitalization of existing communities through transit-oriented, mixed used development? |
| | Does the project enhance the unique characteristics of the community by investing in healthy, safe, and walkable neighborhoods? |
| Does the project protect and enhance the environment? | Does the project encourage the efficient use of energy resources and renewable alternatives? Examples are:<br>• Energy and Atmosphere—reduce petroleum consumption and air emissions by improving traffic flow through coordinated signal systems, installing of a transit express system, and limiting access points along a highway.<br>• Electrical consumption—use LED street lighting and LED traffic lights |

(continued)

**Table 10.3** (continued)

| Goal | Criteria (*1 point is awarded for each "yes" answer*) |
|---|---|
| | • Petroleum consumption—reduce petroleum consumption by providing new Park and Ride lots; increasing bicycle amenities at Park and Rides and transit stations; incorporating ITS technology to improve traffic flow |
| | Does the project consider aesthetics in design—context-sensitive design, landscaping, visual easements, etc.? |
| | Does the project include Ecology and Habitat Enhancements, such as species protection, wetlands protection, and native communities? |
| | Does the project involve redevelopment or reuse of Brownfields? The redevelopment of Brownfields leads to public benefits through the removal of hazardous wastes |
| | Does the project contribute toward reducing Greenhouse Gas Emissions (GHGs)? |
| Does the project support the economic vitality of the affected area, and at the same time, minimize adverse environmental impacts? | Does the project enhance the region's attractiveness to new/existing businesses? |
| | Does the project support use of or reinvestment in high density mixed use urban areas or villages? |
| | Does the project avoid previously undeveloped land (open spaces or greenfields)? |
| | Does the project avoid or minimize impacts to social/environmental resources (parklands, wetlands, historic sites, farmlands, and viewsheds)? |
| Does the project contribute toward increasing accessibility and mobility options? | Does the project improve bicycle and pedestrian facilities, such as shoulder widening to provide for on-road bike-lanes, new pedestrian signals, new or extended sidewalks, etc.? |
| | Does the project improve access to transit facilities for multiple users? This may include new/expanded transit infrastructure, such as platforms, stations, parking, and rail lines |
| | Does the project enhance accessibility for persons with disabilities and meet ADA requirements? |

(continued)

**Table 10.3** (continued)

| Goal | Criteria (*1 point is awarded for each "yes" answer*) |
|---|---|
| Does the project employ unique financing arrangements? | Does the project uses Public/Private partnerships to finance the initial cost, or some aspect of this project (operating costs)? |
| | Is the project located in a special assessment district, and is it being financed through taxes or fees collected from developments in the district? |
| | Does the project use other innovative financing arrangements? |
| Other considerations—Does the project address other sustainable transportation practices that are not included in this guidance? For example, does the project employ methods that will lead to a longer life of that facility (i.e., life cycle cost savings)? | |

*Source*: NYSDOT (2011)

to start a conversation about the merits of a proposed project, rather than provide an assessment framework for project selection.

While the GreenLITES project solicitation tool is likely to help identify those projects that align with the concept of sustainability, it does not provide a mechanism to address difficult trade-offs that may exist, for example, between the objectives of economic development and environmental protection. We might also question the connection between "unique financing arrangements" and sustainability. A close read of the questions in Table 10.3 indicates that the GreenLITES project solicitation tool is more of a project rating tool than a planning tool. However, the existence of a standard set of questions is likely to influence the attention MPOs give to sustainability considerations in the development of their portfolio of projects.

As a comparison, the interested reader is referred to the Sustainable Transportation Analysis and Rating System (STARS) planning tool (STARS-Plan), developed by the Santa Cruz County Regional Transportation Commission and Portland Bureau of Transportation (STC 2012).[8] STARS-Plan was created to help transportation planners and decision-makers make substantive progress toward the triple bottom line (TBL) through regional and local transportation plans. A key feature of the STARS-Plan tool is the requirement that a "constrained set" of goals and measurable objectives (TBL targets) are established to guide the development of projects, plans, and programs. The STARS-Plan tool uses a "backcasting" approach to identify the most cost-effective way to achieve the TBL targets. The requirement

[8] The STARS-Plan tool is accompanied by a STARS-Project tool (focused on the design of road, transit, bicycle, and pedestrian projects) and STARS Safety, Health, and Equity tool (focused on improving the health outcomes from transportation projects) (STC 2012).

that substantive TBL progress needs to be made is a unique feature of the STARS-Plan tool. Since the GreenLITES project solicitation tool does not establish any specific targets, there is likely to be considerable variation in how it is applied by MPOs. While NYSDOT has explored other ways to integrate sustainability into the planning process, such as "incorporating sustainable goals in long-range plans and in the development of the Department's capital program,"[9] there are currently no formal requirements for such actions.

Collectively, NYSDOT's GreenLITES programs are designed to embed sustainability thinking into the agency's core functional areas. The self-certification programs and solicitation tool provide a broad array of indicators that NYSDOT can use to track performance in terms of planning, project design and construction, and operations and maintenance. While the specific project and activity focus of these tools are valuable, by themselves they do not provide a holistic sense of the regional performance of transportation services. To address this situation, the GreenLITES regions program was developed to align data from GreenLITES initiatives with regional TBL goals (economic, environmental, and social) captured in the sustainability assessment table (Table 10.4). Since the completion of Table 10.4 is voluntary, there is currently no complete dataset available for comparison purposes (NYSDOT 2014). As the regional assessment tool evolves, attention should be paid to whether NYSDOT has the ability to control/manage those factors that contribute to each indicator. For example, while NYSDOT can directly influence the GreenLITES programs, maintenance backlogs, and system connectivity, it may be more difficult to influence indicators measuring generational equity and community cohesion. These latter indicators may fall under the remit of local government, which raises a question of scope and whether NYSDOT should attempt to change these metrics. However, this broad scope does align well with NYSDOT's leadership role in the promotion of a sustainable society.

Since 2008, all new capital projects have been assessed using GreenLITES project design certification program, but the percentage of projects receiving some form of certification has not materialized as expected. In 2013, only 38 % of all the 971 projects reviewed since 2008 had received some level of GreenLITES certification (Fig. 10.4). Just under one-third (28 %) of these projects were *Certified* and only 10 % were classified as having a high (i.e., *Silver*, *Gold*, or *Evergreen*) level of environmental performance. Viewing these data over time reveals that while the total number of projects being granted some level of certification appears relatively stable at around 70 per year, the total number of projects evaluated is increasing annually (Fig. 10.5).

The "Preservation First" principle was described by NYSDOT as the main reason for the increase in the number of projects not receiving some level of certification (NYSDOT 2014). By prioritizing the preservation of the existing

[9] Source: NYSDOT, GreenLITES for Sustainable Planning, https://www.dot.ny.gov/programs/greenlites/GreenLITESplanning (accessed on May 24, 2015).

**Table 10.4** Regional GreenLITES sustainability assessment table

| Assessment items | Current state (Base line—2010) | Desired state (*Place target year here*) | Action plans—Getting from current to desired state | Accomplishments/ status Year (*Insert year #*) | Next steps |
|---|---|---|---|---|---|
| *Economy* | | | | | |
| Access to jobs and labor | | | | | |
| Access to non-work activities (Quality of life: Recreation, schools, etc.) | | | | | |
| System connectivity | | | | | |
| Transportation preservation (Maintenance backlog) | | | | | |
| Competitiveness (All modes) | | | | | |
| Reliability | | | | | |
| Timely | | | | | |
| Predictability | | | | | |
| Attractiveness to business | | | | | |
| Transit passenger miles | | | | | |
| Other | | | | | |
| *Environment* | | | | | |
| Petroleum consumption reduction | | | | | |
| Air quality—$CO_2$ emissions | | | | | |
| Water quality | | | | | |
| Groundwater | | | | | |
| Surface water | | | | | |
| Habitat | | | | | |
| Terrestrial | | | | | |
| Aquatic | | | | | |
| Visual/Aesthetics | | | | | |
| Electrical energy reduction | | | | | |
| Noise reduction | | | | | |
| Other | | | | | |

(continued)

**Table 10.4** (continued)

| Assessment items | Current state (Base line—2010) | Desired state (*Place target year here*) | Action plans—Getting from current to desired state | Accomplishments/status Year (*Insert year #*) | Next steps |
|---|---|---|---|---|---|
| *Social equity* | | | | | |
| Fatality and injury reductions per VMT | | | | | |
| Improved mobility for all including the disadvantaged and disabled | | | | | |
| Improved mobility options and choices | | | | | |
| Generational equity | | | | | |
| Access to affordable transportation | | | | | |
| Incorporate community cohesion, long-range land use plans, and smart growth principles | | | | | |
| Progress environmental justice and ADA | | | | | |
| Other | | | | | |

*Source*: Adapted from NYSDOT (2010b)

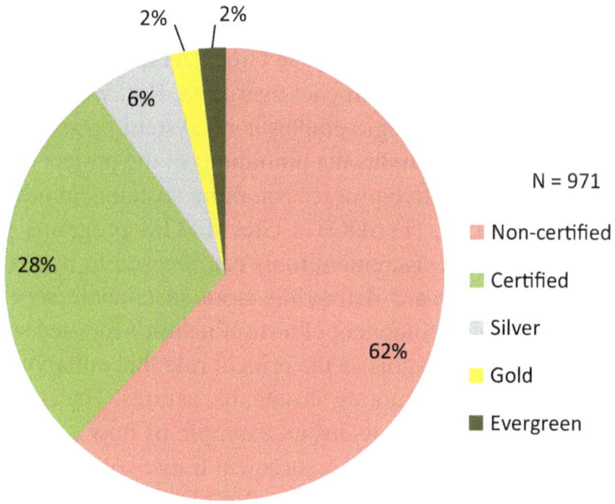


**Fig. 10.4** GreenLITES project certification by percent. *Source*: NYSDOT (2013)

infrastructure over new construction, project designs have focused increasingly on replacing or rehabilitating existing system components. Since these types of projects do not score well in the GreenLITES project design certification program,

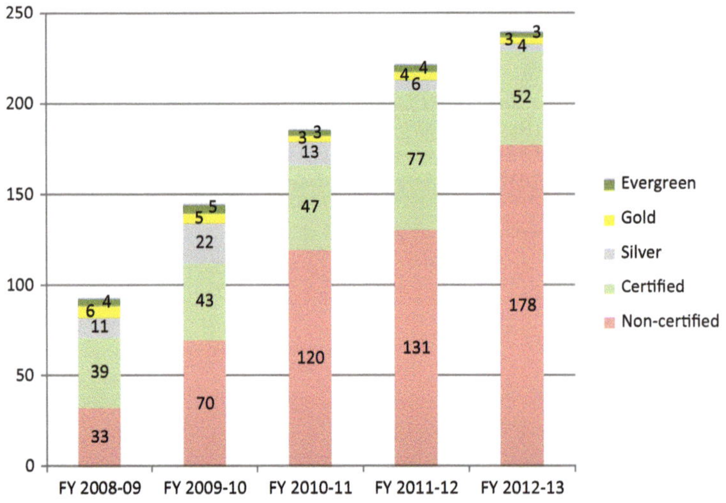


**Fig. 10.5** Number of projects receiving GreenLITES certification by year. *Source*: NYSDOT (2013)

they are less likely to be certified. For example, from 2009 to 2013, around 15 % of the projects relating to pavement resurfacing, preventative maintenance on bridges, and the repair of safety appurtenances were certified (NYSDOT 2013). These projects accounted for two-fifths (38 %) of all the projects evaluated using the GreenLITES project design certification program. This finding highlights an interesting tension between the objectives of a performance measurement program and the broader strategic goal of system preservation, both of which can be considered as key elements of NYSDOT's sustainability strategy. If Fig. 10.5 is taken at face value, it would appear that the GreenLITES project design certification program is having relatively less impact over time. But, when these data are considered in the context of the strategic emphasis on system preservation, a more complex picture emerges that highlights the limitations of the project design tool rather than a failure to make substantive progress on more sustainable project designs.

In summary, NYSDOT's GreenLITES programs provide an example of how performance measurement tools can be used to help a transportation agency track its progress toward delivering more sustainable services. In effect, they are the "operational" component of a sustainability-focused strategy. What is evident from the GreenLITES story is the critical role that collaboration and the communication of data play in promoting sustainable actions. The creation of the GreenLITES suite of programs also provides an example of how organic the process of creating an effective performance measurement framework can be. Rather than architecting the entire framework in advance, NYSDOT grew the framework by piloting new programs over a period of several years. However, as the above discussion reveals, the real-world policy environment in which these performance measurement tools are applied is continually evolving. Thus, what might at first seem like the most appropriate way to measure progress may require refinement as priorities and policies shift. Further, the data provided by the comprehensive set of GreenLITES tools is only one group of information that is considered in NYSDOT's decision-making process. Other types of data relate to issues such as safety, mobility, asset management, and the tactical operation of the transportation system. These data can be linked with groups or teams within NYSDOT who each have priorities and agendas guided by their specific institutional missions. The following section takes a closer look at the institutional setting in which GreenLITES influences capital infrastructure programming.

**Discussion Topics**

- Think of a program like the GreenLITES project design certification program and identify its key data/indicators. How are these data/indicators being used by its creators and by other groups/organizations? Can you identify more than one use of the *same* data/indicator—e.g., to communicate progress that has been made, to support a new policy/decision, etc.?
- In looking at the history of the program you selected, is there evidence that it has changed over time? If so, what factors led to this change? How can the changes you identified be characterized? For example, were they focused on correcting a technical problem or targeted at enhancing the relevance of the program? What does this characterization tell you about the program's evolution?

### 10.2.3 Leveraging Data to Champion Change with NYSDOT

In 2011, NYSDOT updated the way it develops its transportation programs and funds infrastructure investments by shifting its focus to asset management and system preservation (NYSDOT 2012b). One outcome of this shift was the "Forward Four" guiding principles (see Fig. 10.1 shown earlier) that must now be reflected in the fiscally constrained TIPS and the STIP. The objective of this new approach is to maximize the value and performance of the system as a whole based on sound engineering principles.

> NYSDOT's overall asset management strategy is to invest in the infrastructure with the right treatment, at the right time in the life of the investment, and in a location that considers the overall travel system. Recognizing the age, condition and utilization of the transportation infrastructure as a whole, this will require consideration of and investment in all modes of transportation, including facilities owned by entities other than NYSDOT. Customers do not view the transportation system from an ownership perspective, but rather from their ability to get from Point A to Point B. NYSDOT is responsible for the

transportation system—all modes, so it is important that we make investments that best meet the overall needs of this integrated system today, while optimizing transportation for future generations to meet the needs of our customers and to move people and goods in support of the economy (NYSDOT 2012b, p. 5).

The "make it sustainable" Forward Four principle requires NYSDOT to find ways to integrate sustainability objectives with traditional asset management approaches. This challenge falls to the Sustainability team—one of four statewide and regional asset management teams. The other three teams cover the more traditional areas of Safety, Pavements, and Structures (NYSDOT 2014). The co-leads of the four asset management teams sit on a CPT, a senior management team that advises the Capital Program Delivery Committee (CPDC) as they review which projects to include in the capital program. Given its position, the CPT makes many of NYSDOT's strategic decisions based on data and indicators developed and managed by the four asset management teams.

The data developed by each of the four asset management teams can be considered as "boundary objects" (Star and Griesemer 1989) that enable a variety of actors to communicate, understand, and engage with one another to promote a desired change. In this context, the Sustainability team shares and leverages the GreenLITES and other sustainability data and concepts to advance NYSDOT's sustainability agenda. Similarly, the Safety, Pavements, and Structures teams leverage their well-established data and information to promote objectives such as safety and operational performance. Given that sustainability is a relatively new strategic objective, the challenge of integrating the concept into NYSDOT's capital program is described as "daunting," but "incremental" progress is being made largely due to the success of the GreenLITES programs (Nelson and Krekeler 2013). Interestingly, while NYSDOT is clearly a leader in its efforts to integrate sustainability into its activities, the fact that the Safety, Pavement, and Structures teams can operate independently from the Sustainability team implies that sustainability has yet to become an all-encompassing framework for decision-making. In the current decision-making environment with NYSDOT, sustainability could be viewed as one of several critical factors to be considered in the decision process.

The following sections consider how the Sustainability team uses data from the GreenLITES programs in the context of the Learn, Decide, Forecast, and Communicate indicator application areas outlined in Chap. 6.

## 10.3 Indicator Applications

### 10.3.1 Learn

The GreenLITES programs and the data they generate have played an important role in changing the agency's culture by informing Department initiatives and other evolving programs to promote change within NYSDOT. The GreenLITES programs were created to promote a circular knowledge cycle, whereby learning that occurs through the certification programs is shared and incorporated back into the certification tools and policy decisions.

The GreenLITES tools can be considered as *tactical*, in that they improve the design of a project or operation and support the "delivery" of transportation services (see Chap. 5). They do not, by themselves, advance the strategic actions of an agency that fall into what is characterized in this book as the "planning" stage of the transportation process. However, by sharing the GreenLITES data and concepts with the CPT, the Sustainability team is able to leverage these data and sustainability principles to promote *strategic/planning* decisions that could impact the entire transportation system. In leveraging these concepts, the Sustainability team is, in effect, promoting learning across NYSDOT's executive management team. For example, the Sustainability team can present data/information to demonstrate how incorporating sustainability-related approaches into a bridge program or pavement project could not only improve the technical performance of these assets but also address a broader range of considerations.

### 10.3.2 Decide

The traditional approach to transportation decision-making tends to rely on "tangible" data, such as the condition of an asset. Restricting a decision-making process to such data can be limiting (Holden 2013). The process NYSDOT has adopted is not just about the hard data brought into the decision-making process, but about how these data are interpreted and used to advance new practices and ideas. Each of the asset management teams has their own datasets that vary in quality and size. The Sustainability team is still developing its core or baseline data that are continually evolving with the changing GreenLITES programs. In contrast, the Safety, Pavements, and Structures teams have more established datasets that have been measuring the safety and engineering properties of the transportation system for decades. The challenge facing the Sustainability team is to encourage the other asset management teams to think beyond the traditional types of data, to consider more "intangible" factors such as the perceived quality of an intermodal hub or a transportation corridor.

One tool developed to address this challenge was the "Strategic Transportation Enhancement Program (STEP) Project Overview and Context Submission Form" that augmented the 2011 Capital Program Update process. The STEP form helped the Sustainability team discern the more sustainable projects by looking at how a specific project balances economic, social, and environmental considerations within the context of the project's location. The form also documents the corridor type, coordination efforts, and partnerships that have been formed in support of the project, the expected modal/mobility benefits, and other measures that would indicate the success of a project. By requiring the written documentation of these items, the Sustainability team is able to elevate the consideration of more "intangible" qualitative factors in the decision-making process.

The above discussion reveals the complex environment that needs to be navigated when trying to embed the concept of sustainability into NYSDOT's decision-making processes. Rather than being the primary source of data supporting decision-making for sustainability, the GreenLITES data should be viewed as one of several sources that are leveraged by the asset management teams in the decision process.

### 10.3.3 Forecast

The ability to forecast is a critical component of the transportation planning and decision-making process. To this end, NYSDOT is developing a GIS (geographical information system) to support its internal Comprehensive Asset Management and Capital Investment (CAM-CI) decision-making process. The GIS platform, known as the "CAM-CI viewer," provides a common platform where the asset management teams can collectively view their indicators relating to condition, safety, etc. Prior to the development of the GIS platform, each asset management team managed its own structural, safety, or environmental data that were not connected. By combining these data in one platform, the teams can begin to evaluate difficult questions such as how to enhance mobility along a corridor while improving safety and asset condition, reducing environmental impacts, and supporting livable communities. NYSDOT is now working to integrate travel demand models into the platform to assess the impacts of future demand on, for example, pavement condition.

The GIS platform should provide the CPT with the ability to study the multiple impacts of future demand in a way that can be more easily understood across the asset management teams and to the CPDC. Interestingly, the sharing of data in the GIS platform has revealed a number of "cross-team" performance focus areas that integrate the work of the four asset management teams. One of the most important focus areas discovered is mobility (NYSDOT 2014).

In addressing mobility, the agency needs to consider how current and future travel demand is managed in the most environmentally sound, efficient, and safe way. Further, mobility can only be realized if the roads and bridges are well maintained. The more "intangible" factors associated with mobility relate to the creation of places where people want to live due to its aesthetic appearance and the efficient and safe access to basic goods and services. Since each of the asset management teams can directly contribute to the improvement of mobility, the mobility focus area is elevated in the decision-making process. This led to improving the CAM-CI viewer by including the incorporation of a pedestrian and bicycle demand model, a flood vulnerability list, and the addition of employment and population data, which will equip the asset management teams to incorporate economic impacts into infrastructure decision-making.

### 10.3.4 Communicate

The GreenLITES programs have greatly improved the communication of sustainability-related indicators/performance within NYSDOT. They have also enabled the agency to communicate with the public on how it is incorporating sustainability into its various activities. For example, each Earth Day, NYSDOT releases descriptions of those transportation projects that received an "Evergreen" or "Gold" certification.[10] Information is also provided on best practices and innovation in the area of transportation operations. Indicators are frequently used in the qualitative descriptions that summarize project design components or operational activities that are highlighted as promoting environmental sustainability. However, since each project or activity is highly context specific, there is no common set of indicators or descriptions that can be used for comparison. Further, there is no overall performance context within which these summaries can be viewed. Thus, NYSDOT should consider publishing summary data from its GreenLITES programs to enable the public and stakeholders to assess how the agency is performing overall.

## 10.4 Discussion

Two important themes emerged from this case study. First, the application of GreenLITES indicators was found to be influenced by the rich organizational and institutional context in which they are applied. Second, rather than being established in one go, the GreenLITES programs were each piloted and then revised over time to respond to identified problems and to implement best practices. Thus, rather than being a static performance measurement system, the GreenLITES programs can be described as continually evolving. This approach has enabled NYSDOT to tailor the GreenLITES programs to their unique institutional, organizational, and operational environment.

As with any large agency there are multiple factors that shape the decision-making process. In the case of NYSDOT, this situation is captured well by the four asset management teams who are charged with promoting investments that ensure sustainability and public safety, while keeping pavements and structures in a good working order. Each team relies on its own data to advance its concerns in the decision process. Thus, while the Sustainability team can leverage the GreenLITES data and concepts to promote change, these data are considered alongside other, more traditional/hard data in the decision process. The more "intangible" nature of the GreenLITES data means that without a cultural shift toward sustainability concerns across the agency's functional areas, these data may be more easily ignored. Thus, the emphasis that NYSDOT put on changing its internal culture

[10] See GreenLITES Awards, https://www.dot.ny.gov/programs/greenlites/awards (accessed on May 24, 2015).

toward sustainability cannot be overstated and the progress that has been made in this area is largely due to the GreenLITES programs (NYSDOT 2014). Further, whereas we have emphasized the need to adopt a cross-sectoral approach to address sustainability (see Sect. 4.3), the NYSDOT case study reveals the importance of cross-team or cross-division coordination *within* an agency.

The GreenLITES programs were frequently described as needing to be agile and adaptable to a changing environment. NYSDOT's strategic shift toward system preservation due to the declining condition of the transportation system and availability of financial resources provides a good example of how a change in the decision-making environment can undermine the perceived performance of a program. While the GreenLITES project design and certification program is arguably still effective, the emphasis placed on preservation means that fewer projects are being certified. In this situation, either the expectations of the program need to be changed or the scope of the program may need to be revised to enhance its saliency in the decision process.

The NYSDOT case provides some insight into how the frame of reference (or context) can have an important role in how data/indicators are applied. For example, the same data/indicators can be used to promote learning, communicate an idea, or support decision-making. Further, the indicators used in the GreenLITES programs are a mixture of outcome, output, and process indicators. Clearly identifying the types of indicators selected should provide a sense of whether a performance measurement system is leaning toward the achievement of targets (using outcome indicators) or changing how the planning and delivery of transportation services is performed (using process indicators). The challenge now facing NYSDOT's Sustainability team is how to package the GreenLITES data so that it provides valuable/influential indicators in the decision process. The more intangible and evolving nature of these data makes this a difficult task, but the collaborative environment within NYSDOT is likely to support their efforts.

Finally, this case study describes how a large public works agency can take a lead role in promoting a concern for sustainable development across a region. In addition, the actions taken by NYSDOT have inspired other initiatives in the USA and overseas. For example, the GreenLITES programs, along with other initiatives such as the Greenroads rating system, helped lay the foundation for the US Federal Highway Administration's INVEST (Infrastructure Voluntary Evaluation Sustainability Tool) program.[11] In Europe, the GreenLITES programs were studied alongside other sustainability rating systems to inform the development of the Sustainability for National Road Administrations (SUNRA) system.

In summary, the NYSDOT case highlights two important aspects relating to the implementation of a performance measurement system. First, the data generated needs to be fully integrated into the planning and delivery of transportation services. Second, to utilize these data, an agency needs to move toward a

[11] For more information about FHWA's Sustainable Highways Self-Evaluation Tool, see the INVEST website, https://www.sustainablehighways.org/ (accessed on May 24, 2015).

decision-making approach where all decisions are viewed through a sustainability lens. NYSDOT has made substantive progress on the first aspect and is now searching for new ways to elevate sustainability in the decision process. The agency's willingness to experiment with new ideas and approaches demonstrates that moving toward a sustainable transportation system is a continual process of improvement that may take years to perfect.

**Acknowledgment** We would like to thank Debra Nelson and Paul Krekeler from NYSDOT for the time, information, and insight they shared with us during the preparation of this case study.

## References

Bureau of Transportation Statistics (BTS). (2012). *State transportation statistics 2012*. U.S. Department of Transportation (US DOT).

European Council. (2001, April 4–5). *2340th council meeting—Transport/telecommunications—Luxembourg*. Belgium: Council of the European Union.

Holden, M. (2013). Sustainability indicator systems within urban governance: Usability analysis of sustainability indicator systems as boundary objects. *Ecological Indicators, 32*, 89–96.

McVoy, G. R., Nelson, D. A., Krekeler, P., Kolb, E., & Gritsavage, J. S. (2010, November 14–17). Moving towards sustainability: New York State Department of Transportation's GreenLITES story. In *Green streets and highways 2010: An interactive conference on the state of the art and how to achieve sustainable outcomes*, Denver, CO.

McVoy, G. R., Sengenberger, M., & Novak, E. (2000). New paradigm for state department of transportation environmental initiatives: New York State's guidelines and procedures. *Transportation Research Record, 1702*(1), 92–96.

Nelson, D. A., & Krekeler, P. (2013, June 23–27). *Building coalitions and leaders in sustainability: NYSDOT's actions to interweave sustainability and asset management*. Paper read at the 2013 International conference on ecology and transportation (ICOET), Scottsdale, AZ.

Nelson, D. A., Krekeler, P., & Rossi, M. (2011). *Incorporating sustainability into NYSDOT's decisions*. New York: New York State Department of Transportation.

Nelson, D. A., McVoy, G. R., & Greninger, L. (2002). Promoting environmental stewardship in transportation: Maintenance and operations at the New York State Department of Transportation (Paper No. 02-2811). *Transportation Research Record, 1792*, 89–92.

New York State Department of Transportation (NYSDOT). (1999). *Environmental initiative guidelines and procedures. EI 99-026*. New York: New York State Department of Transportation.

New York State Department of Transportation (NYSDOT). (2006). *Strategies for a new age: New York State's transportation master plan for 2030*. New York: New York State Department of Transportation.

New York State Department of Transportation (NYSDOT). (2007). *Multimodal investment needs and goals for the future*. New York: New York State Department of Transportation.

New York State Department of Transportation (NYSDOT). (2008). *Multimodal transportation program submission: 2009–2014*. New York: New York State Department of Transportation.

New York State Department of Transportation (NYSDOT). (2010a). *GreenLITES project design certification program. Version 2.1.0*. New York: New York State Department of Transportation.

New York State Department of Transportation (NYSDOT). (2010b, July). *Region X GreenLITES sustainability assessment table*. New York: New York State Department of Transportation.

New York State Department of Transportation (NYSDOT). (2011). *GreenLITES for sustainable planning*. New York: New York State Department of Transportation.
New York State Department of Transportation (NYSDOT). (2012a). *Draft operations certification program: Recognizing outstanding leadership in transportation and environmental sustainability*. New York: New York State Department of Transportation.
New York State Department of Transportation (NYSDOT). (2012b). *TIP/STIP policy guidance and instructions. October 2013 to September 2017 STIP. October 2013 to September 2018 MPO TIPs*. New York: New York State Department of Transportation.
New York State Department of Transportation (NYSDOT). (2012c). *A transportation profile of New York State*. New York: New York State Department of Transportation.
New York State Department of Transportation (NYSDOT). (2013). *Score card tracking data*. New York: NYSDOT.
New York State Department of Transportation (NYSDOT). (2014). Personal Communication with Ralph P. Hall (2/10/2014).
Star, S. L., & Griesemer, J. R. (1989). Institutional ecology, 'translations' and boundary objects: Amateurs and professionals in Berkeley's Museum of Vertebrate Zoology. *Social Studies of Science, 19*(3), 387–420.
Sustainable Transportation Council (STC). (2012). STARS sustainable transportation analysis & rating system: A framework for integrating sustainability into transportation plans. Pilot plan. Application manual version 1.0. North American Sustainable Transportation Council, Portland.
Venner Consulting and Parsons Brinckerhoff. (2004). Environmental stewardship practices, procedures, and policies for highway construction and maintenance. NCHRP Project 25–25 (04). Washington, DC: Transportation Research Board.

# Japan's "Eco-Model City" Program

# 11

This case focuses on the Eco-Model City (EMC) program created by the Japanese Government in 2008. The EMC program was created to demonstrate how cities could radically transform themselves toward a low-carbon future. The program uses a range of indicators and other evaluation tools to monitor a city's progress and performance, which are discussed throughout the case study.

The case uses the example of the regional capital city of Toyama, which has adopted a range of transportation and land use measures to reach its climate goals as part of its involvement in the EMC program and other government-supported programs. The focus of this case study is on how the progress and results are evaluated within the EMC program, and the type of information it generates with regard to transportation planning targeted at lowering greenhouse gas (GHG) emissions. A special situation concerns the effects of the 2011 Tohoku earthquake that has had a profound impact on Japan's policies in several areas and has also influenced the reporting of progress toward the goal of realizing a low-carbon society. The case mainly illustrates the use of a communicative, results-oriented framework involving central and local government as well as independent experts that use indicators in applications such as ex post "review," "diagnose," and "learning."

## 11.1 Background

The transportation systems of Japan have been developed in accordance with the nation's particular history and special geographic conditions (Enoch and Nakamura 2008; Black and Rimmer 1982). Among the unique characteristics are Japan's mountainous geography and high population density in urban areas along the pacific coast. The high concentration of population has supported an extensive and globally unparalleled development of public transportation systems (Fig. 11.1), which jointly support relatively low levels of car use and $CO_2$ emissions (Taniguchi et al. 2008).



H. Gudmundsson et al., *Sustainable Transportation*, Springer Texts in Business and Economics, DOI 10.1007/978-3-662-46924-8_11

**Fig. 11.1** Transportation in Japan can be intense. Photo: Henrik Gudmundsson

However, strong economic growth and the successful Japanese car industry in the postwar period have also contributed to significant levels of car-oriented development. The growing concerns over negative environmental effects of traffic from the late 1960s onwards have led to the introduction of several policies to limit pollution and control other undesirable impacts (Suzuki et al. 2011).

The predicament of Japan's full dependence on imported oil for automobile transportation became evident in connection with oil price shocks in the 1970s. Later, Japan was among the leading countries behind global initiatives to address climate change, which also influenced urban planning and transportation policy (Suzuki et al. 2011). Today, Japan has one of the most energy efficient transportation systems in the world (Lipsey and Schipper 2013).

Even though Japan was among the first developed countries to report a decline in transportation-induced GHG emissions from around 2001, the country still has a long way to go to ensure significant lasting emission reductions in accordance with general policy goals for climate change and sustainable transportation. Moreover, Japan faces several challenges that significantly constrain the rapid adoption of sustainable transportation solutions and mobility patterns, including factors such as an aging and decreasing population, urban sprawl, decades of restrained economic development, mounting public debt, and devastating natural disasters. The Tohoku earthquake in March 2011, for example, resulted in significant loss of life as well as extensive damage to infrastructure. Moreover, it struck a severe blow to the country's policy to increasingly rely on non-fossil fuel (i.e., nuclear-based) electricity.

Despite such challenges, consecutive Japanese governments have made several commitments to promote a sustainable low-carbon society and transportation system (Suzuki et al. 2011; MoE and MLIT 2009; Government of Japan 2007). Such efforts have been promoted by governments led by the Liberal Democratic Party (LDP) as well as by the Democratic Party of Japan (DPJ), and now again since 2012 the LDP, even if the direction of the programs has shifted with the political changes.

This chapter focuses on one of these initiatives, namely the so-called Environmental Model City (EMC) program launched by the LDP-led government in 2008. The aim of the EMC program is to transform Japan into a low-carbon society by supporting selected model cities that are prepared to take pioneering initiatives with ambitious targets for GHG emission reductions. The approach involves "simultaneous pursuit of a low-carbon society and sustainable development exerting united efforts and potentials of local communities" (Government of Japan 2011, p. 1). An OECD review of Japan's environmental policies in 2010 noted this program to be "a promising initiative for both technical and social innovation," but found that it was too early to assess results in terms of actual GHG reductions (OECD 2010, p. 144). This chapter looks at the indicators that have since been applied to support such an assessment.

The EMC program was launched as an element in the updated national climate policy strategies in 2008 and was also a part of development policies aiming to revitalize urban areas facing economic decline (Kamata 2010). In this way, economic and environmental aspects of sustainability were both combined, even when the overarching objective was to find ways to reduce GHG emissions. Urban planning and decision-making are targeted since cities are large energy consumers, and because city governments are seen to be able to directly influence citizens' behavior in various ways (Murakami 2008).

A number of other "green cities" initiatives have been initiated in Japan, some before and some after the EMC program. In 2010, the DPJ Government, for example, initiated the "Future City" initiative which has a broader scope to promote growth through developing cities where "everyone wants to live and has vitality, by creating environmental, social, and economic values" (Government of Japan 2011, p. 12).

The EMC and Future Cities programs are both maintained by the current LDP government and now integrated so that the EMC program forms an entry point for some cities to also join the more exclusive group of Future Cities, which obtain additional governmental support as international showcases for integrated solutions to problems related to the aging population and the environment (see Fig. 11.2). Cities aspiring to join or learn from the selected cities are invited to the Promotion Council for the Future Cities initiative.

In the first round in 2008, 89 cities applied to the EMC program and 6 were selected. More cities have since been included. As of November 2013, 23 cities have been included in the EMC program, of which 11 are also "Future Cities." Box 11.1 shows key steps in the process up to December 2014.

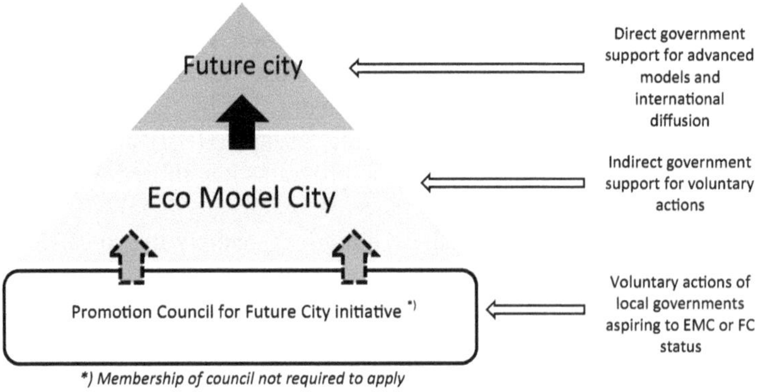


**Fig. 11.2** Relations between programs and types of support. *Source*: Adapted from Edahiro (2014)

**Box 11.1: Timeline of the "EMC" and "Future" City Programs**

| | |
|---|---|
| Jan. 2008 | Japanese government announces EMC program |
| May 2008 | 89 cities submit applications |
| July 2008 | Six cities selected |
| Jan. 2009 | Seven additional cities included |
| April 2009 | Initiatives in each city announced |
| *Sept. 2009* | *(Change in government to DPJ)* |
| May 2010 | First Progress evaluation report of the EMC program cities |
| June 2010 | Japanese government announces "Future City" Initiative as part of Growth Strategy |
| Feb. 2011 | Second Progress evaluation report of the EMC program cities |
| Dec. 2011 | Selection of 11 model cities of the "Future City" Initiative |
| Feb. 2012 | Third Progress evaluation report of the EMC program cities |
| April 2012 | Seven new cities included in the EMC program |
| *Dec. 2012* | *(Change in government to LDP)* |
| Feb. 2013 | Fourth Progress evaluation report of the "Future City" Initiative |
| Nov. 2013 | Three additional cities entering the EMC program |
| Dec. 2014 | International Forum of "Future City" initiative |

The cities originally applying to the EMC program were selected using the following five criteria (Government of Japan 2011):

- Ambitious target-setting for GHG emissions reduction;
- Excellence in acting as a pioneering model;
- Regional adaptability;
- High feasibility for smooth implementation; and
- Continuous development of new initiatives.

The aim was to include cities that would set clear, yet realistic performance goals for their efforts, while also being able to serve as useful role models for other cities. The cities are tiered in four size classes from "major cities" like Yokohama with over three million inhabitants, to "special Tokyo ward," to "regional core cities," to "towns and villages" with a few thousand inhabitants as the smallest category. The EMC and Future City programs are both managed by the Regional Revitalization Bureau, under the Cabinet Secretariat in the Prime Minister's Office of the Japanese Government. This organizational arrangement emphasizes the strategic, cross-sector nature of these programs. A program secretariat has been set up to provide information and logistical support for the applying cities and to help coordinate financial assistance to related projects implemented in each city offered by relevant agencies.

In this chapter, the evaluation within the EMC program will be explored by looking at the example city of *Toyama*, a provincial capital on the Japan Sea coast, with 413,000 inhabitants (Fig. 11.3).

Toyama was among the first wave of cities adopted in the EMC program in 2008. The city falls in the category of "Regional core cities" based on its size. Toyama city's entry to the EMC program focused heavily on transportation and land use initiatives as measures to potentially reduce transportation GHG emissions. Toyama has a recent history as a wealthy city with a relatively high car ownership rate and sprawling suburbs, accompanied by a severe weakening of the traditional urban core. The population in the city center dropped by 50 % over 40 years and retailing has declined by 40 % in just 10 years from 1994 to 2004 (Takami and Hatoyama 2008). From 1990 to 2003, $CO_2$ emissions from transportation in the

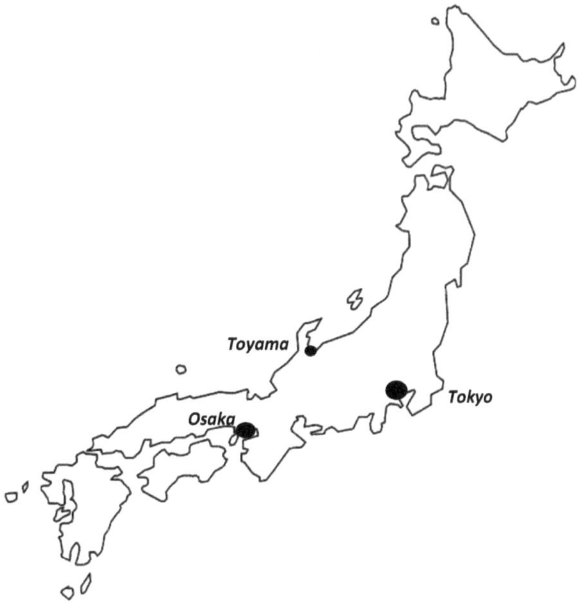


**Fig. 11.3** Location of Toyama City

Toyama region increased by 8 % points more than the national average (Mori 2008). Furthermore, regional cities like Toyama are especially challenged by the rapidly aging population in surrounding areas (Kidokoro 2008). In general, aging in Japan increases the demand for social services such as accessible public transportation, while at the same time undermining the financial basis for them (Sakakibara 2011).

During the 2000s, the city started a process to address these issues through urban and transportation planning policies under strong political leadership of the Mayor Mr. Masashi Mori (Kono 2011). A broad range of investments and other measures have been proposed in Toyama. Toyama's basic urban development policy adopted in 2008 emphasizes development or renovation of rail and other forms of public transport, and the clustering of residential, business, commercial, and cultural facilities in public transit corridors, also known as a form of "compact urban development" (or referred to as "*Kushi-to-Dango*" *[串と団子]* or "Stick and Dumpling") (Mori 2008). This involves concentration in some core nodes in the public transport network, rather than full (one polar) centralization. The strategies of "compact city" and "transit-oriented development" are generally supported by studies showing that high population density and high levels of public transportation infrastructure tend to decrease transportation $CO_2$ emissions (Taniguchi et al. 2008).

Major investments undertaken as part of Toyama city's transportation and urban revitalization schemes include Japan's first light rail scheme opened in 2006 and a large development project to prepare Toyama central station for the Shinkansen high-speed rail service planned to be open in 2015 together with the extension of LRT lines or local rail lines. Other measures adopted in the transport area include an innovative scheme of incentives to encourage citizens to locate in the city center or along public transit corridors. For the EMC program application, Toyama aims to reduce its GHG emissions by 30 % from 2005 to 2030 and by 50 % by 2050. Toyama has also been selected for the "Future City" initiative, but in this chapter the focus is on the EMC program, which has a longer history and more established performance indicators (Fig. 11.4).

**Discussion Topics**

- Are the challenges faced by Japanese cities in terms of aging population and urban sprawl unique or found elsewhere?
- What could be the advantages for the government of focusing on selected model cities and supporting their efforts rather than adopting a uniform strategy with similar goals and measures for all cities in the country?

**Fig. 11.4** Toyama's CENTRAM, opened 2009. Photo: Daisuke Fukuda

## 11.2 Framework for Using Indicators in the EMC Program

The selected cities in the EMC program are monitored and evaluated ex post by the Secretariat of the program that is supported by an independent expert panel. The evaluation assesses performance with regard to each city's implementation of measures, the progress toward their goals, and their performance with regard to the program's broader objective to serve as models for other cities. Each city determines their own performance targets, but the performance measures are defined by the evaluation panel. Conceptually, the focus is on ways to reduce climate change impacts and GHG emissions from all sectors.

The evaluation uses a combination of quantitative indicators and a qualitative assessment of progress with the implementation of measures and initiatives, both of which are based on information provided by the city. It is structured by the following five categories that are based on the EMC program's selection criteria (in unofficial translation):

A: Implementation of efforts;
B: GHG (absorption and emissions);
C: Inducement of community vitality;
D: Stimulating citizens' ideas; and
E: Spread and extension of ideas.

The evaluation panel scores cities from 1 to 5 for each of these five criteria, with 5 as the maximum value. The overall results are illustrated in a "spider diagram" with results here shown for Toyama in the two Fiscal years 2010–2011 and 2012–2013 (Fig. 11.5) by combining data from the two separate reports. As can be seen, Toyama has improved over the period for three of the criteria. No further

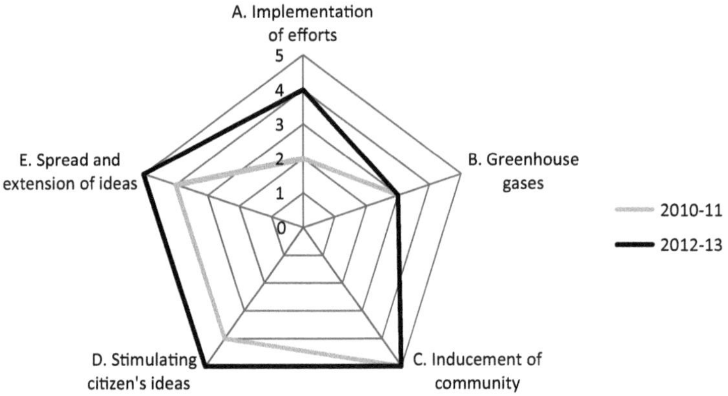


**Fig. 11.5** Overall results for Toyama, fiscal years 2010–2011 and 2012–2013. Based on: Joint Bureau for Regional Revitalization (2010a; 2013b)

aggregation to a unified scale is provided. The details of the evaluation are discussed in Sect. 11.4.

A set of methodologies termed as "CASBEE,"[1] developed by the Chairman of the Evaluation panel Professor Murakami, have been proposed to assist cities in managing the planning and implementation of strategies and to support the evaluation work. Though CASBEE mainly focuses on the evaluation of the built environment, similar to rating tools used in other countries (see Chap. 10 on the New York State Department of Transportation's GreenLITES programs), an extended version called "CASBEE-City" has been developed with a wider scope in mind (Murakami 2008; Murakami et al. 2011). It is especially designed as a comprehensive tool for city-wide assessment. However, the tool is not formally used for the evaluation process. It should be noted that the city itself tracks additional indicators as part of their own planning, which are not reported as part of the EMC program (Personal Comment 2011).

The framework applied for the evaluation of EMC cities does not seek to regulate the cities or to allocate funding for them, although some coordination with funding schemes (e.g., the so-called "Comprehensive Special Zones System") takes place. The framework primarily aims to help cities measure and manage progress and to inform and inspire other cities willing to learn from the experience of the EMC cities. Performance is measured and reported once a year (as of 2013 four years of reports had been issued). There are individual reports for each city as well as an annual summary report for all cities. However, as the context of each city is recognized as unique and as performance measurements are diverse across cities, there are no attempts to directly compare cities or make them "compete" with each other for superior performance.

[1] CASBEE: "Comprehensive Assessment System for Built Environment Efficiency."

## 11.3 Indicators

This section looks closer at the indicators and methods that are used in three of the five evaluation categories—"A: Implementation of efforts"; "B: GHG emissions"; and "E: Spread and extension of ideas." The evaluation of Toyama city's performance will be used as an example, with emphasis on the reports from December 2013 that cover the fiscal year 2011–2012 (Joint Bureau for Regional Revitalization 2013a, b).

### 11.3.1 A: Implementation of Efforts

In this assessment, the indicators used refer to the number of implemented initiatives compared with what was planned and the progress with their implementation. Toyama City has around 70 projects in total, 18 of which are related to transportation. According to the first 3 years of evaluation reports, most of the initiatives are being implemented according to plan or even ahead of plan.

The assessment of implementation progress is rigid, using a normative indicator approach. Each initiative can be implemented ahead of time, on time, delayed, or not at all (four categories). The number of initiatives in each category is counted. Then a weighting is applied, which gives added reward for initiatives that are ahead or more ambitious than planned. Table 11.1 shows this assessment for all of Toyama's projects. The aggregate calculation of "121" expresses the degree to which Toyama is ahead of its own schedule (100 would correspond to delivery as planned on average).

The calculated aggregate value is used as a basis for scoring the city's performance on a 1–5 scale for this criterion using the score bands shown on the right side of Table 11.1. Toyama is placed in category 4, due to many projects that exceed the planned implementation schedule. This evaluation does not assess how effective the measures are for reaching targets, only if they are implemented or not (see the following sections). It is not possible to distinguish how well the transportation projects are performing compared to others, but several of them, including the extension of a commuter rail line, introduction of smart cards, and setting up Park and Ride facilities, are mentioned as making good progress in the written descriptions.

**Table 11.1** Counting and scoring of Toyama's performance for indicator area A

| | Weight | Number (1) | Score (2) | Calculation | Evaluation categories | |
|---|---|---|---|---|---|---|
| Ahead | 2 | 28 | 56 | $(2)/(1) \times 100$ | 5 | 130– |
| On time | 1 | 30 | 30 | | **4** | 110–129 |
| Delayed | 0 | 13 | 0 | | 3 | 190–109 |
| Not active | −1 | 0 | 0 | | 2 | 70–89 |
| Total | | 71 | 86 | 121 | 1 | –69 |

*Source*: Joint Bureau for Regional Revitalization (2013b)

### 11.3.2 B: Greenhouse Gas Emissions

The evaluation of GHG emissions includes all $CO_2$ emissions from the city as well as the net uptake of $CO_2$ by vegetation. Emissions are calculated for five sectors—industry, business, households, transportation, and energy conversion. The results are documented in a detailed companion report based on data submitted by the city and shown here in summary in Figs. 11.6 and 11.7. The figures show 4 years of data plus the base year, with the evaluation focused on the 2011 data. The target for the city is not directly inserted, although the desired direction of change (reduction) is obvious.

The $CO_2$ emissions are calculated using two different methodologies, which is why two figures are needed to present the data. After the 2011 Tohoku earthquake, the shutdown of nuclear reactors in Japan necessitated a significant shift in the country's energy supply toward imported fossil fuels with higher $CO_2$ factors. Figure 11.6 accounts for the actual emissions, whereas Fig. 11.7 reflects what the emissions would have been if the energy mix and emission factors had been unchanged. The use of two methodologies allows the results of local initiatives that are not related to the overall energy shift to be isolated. The significant increase in actual $CO_2$ emissions from 2005 to 2011 (Fig. 11.6) is not seen once the adjustments have been made (Fig. 11.7). This result suggests—assuming all else is equal—that the initiatives in Toyama have not yet been effective in reducing actual $CO_2$ emissions. Thus, despite the rapid implementation of measures (as shown by the indicators for category A), there has yet to be a significant reduction in $CO_2$ emissions.

Figures 11.6 and 11.7 both show that transportation emissions have stabilized. Transportation emissions are not significantly influenced by the shift in the energy

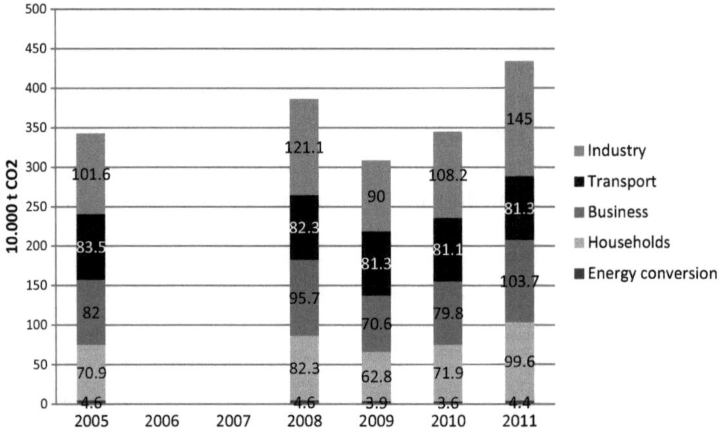


**Fig. 11.6** Actual $CO_2$ emissions from activities in Toyama City (2005–2011). Based on: Joint Bureau for Regional Revitalization (2013a).

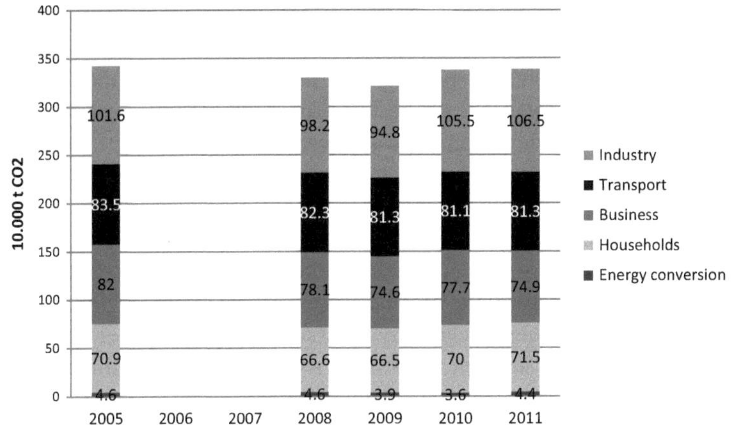


**Fig. 11.7** Adjusted $CO_2$ emissions from activities in Toyama City (2005–2011)—Assumes an unchanged energy mix after the 2011 nuclear accident. Based on Joint Bureau for Regional Revitalization (2013a)

production in other sectors. A small decline from the base year, which was already observed by the evaluation panel in its first report (Joint Bureau for Regional Revitalization 2010b), has been maintained but not further enhanced.

Toyama received a score of "3" on a 1–5 scale for category B, as was shown in Fig. 11.5. When compared to indicators used in category A, the evaluation panel has not defined a clear categorization procedure for the GHG indicator. While this indicator is strictly quantitative, which makes it easy to interpret the direction of change, it has obviously been more difficult for the panel to define criteria for categorizing the levels of $CO_2$ change into bands—i.e., what would count as a significant change? That no progress toward the target is assessed with the middle grade of "3" and not a lower score could suggest that changes are not expected to follow a linear path, and it is hoped that improvements will start to snowball when implemented measures begin to take effect.

### 11.3.3 E: Spread and Extension of Ideas

A key purpose of the EMC program is to disseminate experience and inspiration to other Japanese cities. Therefore, efforts are made to also evaluate cities based on their ability to serve as role models.

The indicators used to evaluate this category are relatively few, diverse across the cities, and quite scattered. For Toyama, indicators are divided into usage statistics for novel transport services such as light rail and bike sharing, and the number of visitor groups coming to learn about Toyama's plans and initiatives. For the former, a three-and-a-half fold increase in the patronage of the light rail system by elderly above 70 years of age is, for example, noted. For the latter, almost

250 groups came to Toyama in 2011. Many of the groups represented other cities who intend to implement similar measures, while there were also international groups who were interested in studying best practices in the region (these are not distinguished in the report). The interest to learn from the city department responsible for "compact city" development was the strongest (200 groups; 1697 people in total), whereas the department directly responsible for the EMC program had fewer visitors (13 groups; 61 people in total).

The evaluation panel has not presented a systematic methodology for how to translate the data for patronage and visitors into score bands 1–5. In this category, Toyama is nevertheless given the highest score with the grade "5." As the measures are not similar across cities, it can only be assumed that the panel may have been satisfied with Toyama's attempt to provide some form of quantitative measures, besides the level of accomplishments demonstrated.

The two other categories—C: Stimulating citizens' ideas, and D: Inducement of community vitality—are not studied here since they are evaluated using a less rigorous approach to that used for category E.

In summary, Toyama received high scores in the evaluation in most categories in the 2013 report, and in three categories the city improved its results from the previous evaluation. Notably, the score is lowest in the category of $CO_2$ emissions, which is arguably one of the most important, considering the focus of the program. After all, to what avail is strict implementation and a large number of visiting groups if these are not connected to actual emission reductions? However, it could also be too early to expect significant and substantive changes from the implemented programs. We return to this question again in the following section.

**Discussion Topics**

- What are the key differences in the way that each indicator area is evaluated in the EMC program?
- Would it be possible to use a more systematic way of scoring indicator areas such as "spread and extension of ideas"? As a context to this question, see the "SMART" criteria discussed in Chap. 6.
- How could the five different evaluation categories of the EMC program be located in an overarching framework? For example, see the "linkages" type of indicator systems introduced in Sect. 7.4.2.

## 11.4 Indicator Applications

### 11.4.1 Review

The direct role of the evaluation in the EMC program is to show how well each city is performing in terms of its planned activities. The most obvious role the indicators play is therefore as a "review" application, asking "how are we doing?" (Chap. 6). Review in this context is defined as an assessment of progress with regard to established goals for $CO_2$ reduction, the implementation of planned measures to achieve them, and the three other categories evaluated. Spider diagrams showing scores provide a rather instant overview of the relative performance across the five areas, albeit only for 1 year at a time (Fig. 11.5 that contains 2 years of data was constructed by the authors of this book), and it does not answer in detail how well the city is doing. This can be discerned better by looking at the evaluation for each area.

The most rigorous review using a performance standard occurs in category A, where the status of the "implementation" of planned measures serves as the benchmark for reviewing progress. The status of each of the planned measures is assessed and an overall score (on a 1–5 scale) is calculated for the total performance across all measures. This provides both detailed indicators of progress and an overview of a city's performance. The review does not provide a comparison of whether the transportation measures are more or less successfully implemented when compared to other measures, but in principle it would be possible to present these data. In area B, the $CO_2$ emissions could be compared with the targets set by cities themselves. However, these are relatively distant in time and not shown in the figures. The limited progress toward the target could also be shown using a normative indicator if incremental steps toward the targets had been defined and indicated as milestones in the graphs. The approach to evaluating the remaining three areas is less rigorous. Data on measures such as visiting groups are reported, but it is not clear how to interpret, let alone score, such results in a rigorous way. Nevertheless, the evaluation panel is able to use the information and exercise its professional judgment in the scoring. It is noteworthy that no city scores below 3 on the 1–5 scale for any of the categories A–E, while several score more than one 5. Toyama tops the group with three scores of 5, together with the City of Kitakyushu, a city known as a long-standing environmental pioneer city in Japan. The data indicate that all cities are doing fairly well across the board, though a lack of progress is observed with regard to the $CO_2$ emission reductions.

### 11.4.2 Account

The evaluation by the EMC panel has some resemblance to an accounting exercise due to its strong focus on whether city authorities have delivered what they promised in terms of policy measures and initiatives. It could be possible to hold the city representatives to account, especially since they alone are responsible for the measures that are included in the city program and also for the progress of implementation or lack of it. The evaluations are made publicly available through the program website and the program is otherwise broadly communicated. However, the EMC program does not seek to "blame" those involved if targets are not met or performance is less than optimal. Nor does it attempt to directly praise particular individuals or departments for their role in over-fulfilling, for example, the implementation plans in Toyama. The evaluation process as conducted provides more of a supportive than a punitive form of accounting. It is also important in this regard that no government subsidies or investments are directly distributed to cities via the EMC program, even if the involved cities may be receiving central government funding through other programs. The evaluation panel is therefore not charged with the review of how such measures are funded or how resources are spent. It does not have the power to use fiscal accounting indicators, for example, to identify inefficient measures, point to budget risks, or to advise on alternative financing schemes. These types of actions fall outside the scope of the EMC program, which focuses more on measuring progress and identifying good practices.

### 11.4.3 Diagnose

The EMC evaluation requires cities to report quantitative data for GHG emissions. These data could potentially be used together with other information to explain or "diagnose" the reason for the observed trends. First of all, the data are calculated by sector and then added to totals as seen in Figs. 11.6 and 11.7. The former graph suggests that all sectors (except transportation) have seen a significant increase in $CO_2$ emission over the measured 5-year period, whereas the latter suggests that emissions have been relatively stable in all sectors, if the effects of the "forced" nationwide shift away from nuclear power are neutralized. This simple diagnostic procedure reveals not only the massive indirect environmental consequences of the natural disaster but also the extremely limited role of electricity use in the transportation sector in a car-dependent regional city like Toyama. Moreover, it helps to zoom in on the part of the problem that the local level could address. Without this dual approach to calculating $CO_2$ emissions, the transportation sector would appear as the least problematic one, relatively speaking, whereas it is now more clearly seen that all sectors need more attention at the local level.

However, it would be desirable to seek a deeper diagnosis in order to understand the intriguing apparent discrepancy between the successful implementation of planned measures and the disappointing lack of progress toward the emissions target. Does this gap in fact undermine the underlying philosophy of the "compact city" with "transit-oriented development" as an effective strategy to reduce automobile dependence and GHG emissions from transportation; does it rather suggest that the measures taken to promote this strategy in Toyama are still too weak; or does it simply demonstrate that it takes time for such measures to take effect on a scale that is measurable at the aggregate city level? A statistical analysis by Taniguchi et al. (2008, p. 422) of urban density and transportation and emission trends in Japanese cities concludes that a "reduction of $CO_2$ attributable to transportation based on travel behavior of residents will require drastic changes." Unfortunately, the analysis in the evaluation report does not explore this level of diagnosis, although data for changes in travel patterns of the citizens and their response to policy measures encouraging location in dense areas are reported in some of the annual reports. More generally, it can be noted that the five performance areas A–E prescribed for cities do not appear to be selected with a clear causal or linkage model (see Chap. 7) in mind, although this set of areas may have other advantages, such as communication or involvement of citizens.

### 11.4.4 Learn

One of the main objectives of the EMC evaluation is to promote the exchange of knowledge about best practices. Indicators are central to the communication of results. By sharing their initiatives, indicators, and experience with domestic and to some extent international groups, the EMC program participants are promoting learning among themselves and others. Toyama is often referred to in guidance and promotional material to other cities in Japan, not least with regard to its pioneering introduction of LRT and compact city development (see MLIT 2011). The strong interest in conducting professional visits to the city and its "compact city" office also demonstrates willingness among cities to learn from experiences elsewhere and it bears witness to the wide acknowledgment of Toyama City's landmark efforts. However, it remains to be seen in what way Toyama's so far limited success in demonstrating clear GHG emission reductions would influence the interest or actions of other cities, and what to "learn" from this type of mixed observations. That a city under strong political leadership sets forth to shift a negative spiral of development into a positive one, while deploying highly visible, modern transportation systems in the city center is guaranteed to stir some attention. However, what if a key lesson is that such efforts are not sufficient to reach the ambitious GHG reduction goals at the city or national level? What if other impacts benefiting the local environment level such as improved accessibility or urban design were more salient than GHG reductions alone? These "what-if" questions are purely speculative and are not intended to suggest particular conclusions from the EMC program or Toyama City. The point is that the learning to be carried forward to other cities and into the more ambitious "Future Cities" program may yet be difficult to discern. The most interesting findings could emerge from the confrontation of strict emission reporting with further diagnostic analysis of the implemented policy measures. Literature, however, suggests that in a complex area like sustainability in transportation, it is a difficult challenge to combine the rigor of accounting for "frozen ambitions" (van der Knaap 2006) in goals and targets with a need to be responsive and explorative toward emerging issues and local needs (Hezri 2005; Lehtonen 2006). The important question is if it is possible to adapt and develop the indicators and the EMC evaluation setup to reflect an evolving need for knowledge.

## 11.5 Discussion

The Japanese case provides a good example of how indicators can be used to discover, review, diagnose, and communicate efforts to promote low-carbon compact urban development, based on recurring evaluation reports. The evaluation report framework provides a variety of quantitative and qualitative indicators embedded in different methods to address a broad range of issues including urban planning and development for low-carbon transportation. The issues covered span policy implementation, outcomes, and learning. The results are presented in a condensed format where complex information is summarized and the performance of each city is presented in a five-dimensional spider web diagram with five performance levels. The results are presented city by city in the same report (based on more detailed sub-reports), but the cities are not directly compared with regard to their actions or performance, and the results are not further aggregated or ranked beyond the five dimensions. The most recent evaluation report at the time of writing does accumulate some information from previous years, although a change in the aggregate performance for each city can only be seen by looking across the series of annual reports (as partly shown above in Fig. 11.5). The EMC evaluation approach is characterized by the voluntary participation of cities in the program and the central government's soft supporting role. Investments or subsidies are not committed to the program, which would likely invite more rigorous accounting-type applications of indicators.

The case results from the city of Toyama demonstrate that transportation emissions in the city have not been significantly reduced over 5 years, despite the gradually more successful implementation of the city's planned measures, partly ahead of schedule, and despite Toyama City being already widely acclaimed and recognized as a frontrunner in the field of transit-oriented, compact city development. According to international literature, limited results accruing from sustainable transportation plans can often be traced to barriers and failures that occur during program implementation. The European Conference of Ministers of Transport, for example, note that, "The challenge of developing transport policies for sustainable development is to orient the sector toward a compromise that maximizes the economic and social benefits of transport and minimizes associated environmental, social and economic costs. Many of the measures required to achieve this balance are not new, the main difficulty is effective implementation" (ECMT 2000, p. 17; see also comprehensive studies by Tengström 1999; Docherty and Shaw 2004). However, this appears not to be the case for Toyama, where implementation progresses ahead of schedule.

The gap between expected and actual results invites deeper scrutiny into the measures that have been adopted to uncover whether they are ineffective because of the way the emissions are calculated, or because the measures themselves only address a relatively marginal proportion of the demand for automobile travel. The evaluation could therefore be used to ask interesting questions about possible learnings with regard to the lead time, effectiveness, and ultimate sufficiency of the adopted measures for possible exploration via further diagnostic analysis of available information. However, some of the performance areas (A, B) have more rigorous indicators and interpretation frameworks than others (C, D, E) and the full setup is not obviously designed as a causal framework inviting systematic analysis. Whether such a framework could be constructed and populated with real data is to some extent an open question. The so-called CASBEE-city system (Murakami et al. 2011) seem to have been developed with such a purpose, but has not so far been implemented in practice.

An interesting feature of the EMC program evaluation is the use of an independent panel to ensure that results are monitored and reported in a transparent and consistent way over time. This requires cities to collect and report information in a standardized format, but it does not require the central government to react in a certain way if cities are found to perform poorly. The panel has "survived" two shifts in government and thereby proved its independence, while the approach and level of detail in the evaluation have changed somewhat over time. The panel is involved in defining the methodology for evaluation as well as undertaking the review itself. The cities are not directly represented in the panel consisting of leading professors, and it is not clear to what extent the cities have a say in the evaluation design.

A clear limitation of the EMC program is that it evaluates sustainable transportation only by the constrained focus on $CO_2$ emissions. It is likely that several of the measures adopted by the city of Toyama and other cities in the program could have broader social and economic impacts that are not currently measured. It is also important to note the EMC program's voluntary nature. It does not replace or supersede existing urban land use and transportation planning strategy, but rather builds on it. The city of Toyama collects and uses a number of indicators locally, such as public transport patronage, modal split, noise levels, and accidents (Personal Communication 2011). Still the EMC program does not support the measurement of the full set of sustainability dimensions and the approach adopted by the city of Toyama falls within the "weak" model of sustainability. However, the multi-sectoral approach to measuring $CO_2$ emissions does present an opportunity for a more holistic consideration of where actions should be taken to reduce overall emissions. If it is true that "drastic changes" in travel behavior would be needed to reduce transportation-related $CO_2$ emissions, this may shift the burden for reducing emissions to other sectors. Thus, the $CO_2$ evaluation framework provided by the EMC program presents an opportunity for cities to advance a cross-sectoral dialogue about how to realize improved performance while minimizing the impact on citizens and customers.

Finally, it is noteworthy that cities in the EMC program also form the pool of candidates for the more recent, comprehensive, and ambitious "Future City" program, which covers a wider set of sustainability dimensions and provides more extensive central government support. It remains to be studied what types of indicators and applications will be put forward in that program and how these relate to the strong or weak forms of sustainability.

**Acknowledgment** We would like to thank Associate Professor Daisuke Fukuda, Tokyo Institute of Technology, who provided essential input for and coauthored this chapter.

## References

Black, J. A., & Rimmer, P. J. (1982). Japanese highway planning: A western interpretation. *Transportation, 11*(1), 29–49.

Docherty, I., & Shaw, J. (2004). *A new deal for transport? The UK's struggle with the sustainable transport agenda*. Oxford: Blackwell.

ECMT. (2000, May). Sustainable transport policies. In *European conference of Ministers of Transport, Paris*.

Edahiro, J. (2014). Towards cities where anybody would want to live: The "FutureCity" initiative. JFS Newsletter No.139 (March 2014). Accessed March 2014, from http://www.japanfs.org/en/news/archives/news_id034802.html.

Enoch, M., & Nakamura, H. (2008). Transport policy and organization in Japan. *Transport Reviews, 28*(2), 159–180.

Government of Japan. (2007, June 1). Becoming a leading environmental nation in the 21st century: Japan's strategy for a sustainable society (Tentative Translation). Cabinet Meeting Decision, Tokyo.

Government of Japan. (2011, July). The eco-model city project and future directions. Regional Revitalization Bureau, Cabinet Secretariat, Government of Japan, Tokyo.

Hezri, A. A. (2005). Utilisation of sustainability indicators and impact through policy learning in the Malaysian policy processes. *Journal of Environmental Assessment Policy and Management, 7*(4), 575–595.

Joint Bureau for Regional Revitalization. (2010a, May 26). *Progress of main projects in 13 eco-model cities and evaluation by the Joint Bureau for Regional Revitalization in Cabinet Secretariat*. Attachment 2. 14 p 環境モデル都市 (Environment model city) [in Japanese].

Joint Bureau for Regional Revitalization. (2010b). 平成21年度 環境モデル都市フォローアップ, 施策の進捗状況等, 総括, 票富山県富山市 [Summary report, Toyama, in Japanese].

Joint Bureau for Regional Revitalization. (2013a). 富山市の平成23 年度温室効果ガス排出量等について [Greenhouse gas emissions for fiscal year Heisei 23—2011-12, Toyama, in Japanese].

Joint Bureau for Regional Revitalization. (2013b). 環境モデル都市における平成23 年度の取組評価 [Evaluation of the environmental model city efforts, fiscal year Heisei 23—2011, in Japanese].

Kamata, S. (2010, March). Global warming countermeasures and city planning. In M. Yokohari & A. Murakami (Eds.), *Special issue: Sustainable cities and regions*. City Planning Institute of Japan, Newsletter No. 33, Tokyo.

Kidokoro, T. (2008). Urban regeneration and the shift of planning approaches. Chapter 8. In N. Harata & J. Jessen (Eds.), *Sustainable city regions: Space, place and governance*. Tokyo: Springer.

Kono, N. (2011). Urban planning and land use management. Chapter 1. In T. Maede, D. G. J. Premakumara, & I. Kikusawa (Eds.), *Local level innovations towards an environmentally sustainable city. Case studies from Japanese cities*. Kamiyaguchi: Institute for Global Environmental Studies (IGES).

Lehtonen, M. (2006). Deliberative democracy, participation, and OECD peer reviews of environmental policies. *American Journal of Evaluation, 27*(2), 185–200.

Lipsey, P. Y., & Schipper, L. (2013). Energy efficiency in the Japanese transport sector. *Energy Policy, 56*, 248–258.

MLIT. (2011, January). Low carbon city development guidance [outline]. Ministry of Land, Infrastructure, Transport and Tourism. Accessed May 2, 2015, from http://www.mlit.go.jp/crd/city_plan/teitanso.html [in English].

MOE and MLIT. (2009, February 24). Efforts for environmentally sustainable transport in Japan. In *Fourth Regional Environmentally Sustainable Transport (EST) Forum in Asia, Seoul*. Ministry of the Environment, Japan; Ministry of Land, Infrastructure, Transport and Tourism, Japan.

Mori, M. (2008). The CO2 reduction plan provided by the Toyama compact city strategy. Presentation slides.

Murakami, S. (2008, December 14). *Promoting eco-model cities to create a low-carbon society*. Presentation at the International seminar on promoting eco-model cities for the low carbon society, Kitakyushu.

Murakami, S., Kawakubo, S., Asam, Y., Ikaga, T., Yamaguchi, N., & Kaburagi, S. (2011). Development of a comprehensive city assessment tool: CASBEE-city. *Building Research & Information, 39*(3), 195–210.

OECD. (2010). *Environmental performance review. Japan 2010*. Paris: Organization for Economic Co-operation and Development.

Personal Comment. (2011, July). *Results of email interview with Toyama City representatives.*

Sakakibara, Y. (2011). Social change and future transport policy in the Japanese context. *IATSS Research, 35*, 56–61.

Suzuki, M., Hayashi, Y., & Kato, H. (2011). Japanese efforts to solve environmental problems with a focus on the transport sector. Chapter 6. In W. Rothengatter et al. (Eds.), *Transport moving to climate intelligence* (Transportation research, economics and policy). New York: Springer. Accessed May 2, 2015, from doi:10.1007/978-1-4419-7643-7_6.

Takami, K., & Hatoyama, K. (2008). Sustainable regeneration of a car-dependent city: The case of Toyama toward a compact city. Chapter 10. In CSUR (Ed.), *Sustainable urban transport in an Asian context* (pp. 184–200). Japan: CSUR Center for Sustainable Urban Regeneration, Tokyo University, Tokyo: Springer Japan.

Taniguchi, M., Matsunaka, R., & Nakamichi, K. (2008). A time-series analysis of the relationship between urban layout and automobile reliance: Have cities shifted to integration of land use and transport? *Urban Transport XIV. WIT Transactions on the Built Environment, 101*, 415–422.

Tengström, E. (1999). *Towards environmental sustainability? A comparative study of Danish, Dutch and Swedish transport policies in a European context*. Aldershot: Ashgate. 241 p.

Van der Knaap, P. (2006). Responsive evaluation and performance management. Overcoming the downsides of policy objectives and performance indicators. *Evaluation, 12*(3), 278–293.

# Conclusions 12

## 12.1 Introduction

In this book, we set out to address the question of "How do we make sustainability count in transportation?" This chapter offers our reflections on Parts I and II of the book, from the process of bringing together state-of-the-art knowledge in the field to putting this information to the test through the critical examination of four real-world case studies.

The movement of our transportation systems toward supporting more sustainable outcomes is something which can only happen over time, through the cumulative decisions of a multitude of public and private entities. This book is unapologetic in not providing the reader with a normative description of what an ideal sustainable transportation system should look like or what the best approach is to pursue it. Instead, it emphasizes that a vision has to be developed based on the definitions and principles of sustainability keeping in mind the context, operating environment, and other constraints. Through the case studies, we gain insight into the practical implementation of this concept. The tools and processes described in this book can then be applied in a flexible and creative way. We find that *the process is just as important as the product* (i.e., thinking through how the concepts of sustainability can be applied to support transportation planning and decision-making in a particular context often generates insights and knowledge that cannot be obtained through a prescriptive, "one-size-fits-all" approach).

This book emphasizes the key role that indicators play in linking the transportation system with sustainability concepts (i.e., making transportation more sustainable and supporting the broader goal of sustainable development). Indicators that are robust and properly framed can be used to bring this vision to life. The book provides critical knowledge to enable the development of an indicator set to best serve this purpose. Keeping with the underlying philosophy of this book, the process does not involve the selection of indicators from a universal recommended set, but rather outlines a process for how the various frameworks can be used and indicators developed and applied to best address the specific context. The various



H. Gudmundsson et al., *Sustainable Transportation*, Springer Texts in Business and Economics, DOI 10.1007/978-3-662-46924-8_12

case studies show how this process has played out in real life and what lessons can be learned from it.

Part I of this book laid the foundation on which the remainder of the text is built. It discusses the origins of the concept of sustainability, providing a clear description of sustainable development, and explores the contribution of the transportation sector to realizing this development objective. It also provides a study of institutional structures and governance systems that serve as the landscape within which sustainability is operationalized. This section concludes with the nature of the information needed to address sustainability, and how it is organized and applied through the use of indicators and organizing frameworks. Part II of this book explores four case studies that represent the use of sustainability indicators in a range of settings, from multinational policy (EU Transport White Paper), to a national-level initiative for rail projects (High Speed Rail in England), statewide planning initiatives (New York State DOT's GreenLITES rating system), and city-level initiatives (Japan's Eco-City Model Program). In this concluding chapter, we recount some of the main takeaway points and conclusions drawn from this work.

## 12.2 Transportation and Sustainability

While sustainability is a broad and complex concept, current thinking on sustainability continues to be rooted in the Brundtland definition with sustainability principles being centered around the needs of current and future generations, focused on three dimensions—environmental, social, and economic. However, the issue of *strong* versus *weak* conceptualizations of sustainability (i.e., whether trade-offs on the environmental and social dimensions are acceptable for economic returns) is an important consideration and often the center of philosophical debates. It should be recognized that there is a spectrum of definitions on how sustainability can be viewed and applied. There are two important reasons not to get too drawn into this debate. First, we set out in Chap. 4 that transportation is only one part of the picture and not capable of being defined as sustainable in its own right. Second, transportation decisions are rarely, if ever, taken by an agency with a holistic remit to act on all relevant matters. However, sustainability should not be viewed as an infinitely flexible notion since its basic dimensions and principles need to be adhered to. Decisions can be more or less aligned with these dimensions and principles.

The concept of sustainability is therefore overarching, and should be viewed as a living idea—there are only trajectories toward sustainability, and no true end state of sustainability. In other words, sustainability is an idealized state, and sustainable development can be viewed as a means to this end. While a holistic approach to sustainability (considering all three dimensions, taking into account inter- and intra-generational equity) is needed, it can be daunting to try and consider the full scope of sustainability at once. Thus, we recommend taking a first step and making a good-faith effort, realizing that it takes time, leadership, and resources to make a difference. It should also be seen as a long-term process and all levels and sectors should be involved in its pursuit. Finally, the dynamic nature of sustainability should be considered because over time issues will change and evolve.

With regard to sustainable transportation, we see that it can be framed in two ways. It can be considered as a subject in its own right where transportation is the center of attention and an effort is made to make it sustainable, *or* it can be considered based on the transportation system's contribution to sustainable development. We show that transportation cannot be made sustainable in isolation—in this regard, sustainable transportation is to some extent a limiting notion as it can imply that coordination with other sectors is not essential, when the opposite is true. The example of Japan's Eco-City Model demonstrates how important changes in land use can be for transportation development, while the case study on the European Union's transport policy reflects the significance of overarching economic strategies that play into the bigger picture. We have argued there is a need for transportation entities to look beyond their institutional mandates and boundaries in search of collaborative solutions that target the root cause of a problem. Both approaches (transportation-centered and holistic) need to be followed in pursuing transportation sustainability and the optimum approach can be found by considering how best to take advantage of their respective strengths.

While a range of sustainability principles and definitions of sustainable transportation exist, we have found the following list of principles and definition to provide a good starting point for discussions on the subject. With regard to the principles, sustainability can be described as meeting human needs for the present and future, while:

- Preserving environmental and ecological systems;
- Improving quality of life;
- Promoting economic development that includes the creation of meaningful and well-paid jobs; and
- Ensuring equity between and among population groups and over generations (see Table 7.5 and the related set of goals shown in the table).

A definition of sustainable transportation that leans toward the more stringent notion of strong sustainability is provided by the European Council. A sustainable transportation system is defined as one that:

> - allows the basic access and development needs of individuals, companies, and societies to be met safely and in a manner consistent with human and ecosystem health, and promotes equity within and between successive generations;
> - is affordable, operates fairly and efficiently, offers choice of transport mode, and supports a competitive economy, as well as balanced regional development;
> - [in coordination with other sectors] limits emissions and waste within the planet's ability to absorb them, uses renewable resources at or below their rates of generation, and uses nonrenewable resources at or below the rates of development of renewable substitutes while minimizing the impact on the use of land and the generation of noise (European Council 2001, pp. 15–16) (see Table 4.1).

A range of components that could be included in a comprehensive definition of sustainable transportation is also provided in Table 4.3.

## 12.3 Framing and Measurement

Indicators are essential in operationalizing the notion of sustainability in the transportation context and beyond. Indicators can be powerful as they focus attention on key criteria and directly impact decision-making. The saying goes that "what gets measured gets done"; however, this is not a given. The literature is awash with studies that provide idealized lists of indicators for sustainability, and examples of sustainable transportation indicators abound without these necessarily having an impact on practice. This book focuses on understanding what an indicator is, how to use indicators meaningfully, the types of information indicators can communicate, and how they can be organized to support different types of decisions. We conclude that there is no such thing as a "sustainability indicator" in itself, or a set of universal sustainability indicators—only indicators that can *tell us about and help move us toward* sustainability. Additionally, when it comes to sustainability there is a need to move beyond the transportation-centered approach and include indicators either outside of transportation or which make clear the contribution of transportation to the system as a whole.

Even when agencies or decision-makers share the same overarching commitment to sustainability, the ways in which they choose to measure it will be somewhat different, depending on the requirements and opportunities of the situation. The combined questions of "why," "what," and "how" to measure should lead the development of frameworks that are tailored to the specific context in terms of geography, politics, socioeconomics, and beyond. If sustainability measurement is not framed appropriately there is a risk that it could be ignored or misapplied. Moreover, variations in measurement and management approaches should reflect a systematic consideration of indicator applications with regard to the communication and decision-making functions they are intended to support rather than an unstructured/uncoordinated collection of data.

## 12.4 Decision-Making

The decision-making environment in transportation is multilevel, multi-sector, and multi-actor, and indicators can be viewed as a common language for exchange. Because indicators are so powerful and also costly to populate with reliable information, it is imperative to know what they will be used for and to select indicators that are robust. The *application* of the indicators (i.e., what they are to be used for) is very important and should be clearly understood before embarking upon the selection or development of indicators. In this book, the possible applications of indicators are organized into the following categories: Describe, Forecast, Review, Diagnose, Decide, Account, Learn, and Communicate.

Sustainability assessment and decision-making should bring to the fore the details of the trade-offs between different elements. Information is important in framing how decisions are made—however, we need to keep in mind that information is largely used to support decision-making, and not directly for decision-making. Therefore, attention also needs to be paid to the politics and the practice of decision-making and the role of information in supporting, challenging, and developing these practices. In one sense, the rise of sustainability reporting and analysis identified in our case studies is a source of hope for increasingly integrated assessments. However, it remains difficult to point to places that are completely on track with sustainable development which suggests that true sustainability often continues to be sidestepped in favor of the status quo or other agendas. It should also be noted that it is fairly easy to implement change at the macro or policy level, but true implementation occurs at the local level and is significantly more challenging and time-consuming.

## 12.5 Lessons from the Case Studies

The four case studies demonstrate how performance measurement frameworks are shaped by the context in which they are created and applied. Understanding the *context* is therefore critical for the use of indicators and the pursuit of sustainability. From the case studies, we see that an overarching national or regional policy framework for sustainable development is essential for advancing sustainable transportation. Such a framework can stimulate greater local political interest in transportation and facilitate sustainability assessments. However, there is a risk that such frameworks can lack sufficient specificity to ensure implementation, fall out of political favor, or only have a marginal influence on the ultimate decision-making process.

At a more localized level, effective progress can be implemented in the absence of such an overarching framework. This was demonstrated through the GreenLITES example, where the initiative was not driven by a federal-level action or policy, and the high-speed rail example, where a comprehensive assessment system was put in place despite the national discussion on sustainable development stalling.

On the whole, the case studies provide a positive message that progress toward sustainability can be made. There is evidence from each case study that the adoption of a sustainability framework for decision support impacts what gets implemented on the ground. The impacts can be seen at local, regional, and national levels for different types of decisions. The case studies also illustrate how the agencies involved work with, engage, and even influence other agencies in the process. Through the case studies, we see a variety of governance systems in operation across the world. In these systems, indicators provide decision-makers with the information they need to effectively manage and develop infrastructure. Thinking through what sustainable development means and how transportation contributes to it is one way in which organizations can reach out to each other and work across or break down existing silos.

All the case studies, however, also have a rather sobering message that when it comes to making transformative progress toward sustainability, there is no "silver bullet" solution to this challenge. Many elements have to work together to make a significant change, and factors such as politics and broader socioeconomic trends play a major role in determining the rate of progress. Sometimes, the question also arises about how we move beyond the tempting case of mere green washing.

While the case studies focused on developed nations, many of the principles and lessons learned translate to other regions across the developing world. Although there will be several differences such as the focus on accessibility and mobility needs, data availability, data collection methods, and technology, the broader principles and methods are, we believe, transferable and globally relevant. It is expected that the biggest difference will be at the level of the indicators and data, whereas the overall approach will be broadly transferable. This is a contention that merits further exploration.

## 12.6 Who Should Drive Change?

This book focuses on how transportation professionals and agencies can use indicators and performance measurement frameworks to promote change toward sustainability. We argue that indicators should be actionable and address issues that a transportation agency can influence. But what if an agency's actions are constrained and it has a limited scope to influence sustainable development/transportation? It could be argued that many transportation agencies have yet to move beyond a "business as usual" approach, raising the question of who should be driving change? It may also be the case that new forms of collaborative governance are needed, where civil society, public agencies, nongovernment entities, and private actors have a more balanced role in shaping sustainable development and sustainable transportation. Thus, the question is not just how transportation agencies can make sustainability count. It is also how indicator frameworks in general can be used by a broad range of stakeholders to promote transformative change.

We have kept our attention on the more traditional models of transportation planning and delivery. However, we are cognizant that new forms of transportation governance may be needed. An inquiry into how sustainability can transform transportation planning and delivery falls beyond the scope of this work. However, the case studies have explored some models of how public sector agencies can act to promote change. The NYSDOT case study (Chap. 10) in particular demonstrates the benefits and challenges associated with implementing change within an organization where a new sustainability entity had to work alongside well-established groups or teams that may or may not share the same enthusiasm for change. Given that transportation systems cross jurisdictional boundaries and are interconnected with multiple sectors (such as energy and communications), the ability to design comprehensive performance assessment frameworks is likely to remain a valued skill. Thus, we hope the approaches and ideas presented in this work will support transportation professionals who have decided to take a lead role in driving change toward sustainability, regardless of whether they have an institutional mandate.

## 12.7 Looking Ahead

This book reflects on over a decade of significant attempts to address sustainability in transportation, including the development and implementation of sustainability indicators and sustainable transportation decision-support and planning tools. There remain some significant and growing global transportation challenges that the transportation sector can and should take the initiative to address.

Building on the material presented in this book, we offer a few key takeaways—(1) Sustainability matters, and transportation plays a key role in global sustainability; (2) There is a continuum between weak and strong sustainability, and the examples of sustainable transportation initiatives we see in practice lie at different points on this spectrum; (3) Indicators are critical tools in pursuing sustainability; (4) There is no universal set of indicators for sustainability, but comprehensive indicator sets do provide a valuable frame of reference for agencies seeking to develop their own frameworks; and (5) Context is important and should drive the selection of indicators to reflect relevant needs and priorities. In this regard, the process is as important as the outcome, and can pave the way for important changes and true progress.

It is our belief that the concept of sustainability is here to stay. Regardless of what the transportation system looks like in the future, we can confidently assert that the basics will not change, though the applications will. Whether we move to automated vehicles, or car ownership paradigms change, or vehicles no longer run on fossil fuel, the world will still grapple with issues relating to the quality of the environment, the use of natural resources, and issues of social equity, health, and safety. It will therefore remain important to address economic, social, and environmental needs now and into the future. Global challenges such as climate change, poverty, and health will also ensure that sustainability remains important. Sustainability is currently gaining traction within the private sector which sees it as a good marketing strategy, part of corporate social responsibility, and a way to reduce costs and increase profits. Other emerging concepts, such as smart growth, livability, and resiliency, can all fit into a broader sustainability framework. Future work in this area could focus on conducting additional case studies which should include developing nations. Future work will likely also need to address emerging trends such as big data, visualization, greater personalization of information and incentives, connected and autonomous vehicles, electric vehicles, and the use of alternative and renewable sources of energy for transportation. We also suggest that understanding the organizational factors that support a sustainable transportation decision-support framework (see Chap. 10) is essential to advancing institutional reform.

We hope this book will help students, academics, and practitioners develop a deeper appreciation for how sustainability can be approached by the transportation sector. We have attempted to provide practically accessible guidance informed by current research about the processes needed to start making a difference. We intentionally have not recommended specific indicators and solutions as these need to be worked through in a context-sensitive manner. While this may frustrate those looking for an "off-the-shelf" approach, the right solutions are more likely to emerge when basic principles are applied, using the right information, framed in an appropriate manner, to inform and support decision-making processes. Armed with a comprehensive understanding of the factors that support a successful indicator framework, we hope that readers will be able to ask more informed questions when trying to make sustainability count in the transportation sector.

## Reference

European Council. (2001, April 4–5). 2340th Council meeting—Transport/telecommunications—Luxembourg. Belgium: Council of the European Union.

MIX
Papier aus verantwortungsvollen Quellen
Paper from responsible sources
FSC® C105338

If you have any concerns about our products, you can contact us on
**ProductSafety@springernature.com**

In case Publisher is established outside the EU, the EU authorized representative is:
**Springer Nature Customer Service Center GmbH**
**Europaplatz 3, 69115 Heidelberg, Germany**

Printed by Libri Plureos GmbH
in Hamburg, Germany